T0396994

Advances in Growth Regulation of Fruit Crops

Life science has experienced a unique level of growth and development in recent times, as has the area of fruit crop regulation. Hence, the authors have been inspired to write this book entitled *Advances in Growth Regulation of Fruit Crops*. There are limited books with advanced knowledge on the growth and development of fruit crops, and therefore, there is a need for greater information to be made available about basic and advanced concepts of growth and regulation vis-a-vis fruit development. Growth regulation of fruit crops is a multifaceted and dynamic subject that requires simplified form so that the students pursuing UG (B.Sc.) in Horticulture or Life Sciences or PG (M.Sc. and Doctorate) in Fruit Science or Pomology can understand the concepts easily. Our primary target is to upgrade students' knowledge bases by providing the latest information to researchers. We hope it will help further knowledge about advances in the growth regulation of fruit crops. This book has been designed with the dual purpose of being a text cum reference. This book contains 20 crucial topics, including an introduction to the growth and development of fruit crops; eco-physiological influences on the growth and development of fruit crops – flowering and fruit set; phloem transport: source and sink; crop load and assimilate partitioning and distribution; root and canopy regulation of fruit crops; plant growth regulators – structure, biosynthesis and mode of action; plant growth inhibitors and growth retardants – metabolic and morphogenetic effects; absorption, translocation and degradation of phytohormones; growth manipulation through canopy architecture; growth regulation aspects of propagation; embryogenesis; seed and bud dormancy; physiology of flowering; regulation of flowering and off-season production; flower drop and thinning; fruit set and development; fruit drop and parthenocarpy; pre-harvest factors affecting post-harvest fruit quality; fruit maturity, ripening and storage; and molecular approaches in crop growth regulation.

In a nut shell, this book is written with the objective of scientific appraisal of the advances in the growth and development of fruit crops.

Advances in Growth Regulation of Fruit Crops

Edited by
Vishal Singh Rana, Neerja Rana, and
Sunny Sharma

CRC Press is an imprint of the
Taylor & Francis Group, an **informa** business

Designed cover image: Author's own image

First edition published 2025
by CRC Press
2385 NW Executive Center Drive, Suite 320, Boca Raton FL 33431

and by CRC Press
4 Park Square, Milton Park, Abingdon, Oxon, OX14 4RN

CRC Press is an imprint of Taylor & Francis Group, LLC

© 2025 selection and editorial matter, Vishal Singh Rana, Neerja Rana, and Sunny Sharma; individual chapters, the contributors

Reasonable efforts have been made to publish reliable data and information, but the author and publisher cannot assume responsibility for the validity of all materials or the consequences of their use. The authors and publishers have attempted to trace the copyright holders of all material reproduced in this publication and apologize to copyright holders if permission to publish in this form has not been obtained. If any copyright material has not been acknowledged, please write and let us know so we may rectify in any future reprint.

Except as permitted under U.S. Copyright Law, no part of this book may be reprinted, reproduced, transmitted, or utilized in any form by any electronic, mechanical, or other means, now known or hereafter invented, including photocopying, microfilming, and recording, or in any information storage or retrieval system, without written permission from the publishers.

For permission to photocopy or use material electronically from this work, access www.copyright.com or contact the Copyright Clearance Center, Inc. (CCC), 222 Rosewood Drive, Danvers, MA 01923, 978-750-8400. For works that are not available on CCC please contact mpkbookspermissions@tandf.co.uk

Trademark notice: Product or corporate names may be trademarks or registered trademarks and are used only for identification and explanation without intent to infringe.

ISBN: 9781032406367 (hbk)
ISBN: 9781032406398 (pbk)
ISBN: 9781003354055 (ebk)

DOI: 10.1201/9781003354055

Typeset in Times
by codeMantra

Contents

Preface

Horticulture, a branch of agriculture, is broadly concerned with the intensively cultivated plants directly used by man for food, medicinal purposes or aesthetic gratification. More accurately, horticulture deals with the production, utilization and improvement of garden crops like fruits, vegetables, ornamental, spices and condiments, aromatic and plantation crops. Horticultural crops, in general, and fruit, in particular, are highly perishable as these commodities contain higher water content and are mostly used in living states. These crops are highly remunerative and also regarded as protective food due to their medicinal and nutritional properties. Growth and development processes are very crucial to every biotic creature. Life sciences have experienced a unique level of growth and development in recent times, as has the area of fruit crop regulation. Books related to the basic and advanced concepts of growth and development as well as regulation of various vital processes in fruit crops are scarce. Regulation of growth and development in fruit crops is a multifaceted and dynamic subject that requires a simplified form, so that the students pursuing UG (B.Sc.) in Horticulture Life Sciences or PG (M.Sc. and Doctorate) in Pomology deal with the cultivation of fruit crops to get better insight into the subject. Upgrading of student's knowledge by providing latest information is of prime importance. The aforementioned reasons have prompted us to write this book, and we hope it will help further the knowledge about advances in the growth regulation of fruit crops.

This book has been designed with the dual purpose of being a text cum reference. This book proposes 20 crucial topics, namely, introduction to the growth and development of fruit crops; eco-physiological influences on the growth and development of fruit crops – flowering and fruit set; phloem transport: source and sink; crop load and assimilate partitioning and distribution; root and canopy regulation of fruit crops; plant growth regulators – structure, biosynthesis and mode of action; plant growth inhibitors and growth retardants – metabolic and morphogenetic effects; absorption, translocation and degradation of phytohormones; growth manipulation through canopy architecture; growth regulation aspects of propagation; embryogenesis; seed and bud dormancy; physiology of flowering; regulation of flowering and off-season production; flower drop and thinning; fruit set and development; fruit drop and parthenocarpy; pre-harvest factors affecting post-harvest fruit quality; fruit maturity, ripening and storage; and molecular approaches in crop growth regulation.

The literature cited is given at the end of each chapter. The final section also includes a glossary of related terms. Efforts are made to describe all the vital fundamental cultural practices being followed in fruit production in a comprehensive manner. The subject matter is illustrated with appropriate tables, figures and photographs wherever the need is felt. This book will definitely be useful to amateurs as well as students and researchers engaged in the field of horticulture. The help and criticism during the compilation of this book will definitely be acknowledged.

Dr. Vishal Singh Rana
Dr. Neerja Rana
Dr. Sunny Sharma

About the Authors/Editors

Dr. Vishal Singh Rana obtained his B.Sc. (Agri) from CSKHPKV, Himachal Pradesh in 1989 and his M.Sc. (Hort.) in Post Harvest Technology and Ph.D. in Pomology from Dr. YS Parmar University of Horticulture and Forestry, Nauni, Solan, Himachal Pradesh in 1992 (with first division) and 1997, respectively, which is Asia's first horticulture and forestry university. He was awarded ICAR-JRF and ICAR-NET in Fruit Science in 1993 and 1995 respectively. He has contributed significantly to the field of education and research, and guided undergraduate students as a class in charge and postgraduate students as a major advisor (20 M.Sc. students and 6 Ph.D. students). He has been actively involved in the restructuring of UG and PG course curriculum of horticulture. He has prepared 3 practical manuals, more than 70 research papers in reputed journals and 20 book chapters and presented 40 papers in national/international seminars/symposia/workshops. He has also made remarkable contributions to horticulture research and different orchard management aspects like fertigation, pollination, crop regulation, canopy management and rejuvenation of senile orchards. He has handled many state-funded projects as well as externally funded projects. He is also a life member of many societies and a peer reviewer of many reputed horticultural journals like the *Indian Journal of Horticulture*. He has 23 years of research experience on fruit crops and has standardized the production technologies for quality production of fruit crops in the mid-Himalayan region. Besides this, he has also developed rejuvenation techniques for senile peach orchards through judicious use of nutrients and pruning intensities. At present, he is engaged in standardizing the use of organic inputs in kiwifruit and developing new kiwifruit cultivars through selection/hybridization.

Dr. Neerja Rana obtained her B.Sc. in medical subjects from Himachal Pradesh University, Himachal Pradesh in 1991 and M.Sc. in Biochemistry from Dr. YS Parmar University of Horticulture and Forestry, Nauni, Solan in 1994 with first division. She earned a Ph.D. in Forestry (Biochemistry) from the Forest Research Institute, Dehradun in 1999. She was awarded the World Bank Fellowship by the Indian Council of Forestry Research and Education during her Ph.D. program. She joined as an Assistant Biochemist in the Department of Vegetable Science, Dr. YS Parmar University of Horticulture and Forestry, Nauni, Solan, Himachal Pradesh and worked for about 9 years under the All India Coordinated Project on Spices. She worked on the improvement of ginger and turmeric under this project. Consequently, she joined the Department of Basic Science as an Assistant Professor (Biochemistry) in August 2009. She has been actively engaged in UG and PG teaching, research and extension work for the last 21 years. Presently, she is guiding M.Sc. and Ph.D. students in the field of Microbial Biochemistry. She has published about 75 research papers, 15 book chapters and 4 manuals and presented 16 papers in national/international seminars/symposia/workshops. She is a life member of the Indian Society for Spices and an editor of the *Journal of Food Science and Fermentation Technology*. She has handled two research projects under the National Mission on Himalayan Research Scheme and one project with the Himachal Council of Science and Technology. She has also been working on various state projects and on industrially important enzymes and their applications in food industries and eco-friendly technologies for the improvement of horticultural crops/trees.

Dr. Sunny Sharma earned his B.Sc. (Agri) from the School of Agriculture (LPU), Phagwara, Punjab in 2016 and M.Sc. (Fruit Science) from Dr. YS Parmar University of Horticulture and Forestry, Solan, Himachal Pradesh in 2018. He was awarded ICAR-NET in Fruit Science in 2018 and 2019. He joined as an Assistant Professor (Horticulture) at Lovely Professional University, Phagwara, Punjab in 2018. He earned his Ph.D. in Fruit Science from Dr. YS Parmar University of Horticulture and Forestry, Solan, Himachal Pradesh in 2018. At present, he is working as an Assistant Professor (Horticulture) at Lovely Professional University, Phagwara, Punjab. He worked on the project titled 'Assessment and prediction of phenoclimatography of kiwifruit in Himachal Pradesh'. Presently, he is guiding M.Sc. and Ph.D. students in the field of Microbial Biochemistry. He has over 60 scientific publications to his credit which include research papers and review articles, 3 book chapters and 10 popular articles, and he has presented more than 20 papers in national/international seminars/ symposia/workshops. He is a peer reviewer of many reputed journals and a life member of professional scientific societies of India.

List of Contributors

Trina Adhikary
Department of Fruit Science
Punjab Agricultural University
Ludhiana, Punjab, India

Tanuj Bhardwaj
Department of Fruit Science, College of Horticulture
Dr. Yashwant Singh Parmar University of Horticulture and Forestry
Nauni, Solan, Himachal Pradesh, India

Suman Bodh
School of Agriculture
Lovely Professional University
Phagwara, Punjab, India

Subhash Chander
Deapartment of Agriculture
Punjab Agriculture University-Regional Research Station
Abohar, Punjab, India

Vimal Chaudhary
School of Agriculture
Lovely Professional University
Phagwara, Punjab, India

Akriti Chauhan
Department of Fruit Science, College of Horticulture
Dr. Yashwant Singh Parmar University of Horticulture and Forestry
Nauni, Solan, Himachal Pradesh, India

Pratibha Chib
Department of Fruit Science, College of Horticulture
Dr. Yashwant Singh Parmar University of Horticulture and Forestry
Nauni, Solan, Himachal Pradesh, India

Pankaj Das
Department of Agricultural Statistics
ICAR-Indian Agricultural Statistics Research Institute
New Delhi, India

Susmita Das
School of Agriculture
Lovely Professional University
Phagwara, Punjab, India

Komaljeet Gill
Department of Biotechnology, College of Horticulture
Dr. Yashwant Singh Parmar University of Horticulture and Forestry
Nauni, Solan, Himachal Pradesh, India

Debashish Hota
Department of Agriculture
Siksha 'O' Anusandhan (Deemed to be University)
Bhubaneswar, Odisha, India

Mukul Kumar
Department of Agriculture
Krishi Vigyan Kendra
Raisen, Madhya Pradesh, India

Pankaj Kumar
Department of Biotechnology, College of Horticulture
Dr. Yashwant Singh Parmar University of Horticulture and Forestry
Nauni, Solan, Himachal Pradesh, India

Pramod Kumar
Department of Fruit Science, College of Horticulture
Dr. Yashwant Singh Parmar University of Horticulture and Forestry
Nauni, Solan, Himachal Pradesh, India

Sandeep Kumar
School of Agriculture
Lovely Professional University
Phagwara, Punjab, India

Vijay Kumar
Department of Fruit Science, College of Horticulture
Dr. Yashwant Singh Parmar University of Horticulture and Forestry
Nauni, Solan, Himachal Pradesh, India

Pooja Sharma
Department of Fruit Science
Dr YS Parmar University of Horticulture and Forestry
Nauni, Solan, Himachal Pradesh, India

Chetanchidambar N. Mangalore
Department of Fruit Science
Dr. Y. S. R. Horticultural University Banavasi
Sirsi, Uttara Kannada, Karnataka, India

Divya Pandey
Department of Fruit Science, College of Horticulture
Dr. Yashwant Singh Parmar University of Horticulture and Forestry
Nauni, Solan, Himachal Pradesh, India

Stuti Pathak
Department of Agriculture
Maharishi Markandeshwar (DEEMED TO BE UNIVERSITY)
Mullana, Ambala Haryana, India

Vikrant Patiyal
Department of Fruit Science, College of Horticulture
Dr. Yashwant Singh Parmar University of Horticulture and Forestry
Nauni, Solan, Himachal Pradesh, India

Neerja Rana
Department of Tree Improvement and Genetic Resources, College of Forestry
Dr. Yashwant Singh Parmar University of Horticulture and Forestry
Nauni, Solan, Himachal Pradesh, India

Vishal Singh Rana
Department of Fruit Science, College of Horticulture
Dr. Yashwant Singh Parmar University of Horticulture and Forestry
Nauni, Solan, Himachal Pradesh, India

Kondle Ravi
School of Agriculture
Lovely Professional University
Phagwara, Punjab, India

K. Ravindra Kumar
Department of Horticulture
Dr. YSRHU-Horticultural Research Station
Kovvur, Andhra Pradesh, India

Rehan
School of Agriculture
Lovely Professional University
Phagwara, Punjab, India

Anindita Roy
Department of Fruit Science
Centurion University of Technology and Management
Paralakhemundi, Odisha, India

Chandrika Roy
Department of Fruit Science
Odisha University of Agriculture and Technology
Bhubaneswar, Odisha, India

Ankita Sharma
Department of Horticulture, College of Agriculture
Jawaharlal Nehru Krishi Vishwa Vidyalaya
Jabalpur, Madhya Pradesh, India

C. L. Sharma
Department of Seed Science and Technology
Dr YS Parmar University of Horticulture and Forestry Nauni
Solan, Himachal Pradesh, India

Naveen C. Sharma
Department of Fruit Science, College of Horticulture
Dr. Yashwant Singh Parmar University of Horticulture and Forestry
Nauni, Solan, Himachal Pradesh, India

Shagun Sharma
Department of Biotechnology, College of Horticulture
Dr. Yashwant Singh Parmar University of Horticulture and Forestry
Nauni, Solan, Himachal Pradesh, India

Shivali Sharma
Department of Fruit Science
Rani Lakshmi Bai Central Agricultural University
Jhansi, Uttar Pradesh, India

Umesh Sharma
School of Agriculture,
Dev Bhoomi Uttarakhand University
Dev Bhoomi Campus, Chakrata Road, Manduwala, Naugaon
Uttarakhand, India

Aeshna Sinha
School of Agricultural Science and Technology
RIMT University
Gobindgarh, Punjab, India

Dinesh S. Thakur
Department of Fruit Science, College of Horticulture
Dr. Yashwant Singh Parmar University of Horticulture and Forestry
Nauni, Solan, Himachal Pradesh, India

Sandhya Thakur
Department of Fruit Science, College of Horticulture
Dr. Yashwant Singh Parmar University of Horticulture and Forestry
Nauni, Solan, Himachal Pradesh, India

Shivender Thakur
School of Agriculture
Lovely Professional University
Phagwara, Punjab, India

Vandana Thakur
School of Agriculture
Lovely Professional University
Phagwara, Punjab, India

Pramod Verma
Department of Fruit Science, College of Horticulture
Dr. Yashwant Singh Parmar University of Horticulture and Forestry
Nauni, Solan, Himachal Pradesh, India

Praveen Verma
School of Agriculture
Lovely Professional University
Phagwara, Punjab, India

M. Viswanath
Department of Fruit Science
Dr. YSRHU-Horticultural Research Station
Kovvur, Andhra Pradesh, India

1 Introduction to Growth and Development of Fruit Crops

Praveen Verma and Suman Bodh

1.1 INTRODUCTION

The growth and development are frequently correlated to each other, but they denote distinct phases of a crop life cycle (TABLE 1.1). To gain a precise understanding of the subject matter, it is helpful to consider the example of multicellular organisms such as annuals, monocarpic or flowering plants (angiosperms), which encompass a variety of entities including fruits, vegetables, and flowers. The life cycle of an organism commences with a solitary-cell fertilized ovum, commonly referred to as the zygote. The process of embryonic development involves cellular division and differentiation, whereby the zygote transforms into an embryo. Upon sowing a seed in either a garden or a pot, a juvenile seedling will emerge within a short period. Over time, perennial plants experience a growth in the magnitude of their seedlings and the corresponding increase in the number of leaves. The plant undergoes branching for a limited duration. Following this, the plant initiates the process of flowering, leading to the formation of fruits and seeds. Eventually, the plant undergoes senescence at a specific time. The plant cycle is naturally measured by three discrete processes, namely "growth," "differentiation," and "development" (Sharma and Pratima 2018).

TABLE 1.1
Variation between Growth and Development Process

Particulars	**Growth**	**Development**
Explanation	Irreversible change in size and biomass of the plant	Sum up various sets of changes during the life cycle
How to measure	Height, weight, leaf area, etc.	Different phenophases Flowering, fruiting, maturation stages
Nature	**Quantitative**	**Qualitative**
Progression	Cell division and their enlargement	Morphogenesis, differentiation of cells and tissues
Phytohormones relation	Affected by auxins and gibberellins	Like ABA and ethylene
Nutrition	NPK required in huge amount	Vary with phenophase for example calcium during fruit setting
Energy diversion	Toward new cells, and tissues	Toward reproductive structures as well as organs
Activity of cell	Involves mainly mitosis and cell enlargement	Involves the differentiation of cells into specialized types
Seasonal variations	More in favorable seasons Slow in adverse conditions	Very specific (seasonal)
For example	Expansion of leaf, elongation of stem, root emergence	The transition process from the juvenile phase to the reproductive phase, the maturation of fruit

DOI: 10.1201/9781003354055-1

The process of growth can be demarcated as an irretrievable enlargement in both the mass of a cell and a tissue, concomitant with a rise in its dry mass. To clarify, growth refers to the numerical expansion of the plant structure, including the elongation of the stem and root, as well as the proliferation of leaves. The evaluation of an object can be conducted through various quantitative measurements including mass, length, height, surface area, or volume. The field of plant physiology has a rich and defined past in examining the development of fruits, utilizing metrics such as fresh and dry weight, cell count, and cell dimensions to uncover an array of correlations. Several types of fruits, including tomatoes, apples, avocados, bananas, and bananas, exhibit a unimodal growth pattern. The growth curve of this organism exhibits three distinct phases. The initial phase is characterized by a gradual growth rate driven by cell division. This is followed by a phase of cell expansion, during which there is a notable increase in size, fresh weight, and dry weight. The final phase is marked by a decline in growth rate, which coincides with the ripening process. Certain types of fruits, such as the fig (*Ficus carica*), grape, and olive (*Olea europea* L.), demonstrate a dual sigmoidal growth pattern, characterized by two rapid growth phases that are divided by a slower period. According to Nitsch's research in 1953, the enlargement of cells significantly contributes to the eventual increase in the size of fruit, even though a primary phase of rapid development appears to account for most of the cell division activity in such fruits. The process of differentiation refers to a qualitative transformation that occurs during the life cycle of a crop. Specifically, it involves the specialization of newly formed cells following division. Throughout this process, the cell experiences various alterations in its cell wall and protoplasmic components to carry out diverse functions. The newly formed cells exhibit variations in both their size and morphology in comparison to the parent cell. The expanded cellular structure is provided with nourishment that can be distinguished through diverse means. The phloem has the potential to serve various functions such as accumulating food reserves and transforming into storage tissue, as observed in the tuber of a potato or the root of a carrot. Additionally, it can differentiate into floral initials and integrate into the flower structure. Furthermore, it can generate sex cells and play a crucial role in transmitting genetic traits to the succeeding generation. Each specialized cell in a plant undergoes cell division, followed by a degree of enlargement and subsequent differentiation into its specific shape, size, and function (Raghavan 2000). Development refers to the entirety of the transformations that an individual experiences throughout their lifespan. Stated differently, the process of development can be conceptualized as the combination of both expansion and specialization (Arteca 1996). The optimal growth of plants necessitates a balance of three elementary cellular activities, namely cell division, expansion, and differentiation. Plant development is an uninterrupted procedure that encompasses the establishment of core plant parts, including the embryonic roots and shoots, as well as the roots, and flowers after germination. New cells are continuously generated during a plant's lifespan inside specific areas called meristems. These meristems contain stem cells that self-renew (SCs) and can differentiate into various types of cells during development. The process of differentiation is the phenomenon by which cells acquire specific characteristics and functions as they undergo development in plants. Plants can undergo dedifferentiation and rejuvenate into whole plants, a phenomenon referred to as totipotency (David 2017). The term "growth" refers to an increase in size resulting from cell division and expansion, as well as the generation of new cellular substances and the organization of cellulose organelles.

1.2 PLANT GROWTH REGULATION

Growth and development are influenced by a pair of distinct groups of components: external variables, such as air, light, water, and nutrition, and internal variables, such as heredity and hormones. Nutritional components, such as mineral and organic compounds like carbohydrates and protein, supply the essential building blocks for growth. The regulation of plant development through the use of certain substances is facilitated by chemical messengers known as plant hormones. These hormones are capable of modifying or increasing physiological processes in plants, even in small quantities. The plant's distinct characteristic, which is passed down from its parent, can be attributed to its genetic makeup, making it another internal factor (Hopkins 1995).

Plant hormones are an assortment of organic compounds that exist naturally. They have physiological effects when present in low concentrations. The basic processes that are impacted include growth, differentiation, and development, with the possibility of other processes, such as stomatal activity, being influenced as well. Despite being infrequently utilized, the term "phytohormones" has been employed to refer to plant hormones. (Davies 2010). The concept of plant hormone pertains to a substance synthesized endogenously by the plant, which typically translocates from its origin to its target location and modulates plant physiological functions at a relatively low physiological concentration. Auxin is a type of plant hormone, along with gibberellins, cytokinins, abscisic acid, and ethylene. In recent times, several additional compounds possessing properties for regulating plant growth have been discovered, among which are brassinosteroids, which are commonly, referred to as the sixth group of plant growth regulators (PGR), along with jasmonates, salicylates, polyamines, and other compounds; continue to be the subject of ongoing research and investigation (Basra 2004).

Plant growth regulators (PGRs) are substances utilized to alter the proliferation and growth of plants. These are substances called nutrients that, in small amounts, can either enhance, hinder, or alter different biological reactions. PGR, or plant growth regulator, includes both natural ingredients and additional components. Plant growth regulators (PGRs) are naturally produced chemicals that influence physiological processes. Plant growth and development rely on the activation of many biochemical and physiological processes, which can be initiated by synthetic analogs (Bhatla 2018). Plant bio-regulators (PBRs) encompass a variety of materials, such as plant hormones, artificial and biologically actively growing substances, synthetic enzymes, vitamins, triacontanols, organic acids, ethylene, inhibitors, and other analogous substances. Biochemical substances known as plant bio-regulators (PBRs) have been found to enhance the growth and output of plants when applied at appropriate stages of plant development, even in small quantities. Phosphorus-based fertilizers (PBRs) are extensively utilized in the cultivation of horticultural crops, primarily for augmenting yield (Pasala et al. 2017).

1.3 PHASES OF GROWTH

Plants, regardless of their classification as annual, biennial, and perennial, undergo distinct stages or periods of growth. Throughout these phases, every horticultural plant adheres to a predetermined sequence of growth and progression (Sharma and Pratima 2018).

1.3.1 Vegetative Phase

The initial stage of plant growth, characterized by the development of roots and leaves, is commonly referred to as the vegetative phase.

1.3.2 Juvenile Phase

The juvenile phase of a plant's cycle is marked by an onset of photosynthesis following germination. This period is characterized by rapid vegetative growth, during which a fruit tree acquires the necessary size, framework, and strength to produce fruit. During this phase, the plant utilizes manufactured food to support further shoot, leaf, and root development, increasing in size. Juvenile plants can prevent precocious flowering and seed production while their photosynthetic capacity is still restricted, making this stage of their growth crucial (Sgamma 2017).

1.3.3 Transition Phase

When a plant is young, it transitions from using carbohydrates for growth to accumulating them in storage organs. In certain fruits, the transition period may extend for several years, whereas in some annual plants, it may just last a week or less. In either scenario, the transition from juvenile to productive is not rapid but rather indicates a constant and steady transformation up to blossoming. Throughout this period, a series of gradual modifications take place in the morphology, anatomy, and physiology of the plant, encompassing alterations in leaf properties such as shape, thickness, and epidermal traits, as well as phyllotaxis, thorniness, shoot orientation, growth vigor, anthocyanin pigmentation, photosynthetic output, resistance to pests and diseases, and the capacity to generate adventitious buds, roots, and somatic embryos. Even at the most productive stages of plant growth, some vegetative growth persists, but a balance is established when carbohydrate accumulation is preferred over carbohydrate utilization and the plant is regarded as being in its productive stage of growth. The plant reaches the reproductive stage when it transitions from the vegetative to the floral stage. The gametophytic phase begins with meiosis, and the embryonic phase of the subsequent generation is initiated by the fusing of the gametes during fertilization. Environmental signals control several of these transitions to time growth with a favorable environment and maximize reproductive success (Bäurle and Dean 2006).

1.3.4 Reproductive Phase

The reproductive phase is a pivotal developmental stage in the life cycle of the plant, wherein it attains the ability to generate progeny. The accumulation of food reserves in horticultural plants during the productive stage varies depending on the type of plant, with reserves accumulating in vegetative storage organs, reproductive parts, or fruits. The reproductive phase of a plant usually commences upon reaching maturity, whereby the majority of the apex meristems on the branches discontinue leaf initiation and instead generate floral components based on species-specific characteristics. The specific reason for changing from the juvenile to the maturity phase is not yet fully understood. However, it is hypothesized that the controlling of assimilate distribution in the apical meristem region through hormonal control may be a contributing factor.

1.4 DEVELOPMENT OF FLOWERS

Flower development involves an important change of the photosynthetic apex into a functional structure. The commencement of reproductive development throughout the apical meristem is an essential phase in the entire life cycle of plants. The formation of flower buds has been attributed to the interplay between flower-forming genes and meiosis. While annuals may view flower bud formation as an indication of the completion of their life cycle, perennial plants undergo repeated flowering each season. Flowers may be induced in the previously formed buds at the apex of the main central shoot, lateral branches, or both, either as solitary blooms or an inflorescence. The process initiates with the induction phase, which is succeeded by the differentiation of the expanding point, culminating in the differentiation of flower primordia. According to physiological and biochemical definitions, blooming refers to the modifications that occur throughout the development of one blossom or a group of flowers called an inflorescence. When analyzing different species, it is important to remember a fundamental distinction in the structure and morphology of the ovary and the flower. Unlike an inflorescence, where flowers are produced on a floral meristem that comes from the shoot apical meristem, a single flower arises from a shoot apical meristem and develops into an indeterminate floral meristem. Both forms of floral meristems are used to produce a certain number of floral structures, regardless of whether they are formed directly at the tip of a shoot or through involution on an inflorescence meristem (Raghavan 2000).

1.4.1 Initiation of Flower Buds

After the young stems have grown to a particular diameter and most of their leaves have attained maturity, floral buds begin to form. In temperate fruit species like apples, peaches, pears, cherries, or plums, flower buds begin to form between late spring and summer of the life cycle before bloom.

They then continue to grow for a number of months before entering a dormant state throughout the winter. Even when the plants are given ideal temperature, moisture, and light conditions during the rest time, flower and vegetative buds will not open. Flower bud growth continues even after rest is complete, and blooms appear once the necessary chilling period is complete the next spring, when the temperature rises, and the soil moisture level is enough. Flowers begin to form and bloom all year long in the tropics on citrus fruits like oranges, lemons, and pomelos. For these fruits, a chilling winter temperature is not necessary to induce bud dormancy (Sharma and Pratima 2018).

1.4.2 Flower Bud Induction

Induction of buds in perennial fruit crops takes place after the vegetative phase. The process of induction functions as a trigger for suitably developed buds to progress from the nonsexual stage to the generative stage. The aforementioned phenomenon denotes a significant qualitative modification that occurs during the developmental process of plants, wherein meristems that are strategically located are genetically programmed to initiate the formation of flowers. The establishment of an appropriate balance of endogenous hormones in fruit trees, which ultimately leads to the initiation of the process, may be facilitated by the interaction among physiological and ecophysiological factors like temperature, light exposure, and day length. The development of floral organs is closely linked to, and often overlaps with, the process of evocation, with the meristem playing a crucial role in this process. The conventional angiosperm flower comprises groups of modified foliage, which constitute sterile and fertile segments. The sterile components of a blossom consist of two whorls, namely the outer whorl of sepals and the inner whorl of petals. The sepals, typically green in color, enclose and protect the developing flower bud before it blooms. However, the petals are brightly colored and serve the purpose of attracting pollinators such as bees and butterflies (Raghavan 2000).

1.4.3 Flower Bud Differentiation

Subsequent to the initiation of flower induction, a sequence of physiological mechanisms ensues, resulting in the development of distinct morphological features. The aforementioned phenomenon is distinguished by the structural arrangement of the uppermost part of the plant into rudimentary structures that give rise to either a single flower or a cluster of flowers known as an inflorescence. The first discernible morphological alteration that signifies the shift towards a floral state in the meristems is the widening and curving of the apex, which is subsequently succeeded by its extension into a cylindrical configuration and the emergence of sepal primordia in the apical flower. Laterally differentiated organ primordia follow subsequently. The sequential formation of blossoming organs like calyx, corolla, and pistil are consecutively developed. The development of floral tissues takes place in a sequential manner, with the outer structures differentiating before the inner structures. The initiation of the developmental process in the flower bud is instigated by a biochemical stimulus. The biochemical signal facilitates the change of the tissue from a vegetative to a reproductive form within a predetermined duration. A balance between GA3, auxin, cytokinins, and hormones that act like ethylene causes it to happen. When the C/N ratio is ideal, flowering begins (Buban and Faust 1982).

1.5 DEVELOPMENT OF FRUITS

The process of fruit development is commonly characterized as the transformation of mature ovaries into structures that contain seeds. In numerous cases, additional floral components such as the floral receptacle, calyx, and inflorescence axis may also play a role in the expansion of fruit. The genesis of these organs may transpire concurrently with the gynoecium, culminating in their amalgamation into the fruit morphology, as evidenced by fruits such as pome fruits. According to

Nitsch's definition in 1952, a fruit is an anatomical structure that arises from the maturation of the tissues that provide support to a plant's ovules. The primary purpose of a fruit is to facilitate the growth, safeguarding, sustenance, and eventual dissemination of the seeds it encases.

1.5.1 Pre-pollination Development

As previously mentioned, the process of pollination, subsequent pollen tube elongation, and fertilization serve as stimuli for fruit development. In the absence of pollination, flowers are typically shed, with only infrequent deviations from this pattern. The process of fruit development frequently commences well in advance of the opening of flowers during the spring season. The process of flower primordia initiation can commence up to 6 months earlier to the blooming of an individual flower. The growth of a flower involves the continuous development of the ovary, with the ovary tissues forming during the later stages of this process. It is noteworthy that certain fruits, such as olives, undergo complete development of their ovules just before the opening of flowering buds in the spring.

1.5.2 Post-pollination Development

The process of fruit development commences after pollination. The phenomenon of ovary wall growth augmentation and, in specific cases, the concomitant development of associated receptacle tissues has been observed. Concomitant abscission of floral organs occurs, although, in specific instances, the calyx may endure with the fruit until it reaches full maturation, such as in the apple, citrus, tomato, and guava. The pistil is activated by pollination. It appears that this organ undergoes a variety of modifications that support the development of male gametophytes and aid in fertilization. The process of pollination facilitates the secretion of stigma and the discharge of sugar from the transferring tissue and enhances the potential for survival of the embryo sac. It appears that even non-fertilizing pollen grains exhibit a supportive role in facilitating the successful fertilization of others. The development of the pollen tube is also influenced by the pistil. Pollen tube development along the pistil is not constant; it accelerates and slows down following the various tissues they pass through. The pistil has a function in regulating the kinetics of pollen tube growth since pollen tube growth is heterotrophic and consumes pistil reserves, which are not continually generated (Herrero 1992). Parthenocarpy, which is the ability of fruit to develop without the need for pollination and fertilization, can occur in certain cases.

1.6 SEED DEVELOPMENT

A seed is a reproductive structure that is derived from an ovule, usually after the process of fertilization. Monocotyledonous and dicotyledonous seeds share common structures, including a seed coat, cotyledons, endosperm, and an embryo. The process of fertilization is the initial phase in seed development, where male and female gametes fuse. Fertilization, also known as syngamy, occurs when the reproductive organs of both the male and female are fully formed. This phenomenon commonly occurs during double fertilization, which is a procedure involving the merging of two gametes. Pollination initiates the development of a long and narrow tube called a pollen tube that develops from a pollen grain upon coming into touch with the stigma. The elongated component, referred to as the pollen tube, advances via the micropyle and style and ultimately reaches the embryo sac. The tube nucleus accurately tracks the tube apex as it descends. The tube nucleus undergoes quick destruction, however, the two pollen sperm cells demonstrate resistance and successfully pierce the embryo sac. One of the sperm cells combines with the diploid (2N) polar nucleus, resulting in the creation of a triploid (3N) endosperm nucleus. The other sperm cell fuses with the egg cell, resulting in the creation of a diploid (2N) zygote, commonly referred to as a fertilized egg (Copeland and McDonald 1999). In dicotyledonous plants lacking endosperm, the cotyledons function as the primary storage organ.

The process of seed development occurs concurrently with that of fruit development. The latter stages of fruit development are contingent upon the plant growth regulators generated within the seeds. The size of the fruit in certain species, such as apples and pears, has shown a strong correlation with the number of seeds present within the fruit. Fruits that contain fewer seeds tend to be smaller in size, frequently exhibit irregular shapes, and have a higher tendency to undergo premature detachment from the plant.

1.7 FRUIT MATURITY AND RIPENING

Ripening is the outcome of another developmental shift that mature fruits go through that involves coordinated alterations in a variety of catabolic and anabolic processes. The phrase "fruit ripening" is deceptive since it primarily refers to the deteriorative parts of the process that cause spoiling and loss as a developmental phase. Fruit ripening, as per current standards, represents the final phase of growth and is distinguished by harmonious modifications in numerous metabolic processes. Respiration, ethylene generation, carotenoid synthesis, chlorophyll degradation, cell wall hydrolase production, and softening are a few of the alterations investigated. Some fruits not only produce ethylene while they ripen, but external ethylene also unintentionally speeds up the ripening process. Alterations in the color of fruits are attributed to modifications in the manifestation of genes responsible for the biosynthesis of pigments. Apart from its substantial economic implications, fruit ripening maintains a wider significance as it serves as an illustration of numerous biological and biochemical mechanisms, including the regulation of gene expression during cellular maturation (Raghavan 2000).

Alterations in pigmentation may result from the degradation of chlorophylls catalyzed by chlorophyllase or the biosynthesis of anthocyanin, carotenoid, or lycopene. Ethylene shows a substantial character as a regulator in the process of ripening. According to scientific literature, this particular phytohormone is known to regulate the expression of multiple genes that are implicated in the ripening of fruits. The primary role of this entity is to regulate the enzymatic activity of multiple enzymes that participate in the physiological process of fruit maturation. The enzymes in question facilitate the process of fruit skin softening and the conversion of intricate polysaccharides into easily digestible sugars. Rapidly growing plant tissues are known to produce and release ethylene. The substance is emitted from the apical meristems of roots, blossoms, injured tissue, and maturing fruit.

1.8 THE GROWTH PATTERN OF FRUITS

The field of plant physiology has a rich and distinguished past in investigating the development of fruits, utilizing measurements such as fresh and dry weight, cell count, and cell dimensions to uncover diverse associations. As previously mentioned, there is a significant increase in fruit mass or volume, exceeding one hundred times, from the point of fertilization until maturity. Two discernible categories of growth curves can be observed when plotting the rise in variables such as fruit volume, fresh weight, dry weight, and diameter over time following anthesis. Many fruits, such as tomatoes, apples, avocados, and bananas, grow along a single sigmoidal growth curve. The growth curve in challenging is characterized by an initial phase of slowing growth, marked by a prevalence of cell divisions. This is succeeded by a period of cell expansion, during which there are substantial increments in size, fresh weight, and dry weight. Ultimately, the curve culminates in a final phase of growth rate deceleration, which is linked to the ripening process. Certain types of fruits, such as the fig, grape, and olive, demonstrate a biphasic growth pattern characterized by two periods of rapid development separated by a slower phase. The process of development can be divided into three clearly defined stages. At first, an ovary's wall and its components, excluding the embryo and endosperm, observe a rapid rise in size (FIGURE 1.1). In stage II, the embryo and endosperm experience rapid development, while the endocarp layer undergoes lignification, resulting in the hardening of the pit in stone fruits. Furthermore, there is a slight rise in the thickness of the ovarian wall. In the next developmental phase, there is a known raising of the level in the mesocarp's

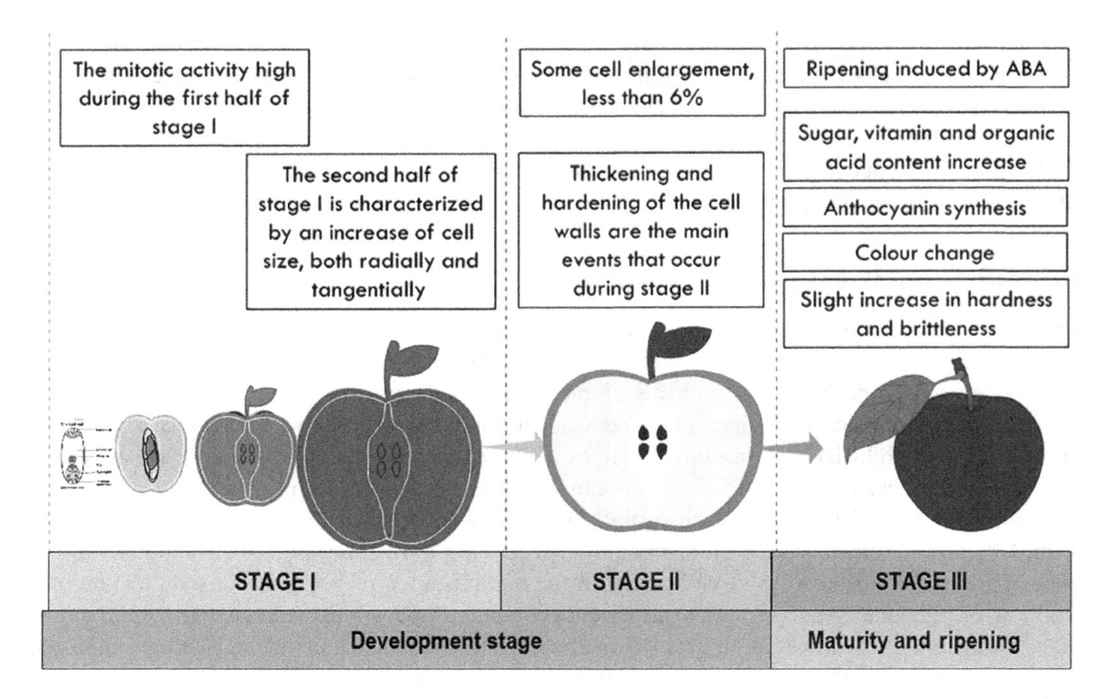

FIGURE 1.1 Steps of development of fruit.

growth rate, resulting in the eventual expansion of the fruit. This is subsequently followed by the maturation process (Raghavan 2000; Sharma and Pratima 2018). These essential characteristics of the fruit lifestyle may also be observed by the in vitro development of fertilized flower ovaries in a rather straightforward media (Nitsch 1951).

1.9 SENESCENCE AND DEATH

The growth, reproduction, and senescence of primary organs—leaves, flowers, and fruits—occur during the complicated juvenile/maturity transition. The growth and senescence of plant organs are governed by multiple genetic systems. A specific system that operates in plants involves the phytohormone ethylene, which collaborates with other hormones to integrate diverse signals and facilitate the initiation of conditions that promote phase advancement, fertility, and organ longevity. The perception of ethylene, its amount, and the hormonal interactions all directly or indirectly affect how long plants survive. Senescence refers to the biological process of degradation and aging that occurs in plants. Senescence is a typical developmental process that relies on energy and is regulated by the plant's inherent genetic program. The demise of the plant or its constituent parts resulting from senescence is referred to as programmed cell death (PCD). The last stage of plant development, known as senescence, is marked by whole-plant degradation and eventual death. Senescence is a process that affects in addition to the entire plant life. This phenomenon may only affect certain plant parts, like leaves and flowers, or certain cell types, like phloem and xylem, or even certain cell organelles, like chloroplasts and mitochondria (Iqbal et al. 2017). Typically, there have been identified four distinct senescence patterns in the realm of plant biology.

i. **Overall senescence**: This form of senescence is observed in annual plants, wherein the entire plant is impacted and ultimately perishes.
ii. **Top senescence**: The phenomenon being described pertains to the enduring nature of perennial herbs, wherein the process of senescence is limited to the aerial portions while the subterranean components, including the root system, remain viable.

iii. **Deciduous senescence**: This form of senescence occurs in woody deciduous plants and is comparatively less severe. Simultaneous senescence of all leaves is observed, while the majority of the stem and root system maintains viability.
iv. **Progressive senescence**: The gradual loss of physiological capacity known as "progressive senescence" leading to ultimately death of the entire plant.

During the plant's living state, the presence of ethylene and its interactions with other plant hormones play a crucial role in the shift from the vegetative stage to the reproductive stage and in the process of senescence. This networking influences the sensitivity of the tissues and the concentration of ethylene. The process of senescence in plants is characterized by a reduction in various biochemical components such as photosynthesis, starch, chlorophyll, DNA, RNA, proteins, gibberellins, and auxins. The study of senescence in plants is most effectively conducted through the examination of plant organs such as leaves, cotyledons, sepals, and petals, as well as isolated chloroplasts at the cellular level. During the process of senescence, there is a high level of metabolic activity in both cells and tissues. This is accompanied by a series of organized cytological and biochemical events.

According to research, plant hormones have an impact on the mechanism of regulated senescence in plants. Abscisic acid, ethylene, jasmonic acid, and salicylic acid all contribute to the promotion of senescence; cytokinin, on the other hand, has been discovered to slow the process down. Various environmental factors, particularly those that inhibit typical plant development, tend to accelerate the process of senescence. Some of the factors that can contribute to soil nutrient deficiencies include high temperatures, water deficits, and lack of exposure to light.

REFERENCES

Arteca, R.N. 1996. Physiology of Fruit Set, Growth, Development, Ripening, Premature Drop, and Abscission. In: *Plant Growth Substances*. Springer, Boston, MA. https://doi.org/10.1007/978-1-4757-2451-6_10

Basra, A.S. 2004. Plant Growth Regulators in Agriculture and Horticulture: Their Role and Commercial Uses. Food products Press: London. 264p.

Bäurle, I., Dean, C. 2006. The Timing of Developmental Transitions in Plants. *Cell*. 125(4):655–664.

Bhatla, S.C. 2018. Plant Growth Regulators: An Overview. In: *Plant Physiology, Development and Metabolism*. Springer, Singapore. https://doi.org/10.1007/978-981-13-2023-1_14

Buban, T., Faust, M. 1982. Flower Bud Initiation in Apple Trees: Internal Control and Differentiation. *Hort Rev* 4:174–203.

Copeland, L.O., McDonald, M.B. 1999. Seed Formation and Development. In: *Principles of Seed Science and Technology*. Springer, Boston, MA. https://doi.org/10.1007/978-1-4615-1783-2_2

David, K. 2017. *Cell Division and Cell Differentiation in Encyclopedia of Applied Plant Sciences* (Second Edition). Academic Press, New York, pp. 149–154

Davies, P.J. 2010. The Plant Hormones: Their Nature, Occurrence, and Functions. In: Davies, P.J. (eds) *Plant Hormones*. Springer, Dordrecht. https://doi.org/10.1007/978-1-4020-2686-7_1

Herrero, M. 1992. From Pollination to Fertilization in Fruit Trees. *Plant Growth Regul* 11:27–32. https://doi.org/10.1007/BF00024429

Hopkins, W.G. 1995. *Introduction to Plant Physiology*. John Wiley e Sons, New York, 464 pp

Iqbal, N., Khan, N.A., Ferrante, A., Trivellini, A., Francini, A., Khan, M.I.R. 2017. Ethylene Role in Plant Growth, Development and Senescence: Interaction with Other Phytohormones. *Front Plant Sci* 8:475. https://doi.org/10.3389/fpls.2017.00475

Nitsch, J.P. 1951. The Physiology of Fruit Growth. *Annu Rev Plant Physiol* 4:199–236

Nitsch, J.P. 1952. Plant Hormones in the Development of Fruits. *Quart Rev Biol* 27(1):33–57. https://www.jstor.org/stable/2812622

Nitsch, J.P. 1953. The Physiology of Fruit Growth. *Annu Rev Plant Physiol* 4(1):199–236

Pasala, R.K., Minhas, P.S., Wakchaure, G.C. 2017. Plant Bioregulators: A Stress Mitigation Strategy for Resilient Agriculture. In: Minhas, P., Rane, J., Pasala, R. (eds) *Abiotic Stress Management for Resilient Agriculture*. Springer, Singapore. https://doi.org/10.1007/978-981-10-5744-1_10

Raghavan, V. 2000. *Developmental Biology of Flowering Plants*. Springer, New York. https://doi.org/10.1007/978-1-4612-1234-8_13

Sgamma, T. 2017. *Plant Physiology and Development in Encyclopedia of Applied Plant Sciences* (Second Edition). Academic Press, New York

Sharma, N., Pratima, P. 2018. Growth and Development of Fruits. In: *Fruit Science: Culture and Technology Basic Aspects and Practices*, pp. 179–202, New India Publishing Agency, New Delhi

2 Ecophysiological Influences on Growth and Development Like Flowering and Assimilate Partitioning and Distribution in Fruit Crops

Vimal Chaudhary, Kondle Ravi, Sandeep Kumar, and Mukul Kumar

2.1 INTRODUCTION

Horticultural crops, due to their high value, are extensively cultivated. These crops comprise fruits and vegetables that offer crucial nutrients, minerals, and vitamins essential for human nutrition (Kwack, 2007; Jain et al., 2023). Lentz (1998) describes the growing of horticultural crops as a complex and unrestricted process that is influenced by multiple factors, including the environment, soil, cropping system, and their interconnections. Various environmental conditions, both directly and indirectly, have a considerable impact on the establishment and development of fruit crops (Schaffer and Anderson, 1994). Hence, to attain efficient cultivation, it is crucial to possess a thorough comprehension of how these components impact the physiological functioning of plants (Wien 1997). Ecophysiology is a scientific discipline that investigates the interplay between plants and their abiotic and biotic surroundings, as stated by Lambers et al. (2008). Environmental physiology is crucial to understand how abiotic factors affect plant growth and development. According to Tiaz and Zeiger (2006), stress response, acclimatization, and adaptation are important ways by which plants combat stress. Understanding how horticulture crops react to temperature, water, light, and CO_2 concentration helps analyze inadequate environmental conditions and improve crop management to maximize efficiency (Schaffer and Andersen, 1994). Understanding how environmental and physiological elements interact may improve horticultural breeding, production durability, and agricultural zoning strategies (Campostrini and Glenn, 2007).

In contrast to numerous animal species, plants generally exhibit indeterminate growth, whereby they continue to grow throughout their lifespan. Upon fertilization, the zygote undergoes a process of growth and differentiation, ultimately giving rise to a complex, multicellular organism characterized by the presence of specialized structures such as roots and leaves. Upon reaching maturity, the reproductive organs of a plant, namely the flowers, fruits, and seeds, undergo development and expansion. The aforementioned alterations take place as a result of the progression in the growth of flora. Development pertains to the complete sequence of modifications that an organism undergoes throughout its lifespan. Development is closely linked to the processes of morphogenesis and differentiation. Morphogenesis refers to the biological process of shaping cells and organs during development. In contrast, differentiation pertains to the numerical distinctions among cells, tissues, and organs. An instance of cellular differentiation occurs when a cambium cell transforms to generate either a xylem or phloem cell, or when a vegetative bud is converted into a flowering bud. All developmental processes entail a state of progression. Cell division is the mechanism that facilitates

DOI: 10.1201/9781003354055-2

the introduction of new cells during the process of growth. The organism experiences growth, which involves an enlargement in weight and size, in addition to the multiplication of protoplasm. This leads to a permanent and irreversible rise in volume. The term "growth" can be defined as a lasting and unchangeable alteration in the magnitude of a cell, organ, or entire organism, which is typically accompanied by a rise in dry weight. While the provided definition for the intricate process of growth is somewhat satisfactory, it fails to explain the established empirical evidence regarding growth.

2.2 GROWTH CURVE

2.2.1 Lag Phase

In this phase, the amount of growth is moderately slow as it represents an embryonic stage of development.

2.2.2 Log Phase

During this period, the growth rate is at its peak because of the rapid division of cells and physiological functions.

2.2.3 Senescence Phase

During this phase, the growth reaches its maximum and then stabilizes. Therefore, the rate of growth reaches a value of zero.

2.3 FLOWERING

2.3.1 Factors for Induction of Flowering in Fruit Crops

Major factors regulating the initiation of flowering could be broadly classified into five major groups, viz.

i. **Environmental factors**: Photoperiod, temperature, water relations and location, etc.
ii. **Internal factors**: Nitrogen metabolism and carbohydrate status (Nutritional factors), phytohormones.
iii. **Soil factors**: Nutrient status, soil structure, texture, moisture content, etc.
iv. **Horticultural traits**: Age, Crop load, rootstocks, type of shoots, etc.
v. **Cultural practices and use of chemicals**: Smudging, girdling, root, and shoot pruning, use of TIBA, SADH, PBZ, CCC, etc.

2.3.2 Temperature

Temperature plays a vital role in determining the success of seed germination, the growth of plants, the development of flowers, the setting of fruits, and the maturing of fruits in horticultural crops. It regulates the rate of a chemical reaction and consequently regulates the rate of plant growth, fruit yield, and quality parameters. Temperature requirements for each plant growth component are specific, and the temperature above or below that range can adversely affect these components. Flowering in mango is directly influenced by temperature. Shu (1999) observed that day-night temperatures of 19°C/13°C have a maximum duration of flowering (36 days), the life span of flowers (187 days), and a total number of flowers (2969) in 'Haden mango'. The development of floral buds in mangoes is greatly influenced by temperature (Singh et al., 1974). Sukhvibul et al. (1999)

reported that temperature had an inverse effect on the total number of flowers per inflorescence, with 619.6±108 at 20°C/10°C, decreasing to 431.3±80.5 at 30°C/20°C in four cultivars of mango. Sex reversal takes place in papaya at low temperatures, i.e. male plants bear fruits sometimes in cool climates (Storey, 1986).

2.3.3 Light

Sunlight is the primary energy source for the process of photosynthesis and the single most important factor in determining crop output in horticulture. There is no plant that can exist without light. Sunlight has two important effects on plants: one is connected to photosynthesis, and the other to growth and morphogenesis. Regardless of photosynthesis, photomorphogenesis is the term used to describe the influence that light has on growth, development, and differentiation. Light controls most plant processes, including reproduction, tissue and organ differentiation, seed germination, and seedling growth.

2.3.4 Photoperiodism

The duration of daily light and dark phases plays a crucial role in the transition of a green shoot meristem into a floral bud. Photoperiodism refers to the process of inducing blooming based on the duration of daily dark and light periods. It refers to the length of time that light is present. Plants exhibit a physiological reaction in response to the length of light exposure. Plants exhibit a range of photoperiodic reactions, but the most significant one is the triggering of flowering. The initial reception of the photoperiodic signal occurs in the leaf by phytochrome, which subsequently initiates the biological clock. Once the biological clock reaches a specific duration (known as the inductive photoperiod), the leaf begins producing the flowering stimulus. The signal for blooming is transferred from the leaf to the shoot apex, causing a shift in the growth pattern of the inflorescence from vegetative to reproductive. The shoot apical meristem generates primordia of floral organs, including sepals, petals, stamens, and carpels, resulting in the formation of a flower bud.

The discovery of photoperiodism was first reported by Garner and Allard in 1920. Plants are categorized into three groups based on the length of their exposure to light, known as the photoperiod (Table 2.1).

Short-day plant (SDP): For the plants to bloom, they need a light period of around 8–10 hours and a dark period of about 14–16 hours. These plants are commonly referred to as long-night plants. The majority of plants found in tropical regions are classified as short-day plants. e.g.-, Pineapple.

Day-neutral plants (DNP): These plants can flower under a wide range of photoperiods, from as little as 5 hours to continuous exposure for 24 hours. e.g.- Guava.

TABLE 2.1
Response of SDP and LDP to Photoperiods

	Duration (hours)			
Day/Night Length	**Day**	**Night**	**SDP**	**LDP**
Long day short night	16	8	No flowering	Flowering
Long day long night	16	16	Flowering	No flowering
Short day short night	8	8	No flowering	Flowering
Short day long night	8	16	Flowering	No flowering
Continuous day	24	–	No flowering	Flowering

Long-day plants (LDP): These plants thrive when exposed to a longer period of daylight, typically around 14–16 hours within a 24-hour cycle, to promote flowering. These plants are commonly referred to as short-night plants. Many of the fruits found in the temperate zone are classified as long-day plants. e.g.- Banana, Passion fruit.

2.3.5 Water Relations

In the tropics, there are clearly defined periods of rainfall and drought. Water stress can serve as a replacement for chilling. When there is too much rain or irrigation, and plants keep growing late into the summer, it can actually hinder the formation of flower buds. However, this problem is lessened when there is moderate drought, and it is almost completely prevented when there is severe stress. It is commonly believed that plant water stress is the main factor that triggers flowering (Schaffer et al., 1994). There is a hypothesis that suggests the buildup of ammonia during periods of stress may lead to an increased production of arginine and polyamines. This, in turn, could result in a higher rate of cell division once the stress is relieved, ultimately leading to the initiation of floral growth.

2.3.6 Nutritional Factors

Nutrition is recognized as a key determinant of the process of flowering. The impact of nitrogen levels differs among species in photo-periodically induced plants. Nitrogen deficit stimulates flowering in long-day plants (LDPs) but delays it in short-day plants (SDPs). Root trimming reduces nutrient intake, increases the C:N ratio, and stimulates flowering. Plants are often classified into four types based on their carbohydrate supply (Table 2.2).

Flower production can be reduced when nitrogen fertilizer application prolongs the period of extension and shoot growth, and delays terminal bud formation. The act of removing branches during late summer or early winter in mango trees can stimulate flowering during non-flowering years and enhance flowering during flowering years. This also provides evidence that nitrogen and carbohydrate reserve significantly impact the initiation of flowers in mango trees by increasing the ratio of carbon to nitrogen in shoots (Pandey, 1989).

Potassium, in particular, appears to have a positive effect on flowering as its increased level enhances amino acid formation, which in turn stimulates the formation of IAA oxidase, which stimulates flower induction as well as carbohydrate levels. K might be effective through pyruvate kinase, which in turn would determine the levels of several amino acids (Fabbri and Benelli, 2000). Normal flower development requires adequate mineral elements in the proper balance. The lack of essential minerals has a negative impact on the process of flowering. For example, when pear trees do not have enough boron, their blossoms start to wither right before and during the stage of anthesis. Research has shown that solutions containing 1%–2% NH_4NO_3 or 2%–4% KNO_3 are beneficial in stimulating the formation of flower buds in mango trees (Nunez-Elisea, 1985). Maximum flowering was obtained by spraying urea (2.5%) and ethrel (200 ppm) in guava by Chandra and Govind (1994) (Table 2.3).

TABLE 2.2
Effect on Nitrogen and CHO on Crops

1.	Moderate nitrogen and more carbohydrate supply	Flowering
2.	Low nitrogen and high carbohydrate supply	Low growth, few buds
3.	High N and low CHO supply	Unfruitfulness
4.	High N and enough CHO supply	Flowering

TABLE 2.3
Commercial Utilization

S.No.	Fruit Crop	Growth Regulator	Response
1	Grapes	GA_3	Increase fruit set
		CCC@ 2000 mg/L	Increased fruit set, inhibited shoot growth
2	Mango	Paclobutrazol	Improve fruit set and retention
		CPPU (10 ppm) 14 days after bloom	Increase fruit retention, yield, and quality
		Alar, TIBA, CCC	Profuse flowering

2.3.7 Plant Growth Regulators

2.3.7.1 Gibberellins

Gibberellins mainly operate as an inhibitor of floral initiation. GA3 @ 100 ppm in guava increases flowering days (50%) and flowers per shoot (15.67%) (Lal et al., 2013). For 2 years, mango trees treated with GA once or twice before flowering at 0, 50, and 100 mg/L experienced two flowering periods (January to March and April to May) and two harvesting periods (June and July). Control trees only blossomed (January–March) and harvested (June). Non-treated trees produced no mixed panicles, while treated trees generated 16%. GA slowed flowering and harvesting by 90 and 42 days. GA3 did not affect fruit output. Papaya sex expression is also affected by GA.

2.3.7.2 Auxin

Auxin is mostly used to blossom pineapples. Auxin was first used to start pineapple blossom in 1939. At 10–20 ppm, NAA and NAA-based compounds like planofix and celmone induce flowering. Auxin induces pineapple flowering by ethylene biogenesis (Burg and Burg, 1966). NAA @ 100–200 ppm a week after auxin treatments increases mango's perfect-to-staminate blossom ratio (Singh et al., 1965). Ubi et al. (2007) tested three forcible hormones, calcium carbide, NAA, and B-hydroxyethyl hydrazine, on two pineapple cultivars' flowering and fruiting. B-hydroxyethyl hydrazine at 0.50 g/L produced flowers in 8 days and was superior to all other treatments. NAA (10 ppm)+urea (2%) @ 50 mL per plant into the crown induces pineapple flowering. Increasing NAA concentration from 0.25 to 0.50 g/L and 0.50 to 0.75 g/L increased flower production. However, increasing β-hydroxyethyl hydrazine from 0.50 to 0.75 g/L reduced flower output by 51.5% and fruit production by 34.1%.

2.3.7.3 Cytokinins

Chen (1985) described mango shoot bud break and flowering after IBA spraying in early October. The treatment also caused total flowering in 1 month, compared to 3 months on fruitless trees. Nunez Elisea et al. (1990) found similar results with synthetic cytokinins as thidiazuron. Putative trans zeatin and its ribosides peaked during early blooming and full bloom but dropped during vegetative development and resting when it was translocated from the roots. Cytokinin migration in tree sap from the xylem peaks in springtime when the trees are in full bloom. Foliar treatment of 50 ppm BA and 2% Ca^{2+} increased mango hermaphrodite blooms (Singh and Rajput, 1990). The mango flowering response to BA and higher cytokinin levels before and during flowering suggest that cytokinins are involved in mango flowering (Chen, 1985).

2.3.7.4 Ethylene

Ethephon boosts mango flower output, especially in low-latitude tropics, according to multiple studies. Certain environmental circumstances cause this effect. Das et al. (1989). The Philippines induces mango blossoming by exposing them to leaf combustion fumes (Pantastico and Manuel, 1978). Spraying 200ppm ethyl (2-chloroethyl phosphonic acid) on mango cv. Alphonso, during the off-year, stimulated flowering. Ethephon @ 0.25 mL/L increases flowering on cv. Haden ringed and non-ringed mango trees, according to Rabelo et al. (1999). Ethylene causes pineapple, Annonaceae, and bulbous plant flowering. Ethylene, an Annonaceae florigen, induces pineapple flowers economically.

2.3.8 Other Chemicals

Fruit crops use plant growth retardants such as Alar, cycocel (CCC), and triazoles such as paclobutrazol, and uniconazole to induce flowering. Cycocel at 1,000ppm and SADH at 500 and 1,000ppm promote flowering in pears and apples, respectively. CCC and TIBA accelerated blooming in Co-2 papaya by 9 and 4 days, respectively, according to Bhattacharya and Rao (1982). Paclobutrazol is often used to induce blooming among triazoles. Paclobutrazol @ 1.0g a.i./m of canopy diameter (Soil application)+foliar spray of 2% KNO_3 decreased the number of days to first flowering (153.67) while increasing the proportion of hermaphrodite flowers (918.72%) in mango cv. Alphonso (Gopu et al., 2017). Paclobutrazol 3 mL/m canopy sprayed 90 days prior to bud break increased Alphonso mango flowering shoots by 89.9% (Reddy and Kurian, 2014). In mango cv. Chausa, foliar application of paclobutrazol at 500ppm in the month of October led to the earliest full bloom and the highest percentage of flowered shoots (36.20%). Paclobutrazol at 500ppm reduces the average minimum days to flowering initiation in Sardar guava (29.0) and enhances the number of flowers per branch (Jain and Dashora, 2007). Paclobutrazol at 2.5–15 g a.i./tree considerably enhances the proportion of reproductive shoots and lengthens the flowering time in mango (Ferrari and Sergent, 1996). Paclobutrazol at 10 g per tree demonstrated a twofold effect, including diminution of tree size and stimulation of early flowering and cauliflory development (Kulkarni, 1988). Under rain-fed conditions, it was advised that a single treatment of 10g PBZ per tree should be followed by applications every 2 years. According to Rao and Srihari (1996), foliar sprays of 100ppm TIBA or 2000ppm PBZ or soil treatment of 10g PBZ encouraged flowering straight on fruited stems in the "off" year by eliminating the need for a vegetative phase. Paclobutrazol (10g a.i. tree-1) applied to mango soil dramatically improved the proportion of perfect-to-staminate flowers (Kurian and Iyer, 1993). Compounds such as SADH and TIBA can prevent biennial bearing in apples, pears, and cherries by preserving the hormonal balance rather than the C:N ratio. The application of 500ppm CCC to guava led to the quickest blossoming and greatest quantity of blooms (Brahmachari et al., 1996). Paclobutrazol at 2 g per plant increased the quantity of mango flower buds (Ataide et al., 2006). Other compounds like benzoic acid, nicotinic acid, nicotinamide, and L-pipecolic acid induce flowering, but their significance as endogenous regulators of flowering has not yet been established.

2.4 CULTURAL PRACTICES

Several measures that will provide vigor control and initiate blossom bud formation include spreading of branches, trunk or branch girdling, defoliation and de-blossoming, chemical treatments, and pruning.

2.4.1 Spreading, Orientation, Bending and Shoot Type

Horizontal branching slows growth and promotes flower buds. When a branch bends, wood stress increases and phloem growth diminishes. Thus, photosynthetic products gradually migrate from bending branch shoots to other locations, maintaining a high C:N ratio and encouraging blooming and fruit development. Using pressure to create inactive reproductive buds.

Ito et al. (1999) found that straight branches yield fewer flowers and fruits than bending branches. Bending regulates blooming (Mitra et al., 2008). According to Eassa et al. (2012), bending consistently increased lipids, proline, polyphenol oxidase, and peroxidase in leaves, bark, and fruits. However, phenolics decreased. These changes may have boosted flowering and fruit production, increasing crop yield (Bagchi et al., 2008 in guava). Tromp found in 1973 that horizontally positioned shoots produced more blooms than vertically positioned ones. By limiting glucose and auxin movement from the top of the limb to the roots, bending young tree branches 45° from the main stem speeds flowering. Thus, flower buds benefit from glucose accumulation and development slowing beyond the bend. Guava cultivation uses shoot bending, which CISH, Lucknow has standardized.

2.4.2 Girdling

Trunk girdling of mango trees to promote flowering is inconsistently effective (Pandey, 1989). The actual cause of flowering by girdling cannot solely be attributed to carbohydrate accumulation but ethylene produced may also play an important role as it is produced under stress conditions (Leather and Abeles, 1972). Ghadage et al. (2017) revealed that girdling of width 1.50 cm in 'Alphonso' mango minimized the day taken to flowering (117.38) and increased the hermaphrodite to male flower ratio (0.26) while girdling of width 1.25 cm maximized (25.55 cm) the length of the flowering shoot. Urban and Alphonsout (2006) reported that flowering occurred 15 days earlier in the girdled than in the non-girdled treatment in mango trees, cv. Cogshall.

2.4.3 Defoliation

By changing the cytokinin-auxin ratio in buds, tree defoliation may encourage fresh leaf growth. Since apical leaves are the main source of auxin, losing them may cause new shoots by increasing the ratio of cytokinin to auxin. Inactive side buds will create flower clusters if they grow in flower-friendly conditions after flower removal. Defoliating leaves affect glucose supply and cytokinin flow.

2.4.4 Pruning

Pruning exerts a beneficial effect on flowering. Increasing severity of pruning usually decreases flowering and cropping (Forshey and Elfying, 1989), and heading-back methods of pruning may convert potential fruiting spurs into shoots. Inadequate pruning may, however, result in excessive tree shade and inhibition of flowering (Feucht, 1976). Saidha et al. (1983) stated that though pruning had no significant effect on the time of floral bud formation, it enhanced the leaf ethylene level and percentage of flowering in *cvs.* Mulgoa, Neelum, and Bangalore. Similarly, Gill et al. (1998) reported that pruning to a length of 30 cm resulted in the greatest flowering in mango cv. Haden. Lal et al. (1996) observed that one leaf pair pruning in guava cv. Sardar proved superior to other pruning treatments for flower bud initiation. In Ber, blossom initiation can be regulated by the time of pruning. By pruning during spring instead of at the normal time 2–3 months later, flowering was advanced by about 3 months *i.e.* during summer (Pareek, 1983).

2.4.5 Rootstock

Rootstock also seems to influence flowering. Young trees on selected clonal rootstocks often flower earlier than those on seedling rootstocks. Mostly the effect of rootstock is observed in temperate fruit crops.

2.4.6 Crop Load

Heavy cropping in 1 year can inhibit flower bud initiation and so reduce flowering in the following year. It is observed that in mango, in the year of a heavy crop (1 year), the demand for carbohydrate supply is such that few flower buds are formed for the next year.

Application of Unicazol (foliar spray) @ 1.5 g/L increased the percent fruit set (0.76%) and number of fruits per tree (56.00) in mango cv. Alphonso (Gopu et al., 2017). Application of paclobutrazol @ 500 ppm in 'Sardar' guava increased the fruit set (71.17%) and fruit retention (73.16%) with minimum days taken to harvesting (115.33) (Jain and Dashora, 2007).

2.5 CONCLUSION

It may be claimed that understanding how environmental elements and plant physiology interact makes it easier to recognize environmental changes such as a lack of light, excessive temperatures, or a water shortage. For instance, shading horticultural crops can increase flower abortion, decrease transpiration, and stomatal density and conductivity. The viability and germination of pollen, the number of blooms per plant, and the number of fruits per plant can all be impacted by high temperatures. Lastly, ecophysiological knowledge is a tool that may be applied to agricultural zoning plans and breeding programs to create better cultivars, increasing production.

REFERENCES

Ataide EM, Ruggiero C, Oliveira JCD, Rodrigues JD and Oliveira HJ. 2006. Effect of paclobutrazol and gibberellic acid in floral induction of yellow passion fruit in intercrop condition. *Revista Brasileira de Fruticultura,* **28**(2): 160–163.

Bagchi TB, Sukul P and Ghosh B. 2008. Biochemical changes during off-season flowering in guava are induced by bending and pruning. *Journal of Tropical Agriculture,* **46**(1/2): 64–66.

Bhattacharya RK and Rao MVN. 1982. *South Indian Horticulture,* **30**: 137–138.

Brahmachari VS, Mandal AK, Kumar R and Rani R. 1996. Effect of growth substances on flowering and fruiting characters of Sardar guava (*Psidium guajava L.*). *Horticulture Journal,* **9**: 1–7.

Burg SP and Burg EA. 1966. The interaction between Auxin and Ethylene and its role in plant growth. *Proceedings of the National Academy of Sciences US,* **55**: 262–269.

Campostrini E and Glenn D. 2007. Ecophysiology of papaya: a review. *Brazilian Journal of Plant Physiology,* **19**(4): 413–424.

Chandra R and Govind S. 1994. Effect of urea and ethrel on flowering and fruiting in guava under intensive planting system. *Indian Journal of Horticulture,* **51:** 340–345.

Chen WS. 1985. Flower induction in mango (*Mangifera indica* L.) with plant growth substances. *Proceedings of the National Science Council, Republic of China. Part B, Life Sciences Taipei, Republic of China,* **9:** 9–12.

Das GC, Sahoo SC and Ray DP. 1989. Effect of GA and urea either alone or in combination on growth and flowering of some off and on year Langra mango shoots. *Acta Horticultarae,* **231**: 495–499.

Eassa KB, Gowda AM and El-Taweel AA. 2012. Effect of GA3, hand pollination and branch-bending on productivity and quality of banati guava trees grown in sandy soils. *Journal of Plant Production (Mansoura University),* **3**(2): 241–251.

Fabbri A and Benelli C. 2000. Flower bud induction and differentiation in olive. *Journal of Horticultural Science and Biotechnology,* **75**: 131–41.

Ferrari D and Sergent EA. 1996. Promoción de la floración y fructificación en mango (*Mangifera indica* L.) cv. Haden, con Paclobutrazol. *Rev Fac Agron,* **22**: 9–17.

Feucht W. 1976. Fruitfullness in pome and stone fruits. *Extension Bulletin,* 665: 1–32.

Forshey CG and Elfying DC. 1989. The relationship between vegetative growth and fruiting in apple trees. *Horticultural Reviews,* **11**: 781–782.

Ghadage NJ, Patil SJ, Khopade RY, Shah NI and Hiray SA. 2017. Effect of time and width of girdling on flowering and yield of mango (*Mangifera indica* L.) cv. Alphonso. *International Journal of Chemical Studies,* **5**(6): 1580–1583.

Gill MP, Sergent E and Leal F. 1998. Effect of pruning on reproductive growth and fruit quality of mango cv. Haden. *Bio Agro,* **10**: 18–23.

Gopu, B., Balamohan, T. N., Swaminathan, V., Jeyacumar, P., & Soman, P. 2017. Effect of Growth Retardants on Yield and Yield Contributing Characters in Mango (Mangifera indica L.) cv. Alphonso under Ultra High Density Plantation. *International Journal of Current Microbiology and Applied Sciences*, *6*, 3865–3873.

Ito A, Yaegaki H, Hayama H, Kusaba S, Yamaguchi I and Yoshioka H. 1999. Bending shoots stimulates flowering and influences hormone levels in lateral buds of Japanese pear. *Horticultural Science,* **34**: 1224–1228.

Jain M. C., and Dashora L. K. 2007. Growth, flowering, fruiting and yield of guava (Psidium guajava L.) cv. Sardar as influenced by various plant growth regulators. *International Journal of Agricultural Sciences,* **3**, (1): 4–7 ref. 12.

Jain S, Rathod M, Banjare R, Nidhi N, Sood A and Sharma, R. 2023. Physiological aspects of flowering, fruit setting, fruit development and fruit drop, regulation and their manipulation: a review. *International Journal of Environment and Climate Change*, **13**(12): 205–224.

Kulkarni VJ. 1988. Chemical control of tree vigour and the promotion of flowering and fruiting in mango (*Mangifera indica* L.) using paclobutrazol. *Journal of Horticulture Sciences,* **63**: 557–566.

Kurian RM and Iyer CPA. 1993. Chemical regulation of tree size in mango (*Mangifera indica* L.) cv. Alphonso. I. Effects of growth retardants on vegetative growth and tree vigour. *Journal of Horticulture Sciences,* **68**: 349–354.

Kwack BH. 2007. The value of human life with horticultural practices and products. *Acta Horticulturae,* **762:** 17–21.

Lal N, Das RP and Verma LR. 2013. Effect of plant growth regulators on flowering and fruit growth of guava (*Psidium guajava* L.) cv. Allahabad Safeda. *The Asian Journal of Horticulture,* **8**(1): 54–56.

Lal S, Tiwari JP and Misra KK. 1996. Effect of plant spacing and pruning intensity on flowering and fruiting of guava. *Annals of Agricultural Research,* **17**(1): 83–89.

Lambers H, Chapin FS and Pons TL. 2008. *Plant Physiological Ecology*. 2nd ed. Springer, Berlin.

Leather GR and Abeles FB. 1972. Increased ethylene production during c1inostat experiments may cause epinasty. *Plant Physiology,* **49**: 183–187.

Lentz W. 1998. Model application in horticulture: a review. *Scientia Horticulturae,* **74**: 151–174.

Marschner P. 2012. (ed.). *Marschner's Mineral Nutrition of Higher Plants*. 3rd ed. Elsevier, Oxford.

Mitra SK, Gurung MR and Pathak PK. 2008. Sustainable guava production in West Bengal, India. *Acta Horticulturae,* **773**: 179–182.

Nunez Elisea R, Caldeira ML and Davenport TL. 1990. Thiadizuron effects on growth initiation and expression in mango. *HortScience,* **25**: 1167.

Nunez-Elisea R. 1985. Flowering and fruit set of a monoembryonic and polyembryonic mango as influenced by potassium nitrate sprays and shoot decapitation. *Proceedings of the Florida State Horticultural Society,* **98**: 179–183.

Pandey RM. 1989. Physiology of flowering in mango. *Acta Horticultarae,* **231**:361–380.

Pantastico EB and Manuel FC. 1978. The Philippines Recommends for Mango 1979. Report for Philippines Council Agriculture. PCARR. 43p.

Pareek OP. 1983. *The Ber.* ICAR, New Delhi.

Rabelo JDS, Couto FA, Siqueira DL and Neves GCL. 1999. Flowering and fruit set in Haden mango trees in response to ringing and ethephon and potassium nitrate sprays. *Rivista Brasileria de Fruticultura,* **21**(2): 135–139.

Rao MM and Srihari D. 1996. Approaches for managing the problem of biennial bearing in Alphonso mango trees. *Journal of Maharashtra Agricultural Universities,* **23**: 19–21.

Reddy YTN and Kurian RM. 2014. Effect of dose and time of paclobutrazol application on the flowering, fruit yield and quality of mango cv. Alphonso. *Journal of Horticultural Sciences,* **9**(1): 32–36.

Saidha T, Rao VNM and Santhanakrishnan P. 1983. Internal leaf ethylene levels in relation to flowering in mango. *Indian Journal of Horticulture,* **40**: 139–145.

Schaffer B, Whiley AW and Crane JH. 1994. Mango. In Schaffer B and Anderson PC (ed.). *Handbook of Environmental Physics of Fruit Crops*. CRC Press, Boca Raton, FL. pp. 165–167.

Shu ZH. 1999. Effect of temperature on the flowering biology and fertilization of mangoes (*Mangifera indica* L.). *Journal of Applied Horticulture,* **1**(2): 79–83.

Singh AK and Rajput CBS. 1990. Effect of GA, BA and calcium on vegetative growth and flowering in mango (*Mangifera indica* L.). *Research Development Report,* **7**: 1–11.

Singh RN, Majumdar PK and Sharma DK. 1965. Studies on the bearing behaviour of some south Indian varieties of mango (*Mangifera indica* L.) under north Indian conditions. *Tropical Agriculture,* **42**: 171–174.

Singh RN, Majumder PK, Sharma DK and Sjinha GC. 1974. Effect of deblossoming on the productivity of mango. *Scientia Horticulturae,* **2**: 399–312.

Storey WB. 1986. Carica papaya. *CRC Handbook of Flowering*, Vo1, 2. CRC Press Inc. Boca Raton, FL.

Sukhvibul N, Whiley AW, Smith MK, Hetherington SE and Vithanage V. 1999. Effect of temperature on inflorescence development and sex expression of mono- and poly-embryonic mango (*Mangifera indica* L.) cultivars. *Journal of Horticultural Science and Biotechnology,* **74**: 64–68.

Tiaz L and Zieger E. 2006. *Plant Physiology*. 4th ed. Sinauer Associates, Sunderland, MA.

Tromp J. 1973. Flower-bud formation in apple as affected by air and root temperature, air humidity, light intensity and day length. *Acta Horticultarae,* **149**: 39–47.

Ubi W, Ubi M, Mike U and Osodeke VE. 2007. The influence of three hormones on the induction of flowering and yield of two pineapple cultivars (*Ananas comosus*). *Global Journal of Pure & Applied Science,* **13**(2): 151–156.

Urban L and Alphonsout L. 2006. Girdling decreases photosynthetic electron fluxes and induces sustained photoprotection in mango leaves. *Tree Physiology,* **27**: 345–352.

Wien HC. 1997. *The Physiology of Vegetable Crops*. CAB International, Wallingford.

3 Phloem Transport
Source and Sink

Ankita Sharma, Rehan, Shivender Thakur, and Sunny Sharma

3.1 INTRODUCTION

During the gradual evolution of plants adapting to terrestrial environments, their anatomical structures underwent differentiation into decentralized organs. At the macroscopic level of the entire plant, inter-organ transportation facilitates connectivity among these anatomical structures. Sugars, in conjugation with water, are one of the most essential elements that comprise this movement. The primary sugar-conducting tissue in plants is the phloem tissue. The widely utilized system of phloem transport was first proposed by Ernest Münch in 1930, over eight decades ago. In accordance with mass flow theory, the movement of mass in the phloem is propelled by a gradient of pressure that is created through osmosis (Figure 3.1). The transportation of compounds within the sieve tube is aided by a mass flow mechanism that is driven by a pressure or turgor gradient. The existence of sieve pores facilitates the connection between protoplasts in the sieve tubes. The pressure gradient observed in the sieve tubes is attributed to the build-up of osmotic substances, including sugars, at the source regions, and their subsequent unloading at the sink regions. The primary sources of nutrients in plants are typically derived from their leaves, while the energy-demanding or storing tissues serve as the primary sinks (De Schepper et al., 2013). Yield is a consequence of the cumulative influence of nutrient and photon sources and sinks strength throughout seed development. The yield potential refers to the highest possible yield that can be achieved by a type of plant that is well-suited to the environment and is grown under optimal conditions with effective management of both abiotic and biotic stress factors. The above methods facilitate the evaluation of variations in production, particularly in terms of yield sensitivity to controllable stresses, and allow for targeted adjustments. The effort of removing the return gap is subject to certain constraints. The yield gap is an indicator that exhibits a non-linear response to increments in yield potential (Lobell et al., 2009).

The magnitude of photoassimilate materials is influenced by both the rate of photoassimilate recapture from source tissues and the net photosynthetic rate. The leaves of higher plants are essential for resource acquisition as they facilitate the reduction of oxidized forms of nitrogen and carbon into sugars and amino acid materials through the process of photosynthesis. This process results in the application of free energy gained. The carbon acquired via photosynthesis is predominantly conveyed via the vascular system of the plant and apportioned among diverse heterotrophic tissues, prior to being dispensed to the import-reliant organs. Sucrose is considered the chief byproduct of photosynthetic carbon metabolism and is primarily responsible for delivering carbon to heterotrophic tissues in most plants. This makes sucrose the primary metabolite in the resource allocation strategy of these plants. The systemic distribution of photosynthates, known as assimilate partitioning, is a crucial process associated with plant growth and productivity.

The loading and unloading mechanisms can be categorized as either passive or active and can take place through apoplastic or symplastic routes. In the apoplastic system, sugars cross the cell membrane and enter the apoplast at least once. In contrast, in the symplastic pathway, sugars go only through the cytoplasm of interconnected cells via plasmodesmata. The apoplast is comprised of a continuous network of cell membranes and xylem vessels. According to Münch's original proposal, the procedures of loading and unloading only take place in the sources and sinks, respectively.

DOI: 10.1201/9781003354055-3

The process of loading and unloading takes place along the transport phloem channel. The above-mentioned behavior is widely recognized as the leakage-retrieval system. Sieve tubes are considered permeable, contrary to impermeable conduits. Thompson (2006) has recently suggested that pressure differences throughout the phloem route must be minimal or inconsequential to ensure the effective exchange of solutes between SECCCs and the surrounding tissues. The proposal put forth by the individual seems to be physiologically plausible, as per a thorough model analysis. This is because, in a decentralized plant body, the pressure gradients between individual cells cannot be sensed or adjusted, and only the turgor of each individual cell can be regulated. According to Thompson (2006), the regulation of phloem turgor is controlled by the transport phloem rather than the turgor gradient.

3.2 MÜNCH THEORY AND PHLOEM FLOW IN TALL TREES

The inherent difficulty in investigating long-distance transport in plants stems from the intracellular nature of the fluxes involved, as well as the protective mechanisms afforded by physical barriers and biological activities, such as forisomes and p-proteins. Additionally, the significant tensions or pressures under which these processes occur further compound the challenge of studying this phenomenon. Theoretically, the acquisition of these qualities necessitates in vivo techniques, which are accompanied by methodological challenges. Biomimetic methodologies have provided solutions to enduring inquiries concerning water transportation in the xylem and have established principles of optimization that govern the arrangement of veins in leaves, both of which are integral components of the water loss stream (Noblin et al., 2008). According to Münch's theory, the movement of phloem sap (FPc) is influenced by the pressure differential (P) between the source and sinks. Furthermore, the control of phloem route resistance plays a crucial role in regulating the movement of phloem sap (RPc):

$$\text{FPc} = \frac{P_{\text{source}} - P_{\text{sink}}}{\text{RPc}}$$

Various research investigations have demonstrated that the motion of phloem, which is directly proportional to FPc, exhibits a consistent pattern across a diverse range of angiosperm species, with an approximate rate of 1 cm/min (De Schepper et al., 2013). The phloem transport system is regulated and modulated to sustain a uniform and relatively moderate rate of flow. The velocity of phloem in angiosperm trees exhibits an apparent augmentation with the increase in tree height. Thompson's (2006) empirical study highlighted the need to minimize or eliminate the pressure differential inside the phloem transmission ($P_{\text{source}} - P_{\text{sink}}$). According to Thompson and Holbrook (2003a), there is a negative correlation between the time it takes for sucrose to pass down a sieve tube and the square of the dimension of the sieve tube. The authors' model suggests that the phloem resistance (RPc) is mostly governed by the number of sieve plates considered, and this is directly correlated with the total amount of sieve elements and, consequently, the size of the sieve tube. These results lead to a dilemma about the existence of big plants, such as trees. Münch's (1930) hypothesis suggests that sources and sinks are interconnected by the symplastic pathway, leading to a direct correlation between the sieve tube and the axial length of the plant. As the size of a tree increases, the resistance of the phloem also tends to increase dramatically (Figure 3.1).

Münch's theory indicates the pressure in the phloem of trees should be higher than that of herbaceous species. This is because the necessary pressure difference increases as the porosity of the sieve tubes and the length of the plant increase. On the other hand, according to literature statistics, trees exhibit the least amount of phloem pressure as reported by Pritchard in 2007.

The terminologies "source" and "sink" pertain to the conveyance of organic compounds within the phloem of plants.

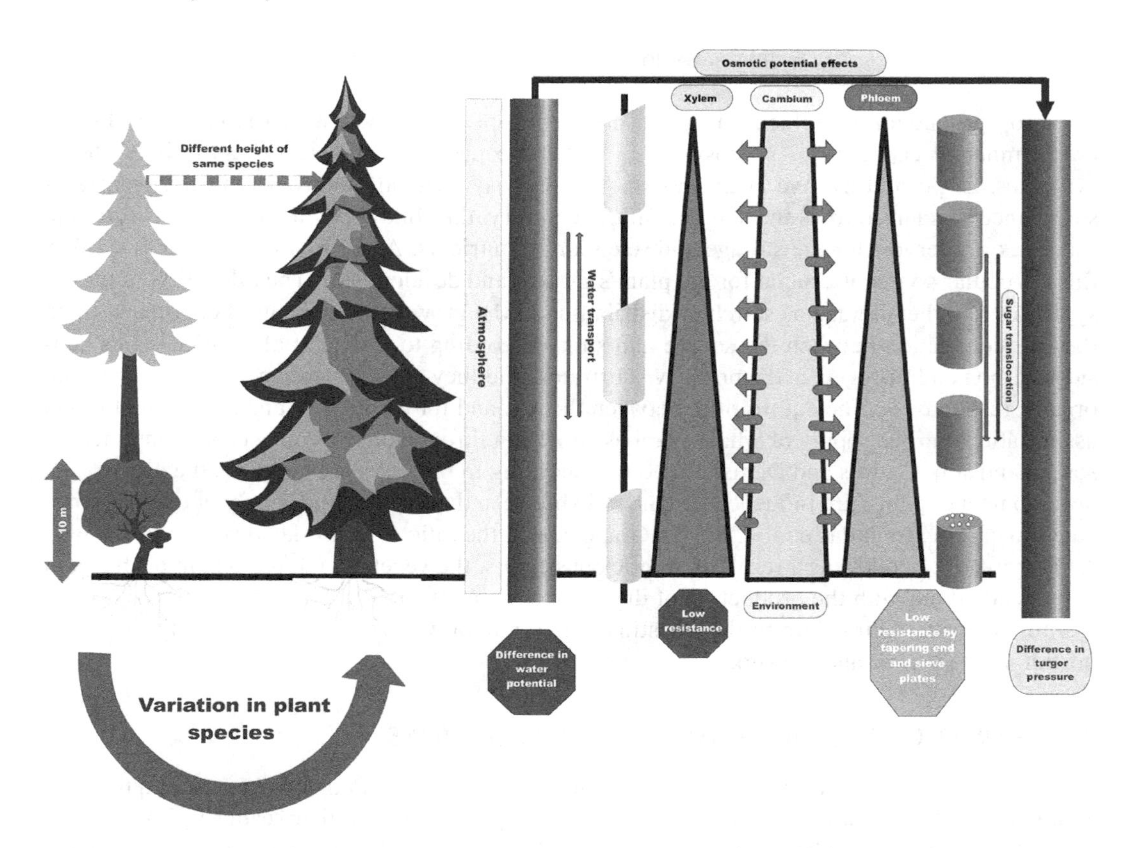

FIGURE 3.1 Phloem sugar transport in tall trees by Münch turgor pressure theory.

A **sink** is any explicit organ that is non-photosynthetic or does not produce adequate photosynthate to meet its requirements. For instance, a root, seed, fruit, or developing root tuber or tuber. A place where organic materials are synthesized is referred to as a source of organic material. Consider a leaf, root, tuber, or tuber at any stage of development. The leaf is considered the principal site for the process of photosynthesis, wherein green plants absorb light energy, primarily through chlorophyll present in their leaves, to synthesize decreased carbon-based substances from CO_2 and water. The process of photosynthesis is responsible for the synthesis of carbohydrates that are utilized for both energy generation and growth. The photosynthates, which are the products of photosynthesis, constitute a significant proportion of a plant's dry matter, with estimates suggesting that they can account for as much as 90%. The processes associated with development and cropping are reliant on a consistent provision of carbohydrates and essential nutrients.

The term "sink strength" refers to the capacity of a tissue or organ to effectively transport photoassimilates. On the other hand, "sink capacity" relates to the ability of a tissue or organ to import or store additional compounds from the source(s). Lastly, "sink activity" is used to describe the rate of respiration of a tissue or organ. Plant growth and development can be limited by photosynthetic resources, also known as being "source limited". The process of leaf photosynthesis is often suppressed by the presence of sinks, such as fruit restrictions or delayed leaf senescence. In the absence of fruit removal, multiple sinks may compete for photoassimilates, leading to the roots becoming the primary sink.

Flowers, petals, fruits, shoots, and roots are all referred to as "sinks". The transition of any of these organs from "sink" to "source" during ontogeny can occur periodically and irreversibly. The hierarchy of sinks in plants is delineated below.

Fruits>>>>>Flower>>>>>>Root >>>>>>Leaf

The crop production is influenced by the production of dry matter, which is regulated by the crop's inherent capacity for photosynthesis and the capacity of spikelets to assimilate photosynthates, as posited by Aye et al. (2020). The photosynthetically active organ that generates sustenance is referred to as the source. On the contrary, the sink is an anatomical structure that serves as a reservoir for the storage and retention of nutrients. According to Garai et al. (2019), this particular source is crucial for the plant's growth and development. The allocation of photosynthates to either storage as starch or distribution to the growth components is contingent upon the interconnection between the source and sink. According to Chang et al. (2017), plants consider carbon and nitrogen as the primary substrates, and they categorize them as source and sink organs, respectively. The equilibrium between the demand for the growth and retention of organ assimilation and the supply of whole-plant assimilation through photosynthesis is maintained by source and sink (Sadras and Denison, 2009). According to Marcelis et al. (2004), the growth and development of plants are influenced by the ratio between the demand and supply of organ assimilates. According to Pallas et al. (2011) in Grapevine, if the ratio of assimilated supply-to-demand is determined to be low as a result of abiotic constraints, the vegetative development of the plant is reduced, along with the production of dry matter. Comprehending the flow and orientation of assimilates from their origin to their destination is imperative in achieving an equilibrium proportion in the source and the sink.

3.3 PARTITIONING OF ASSIMILATES AMONG SINKS

Portioning refers to the distribution of assimilates amongst sinks. Photosynthesis is required for assimilating supply (Marschner, 2012). The volume and pattern of assimilate scattering impact plant development and yield. The determination of translocation is contingent upon the developmental stage of the plant. The determination of transportation direction, as well as magnitude, is dependent upon the position of the sink and the relative intensity of attraction. The majority of solute molecules in sap, approximately 90%, consist of carbohydrates and exhibit a velocity range of 50–100 cm/h. Sucrose is the predominant form of translocation observed in plants. Irrespective of the taxonomic classification or mode of phloem loading, a minor proportion of the saccharides synthesized within a plant are conveyed over extensive distances through the phloem. In addition to sucrose, polyols such as sorbitol and mannitol, as well as raffinose oligosaccharides, can be discerned. Polyols and raffinose are present in the phloem of multiple species. The process of partitioning within a tree is determined by the interplay between different organs that compete for restricted carbohydrates, rather than being predetermined by genetics. The level of competition among sinks is contingent upon the metabolic activity of the organ and its proximity to the source of glucose (Figure 3.2).

The process of assimilate partitioning among competing sinks is primarily influenced by three factors, namely the nature of vascular connections linking the source and sink, the distance that exists between the source and sink, and the strength of the sink. The process of translocation is facilitated by the presence of vascular connections that directly link the source and sink leaves. The vascular tissue of a stem is connected to the primary vascular system of the stem through a vascular trace that extends into the petiole of each leaf. Research studies have demonstrated that photoassimilates exhibit a preference for translocation toward sink leaves that are situated above and aligned with the source leaf, i.e., within the same sink. The sink leaves exhibit the highest degree of proximity to the leaf. The process of assimilation necessitates traversing complex radial pathways amidst the sieve elements, as sink foliage situated on the opposite end of the stem is not equidistantly positioned and therefore, exhibits comparatively less direct connectivity.

Environmental factors, including temperature and hormonal influences, play a crucial role in regulating the growth and differentiation of sink tissue. The majority of photoassimilates originating

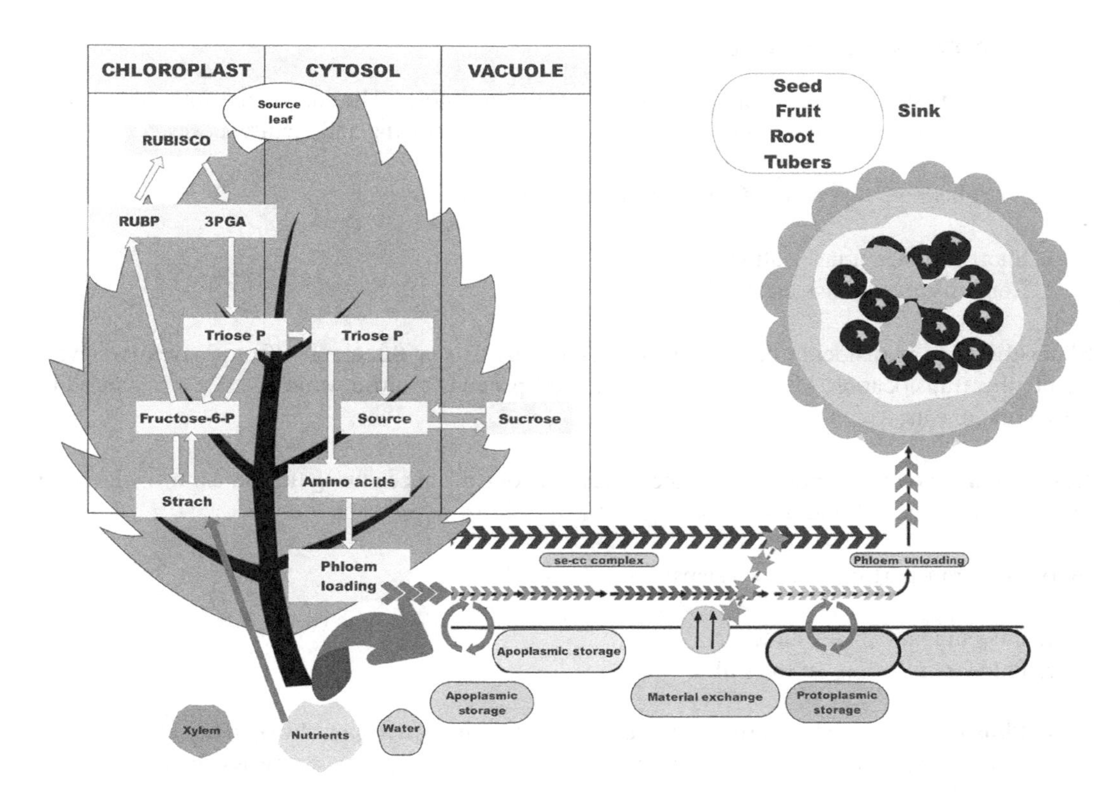

FIGURE 3.2 Partitioning of source and sink in plants.

from source leaves exhibit facile translocation in both upward and downward directions, toward the apex and roots, respectively. In vegetative plants, there is a preference for photoassimilate from juvenile source leaves located at the top of the plant to be translocated toward the stem apex. On the other hand, older leaves that are not senescent and are located near the base of the plant are preferentially supplied to the roots. The magnitude of the hydrostatic pressure gradient within the sieve elements is presumably correlated with the orientation of translocation. Both cell turgor and hormones have been utilized as means to manipulate sink strength.

Hormones produced by plants have been employed for the regulation of long-range transportation, particularly in the context of redirecting assimilates toward novel sinks. The phenomenon of hormone-directed transport may potentially be an indirect consequence of hormonal activity. The challenged significance of hormone-directed transport over long distances notwithstanding, a growing body of evidence suggests that hormones play a more direct role in solute transfer over short distances. A correlation has been observed between the concentration of abscisic acid and the rate of maturation in developing fruits. The application of ABA has been observed to induce the translocation of sugar to the root system of undamaged bean plants. The impact of ABA on the translocation of 14 C-photoassimilates into filled wheat ears has been the topic of conflicting studies. In contrast, auxin suppresses the uptake of sucrose in sugar beet roots while promoting loading in bean leaves. The collective results, in conjunction with other relevant research, suggest that hormones may possess the capacity to impact the mechanisms of loading and unloading. Translocation is a physiological mechanism by which photo-assimilated food and other organic substances are transported from the leaves of a plant to its stem and roots via phloem sieve elements.

3.4 SOURCE-SINK MOVEMENT

The translocation is also considered a source (leaf) to sink (store). The source and sink in interconvertible in developing and germinating seeds and also in developing and mature leaves.

Pathway: There are two types of pathways in translocation

1. Short-distance translocation
2. Long-distance translocation

Short-distance translocation: The solute only travels two or three cells (from leaves to sieves) during this translocation. Mesophyll cells (or leaves) produce sucrose, which is then transported to sieve components.

Long-distance translocation: The solute traverses a long distance during this transit from the sieve to the storage organs (roots and stems).

Steps: The transport involves two steps:

1. Phloem loading (at source)
2. Phloem unloading (at the sink)

1. **Phloem loading**: The sucrose in leaves in activity transported sieve by (i) symplastic pathways (by plasmodesmata that pore in the cell wall) (ii) apoplastic pathways (by cell wall) depending on species.
   - It is an active process that occurs through Sucrose-H symporters.
   - Provides driving force for translocation.
   - The sieve element-companion cell (se-cc) complex is responsible for generating osmotic pressure.
   - Pressure difference results in the bulk flow.
   - Energy used for loading provides a motivating force for translocation.
   - Sucrose is transported from the source in sieve tubes to the sink in a direction that is determined by the gradient of its concentration and water potential. Specifically, it moves from areas of high concentration and water potential to areas of low concentration and water potential.
2. **Unloading**: Photosynthates or sucrose are unloaded from phloem storage. The productivity of the crop is increased by accumulation (storage) at the sink (like sink tissue in cereal grain).
   - The process is facilitated by carrier molecules and is characterized by its dynamic nature.
   - Sucrose unloading at the sink leads to a reduction in osmotic pressure, causing water to move into the adjacent xylem and resulting in a decrease in both turgor pressure and water potential of the phloem.
   - In sink cells, the unloaded sucrose undergoes either starch conversion or consumption to sustain low osmotic pressure and uninterrupted unloading.

3.5 SOURCE-SINK MECHANICS

The ratio between them may allow for growth, or the source and sink may both be able to expand independently. Crop plants can be seen in three different scenarios:

1. The source and sink might be physically closer to one another

2. The sink might be bigger than the source
3. Both could be in dynamic equilibrium.

3.5.1 Condition 1: If the Source Exceeds the Sink

Traditional species of different crops contain more sinks than sources because of their leafy structure. Crop quality improvement decreases leafiness and improves sink size. More photosynthates are produced when the source is higher (more foliage produced by pollination and cultural practices), which raises the photosynthetic rate (Pr). However, because the sink is small, much photosynthesis is unable to leave the leaf, and translocation may be as low as 10%. Some photosynthates that leave the leaf return in the stem or other leaves. Low productivity is the outcome of these circumstances because the leaves become twisted and discolored, and their Pr decreases. This characteristic is present in cereals, pulses, as well as numerous annuals, trees, and other plants. The soil-water system, the climate, and the availability of nutrients can all affect the source-sink balance, even within a single crop or cultivar. If N levels are low, cereals like wheat and rice grow lots of leaves but weak panicles and useless tillers. Usually, nutrient deficits (P, Zn, Mn) and toxins (Fe, Al) lead to enough leaves but low grain production. In each of these situations, the sink is not the source.

3.5.2 Condition 2: If the Sink Exceeds the Source

In fruit trees such as mango, guava, and citrus, as well as in vegetable and cotton, the sink size is greater than the source size. On the other hand, the sinks rapidly decrease as a result of a lack of assimilates. Crop yield can be increased by 50%–100% if the mangoes, guava, and orange fruit set can be enhanced by 1%–2%. The volume of the sink in some grain crops and cereals has increased as a result of breeding. Grain production is poor even though the sink size is adequate. Increasing LAI in rice doesn't boost grain yield; instead, it tends to plateau. The distance between spikelet and grain widens as LAI rises. In other cases, though, functional efficiencies are much lower and the actual leaf area is greater than what is required. The aforementioned crops can perform more effectively with the right care.

3.6 ECO-PHYSIOLOGICAL FUNCTIONS AND PHLOEM LOADING STRATEGIES

Phloem loading is the first step in long-distance sugar transport because the hydrostatic pressure that exists in the sieve tubes increases due to the osmotic accumulation of sugars. Sugar can go from cells of mesophyll to sieve components via an apoplastic or symplastic mechanism. The symplastic mode includes bridging the plasma membrane as opposed to the apoplastic mode. The distinction between active and passive loading can be made in addition to the symplastic-apoplastic difference.

Active loading involves employing metabolic energy to pump assimilates through the phloem despite a gradient of concentration. This idea suggests that there are three alternative loading strategies for the phloem that the minor veins collect:

1. active apoplastic loading,
2. active symplastic polymer trapping, and
3. passive symplastic diffusion.

In many species, loading involves an apoplastic step that is propelled by the proton-motive force and controlled by plasma-membrane transporters (Sauer,2007). According to Turgeon and Ayre (2005), loading in other plants is symplastic and is caused by a gradient of concentration from the mesophyll to the phloem, which necessitates high densities of plasmodesmata. Some synthetic loading plants employ a passive process that is solely driven by diffusion. In some species, intermediate cells, which specialize as companion cells (CCs) in the smaller veins, load through the symplast and turn sucrose from the mesophyll into raffinose and stachyose. These raffinose family oligosaccharides (RFOs) are significantly bigger than sucrose and don't seem to be able to diffuse back to the mesophyll through the intermediate cell plasmodesmata. The RFOs accumulate in the phloem through a process known as polymer trapping to a level of the total content that is equivalent to the sugar in apoplastic loaders. There is a specific partner cell type associated with every loading mechanism in the SECC complex. Only a carrier that is specifically designed for sucrose can pump sucrose from the apoplast into the phloem. Sucrose is intended for export and reaches the apoplast during the active apoplastic loading phase. A proton-motive force, maintained by proton pumps and sucrose carriers (such as the sucrose-symporter SUT4) in the plasma membrane of the SECC complex, drives the apoplastic loading. This active mechanism allows for the creation of a steep osmotic gradient, which calls for an equal inflow of water, leading to a highly localized turgor in the sieve tubes. The companion cell, also known as a transfer cell, is an expert at apoplastic loading. It differs from other cells in two ways: (i) by having numerous cell wall invaginations to increase the plasma membrane surface; and (ii) by having few plasmodesmata that connect to the mesophyll cell. As a result, there is a considerable reduction in the symplastic link between the mesophyll and the SECCC.

3.7 PARTITIONING PRIORITIES AND SINK COMPETITION

- Herbaceous crop plants accumulate photosynthates in their source leaves during the photoperiod, which are subsequently released during the night, resulting in the depletion of the leaf's contents during the day.
- Competition for photosynthates is evident among different organs, such as fruit shoots, and also among separate components of the same type of organ, such as fruit-fruit.
- During their spring flush, citrus trees generate vegetative branches, leafy flowers, and pure, leafless inflorescences. The occurrence of shoot elongation and leaf expansion prior to anthesis and fruit set effectively circumvents direct competition. Moreover, it has been observed that leafy inflorescences exhibit elevated levels of fruit set and persistence, indicating a potential role of leaves in supporting the reproductive organs via the delivery of photosynthate, hormones, or other possible mechanisms.
- The photosynthetic capacity of leaves in close proximity to the developing fruit exhibits a notable increase in comparison to other regions of the tree.
- The outer layer of the calyx exhibits a significant ability to perform photosynthesis. The green calyx of the Cape gooseberry is a crucial factor in carbohydrate production and transmits during the initial 10–20 days of fruit development, as it fully envelops the fruit during this period.
- On the contrary, in instances where the yield is abundant, the nutrient surge during the summer season is inadequate or absent. This implies that the maturing fruit is primarily in a downward position. The delay in root growth observed during the shoot formation phase is attributed to rooted tip competition for photosynthetic resources, with a preference for the tip.

3.8 DISTRIBUTION

The allocation of exported metabolites from a specific leaf is contingent upon the spatial orientation of the leaf. In legumes, a significant proportion of the leaf surfaces are translocated to the root system, while a minor fraction is transported to the upper parts of the plant. The current predominantly travels toward the tip from the higher conductive sheets. Comparable distribution patterns have been documented for grapes (*Vitis vinifera*), cotton (*Gossypium hirsutum* L.), and tea (*Thea sinensis* L.).

In general, the pressure-flow concepts offer an explanation for various observable factors that are implicated in the process of phloem transport, including the distribution of resources. Although there is still a need for unequivocal evidence to support this notion, there is an increasing consensus that the phloem pathway possesses excess transport capability, resulting in a decreased level of scrutiny toward this area of matter. In addition, through the process of pedicel cutting and exudation facilitation, the pressure of the sieve elements located at the sink end of the phloem pathway was reduced to zero. The results of this study indicate that the phloem possesses an additional transport capacity in both monocotyledonous and dicotyledonous plants. Monocotyledonous plants experience considerable selection pressure for additional transport capacity due to the absence of vascular cambial activity that can replace impaired sieve components.

3.9 SOURCE-SINK MANIPULATIONS

Modifying the fruit load is a crucial practice that improves the quality of the fruit yielded in a particular year and guarantees the accumulation of reserves, both of which hold significant importance for the subsequent development of the tree. However, the phenomenon of alternate bearing in fruit production is a crucial issue that can result in considerable financial setbacks for fruit growers. The primary cause of alternate bearing is presumed to be a substantial fruit crop.

The phenomenon of alternate-bearing cultivars during their "on" year, characterized by a substantial crop load, can cause the exhaustion of carbon and mineral reserves, ultimately resulting in the potential collapse of the tree in extreme circumstances. According to Chacko et al. (1982), there is often an inadequacy in the production of photosynthate to fulfill the demands during the stages of fruit setting and fruit development following prolonged and intense blooming. The process of defoliation in trees has been observed to have a positive impact on the level of the photosynthesis process in the remained leaves. This is due to the provision of a larger sink, the magnitude of which is contingent upon the extent of defoliation.

3.10 SOME SPECIAL PRACTICES THAT DIVERT THE ASSIMILATES TO DEVELOPING ORGANS

1. **Root pruning**: Root pruning involves the selective removal of roots located at a distance of 40cm from the plant, e.g., guava.
2. **Ringing**: The act of removing the entire ring of the bark from a limb or trunk is commonly referred to as "ringing," e.g., mango and grapes.
3. **Dehorning**: Dehorning is a horticultural practice that involves the removal of branches to alleviate overcrowding and intermingling.
4. **Notching**: Notching refers to the process of partially ringing divides above a dormant lateral bud, e.g., Poona fig.
5. **Nicking**: Nicking refers to the process of partially ringing branches located beneath a dormant bud, e.g., Poona fig.
6. **Bending**: The phenomenon of bending, which involves the curving or flexing of branches or shoots, is observed in certain plant species, e.g., guava.
7. **Leaf pruning**: Leaf pruning involves the elimination of aged and deteriorated leaves, e.g., date palm.
8. **Skirting**: Skirting refers to the practice of eliminating low-hanging branches from trees, e.g., mango.
9. **Thinning**: Thinning involves the removal of unwanted branches or flowers from a tree, e.g., guava and grapes.
10. **Training and pruning**: Training refers to give framework to the plant during the initial years. Among tropical fruits mainly done in grapes and guava.

3.11 FACTORS CONTROLLING THE RATE OF TRANSLOCATION

1. **Temperature**: The temperature exerts control over the translocation of organic solutes from leaves to their respective destinations. The pace at which materials travel from the soil to the roots increases as soil temperature rises, whereas a decline in soil temperature causes the rate at which solutes pass through the leaves to decrease. Consequently, the

speed of translocation exhibits a direct correlation with the metabolic rate at both the receiving and supplied areas.

In recent times, there has been a significant amount of elucidation regarding the impact of decreased temperature on translocation, to some extent. Both experiments demonstrated that translocation may occur between –1.5° and 2°. The later writers significantly improved our grasp of the conditions governing movement under diverse temperature regimes. They discovered that the pause in translocation caused by low temperatures was merely a transitory situation with a known half-life (or length) and that recovery from it happened spontaneously (without a temperature rise). This indicates that judging a pressure-flow mechanism negatively based only on a temperature response is generally not justified.

2. **Light**: Light has a significant impact on a plant's existence in numerous ways, including growth, blooming, photosynthesis, germination, etc. In addition to these outcomes, it influences the movement of dietary components. Only a limited amount of food items is typically translocated out of the leaves during the day. Out of this modest amount, two-thirds travel to the stem tip and the remaining portion to the root. Contrarily, the majority of the food is moved from the leaves while it is dark or at night. During the nocturnal period, a greater proportion of nutrients is allocated to the roots as opposed to the stem. The action spectrum analysis has revealed that the most effective wavelength of light for a given biological process is in the red spectrum.
3. **Metabolic state of tissue**: Actively dividing tissue requires more energy for its functions, which necessitates additional nourishment sources, which in turn necessitates a greater movement of solutes into these areas. Comparably a winter bud whose growing activity has ceased requires no more energy and hence requires relatively little food. Consequently, the metabolic condition of the cells generates demand, which must be satisfied.

Another critical element of metabolism is the availability of energy-rich molecules such as ATP for translocation. The use of respiratory inhibitors such as KCN, DNP, and others, which impede ATP generation, has a significant influence on the rate of translocation, indicating that translocation is an energy-dependent activity. Anatomical data indicates that partner cells on the side of sieve tube cells are endowed with the cellular machinery required for transport.

3.12 CONCLUSION

Given the current context of climate change and escalating food demand, it is imperative to thoroughly investigate all aspects of the source-sink relationship in order to enhance food production. In order to achieve this objective, it is imperative to develop meticulously structured and mechanistic models of source-sink interaction that can accurately predict crop yield under varying circumstances. This will enable us to optimize resource allocation and enhance crop yield. The Münch concept of phloem transport is being enhanced through the incorporation of active and passive phloem loading and unloading mechanisms in ways to gather and release phloem. Additionally, the transport phloem is being augmented with the introduction of the leakage-retrieval process. The phloem-related data determined on plants and perennial plants exhibited noteworthy distinctions between the two plant types. Despite Münch's initial hypothesis predicting otherwise, there were significant variations observed in loading strategies and determined phloem pressure across different trees and plants. The findings necessitate a reassessment of the transmit system hypothesis, particularly with regard to arboreal structures. The conventional focus on maximizing yield and developing superior genotypes for optimal growth conditions may not be sufficient to address the nutritional requirements of an expanding population in an increasingly unpredictable environment. In order to equip farmers across various stages of development with appropriate resources, it is imperative that breeding techniques incorporate considerations for both resilience and nutritional value. At the level of the plant, it is imperative to focus on factors that enhance plant yields and productivity across diverse environmental conditions and resource availability. In this paper, we present recommendations for enhancing the theoretical framework underlying yield generation.

REFERENCES

Aye, M., Thu, C.N., Htwe, N.M. 2020. Source-sink relationship and their influence on yield performance of YAU promising rice lines. *Journal of Experimental Agriculture International* 42(9):124–135.

Chacko, E.K., Reddy, Y.T.N., Ananthanarayanan, T.V. 1982. Studies on the relationship between leaf number and area and development in mango (*Mangifera indica* L.). *Journal of Horticultural Sciences* 57(4):483–492.

Chang, T.G., Zhu, X.G., Raines, C. 2017. Source-sink interaction: a century old concept under the light of modern molecular systems biology. *Journal of Experimental Botany* 68:4417–4431.

De Schepper, V., De Swaef, T., Bauweraerts, I., Steppe, K. 2013. Phloem transport: a review of mechanisms and controls. *Journal of Experimental Botany* 64(16):4839–4850.

Garai, S., Brahmachari, K., Sarkar, S., Kundu, R., Pal, M, Pramanick, B. 2019. Crop growth and productivity of rainy maize – garden pea copping sequence as influenced by Kappaphycus and Gracilaria saps at alluvial soil of West Bengal, India. *Current Journal of Applied Science and Technology* 36(2):1–11.

Lobell, D.B., Cassman, K.G., Field, C.B. 2009. Crop yield gaps: their importance, magnitudes, and causes. *Annual Review of Environment and Resources* 34:179–204.

Marcelis, L.F.M., Heuvelink, E., Baan Hofman-Eijer, L., Den Bakker, J., Xue, L.B. 2004. Flower and fruit abortion in sweet pepper in relation to source and sink strength. *Journal of Experimental Botany* 55:2261–2268.

Marschner, P. 2012. *Marschner's Mineral Nutrition of Higher Plants*. 3rd ed. Elsevier, Oxford, UK.

Münch, E. 1930. *Die Stoffbewegunen in der Pflanze*. Verlag von Gustav Fischer, Jena.

Noblin, X., Mahadevan, L., Coomaraswamy, I.A., Weitz, D.A., Holbrook, N.M., Zwieniecki, M.A. 2008. Optimal vein density in artificial and real leaves. *Proceedings of the National Academy of Sciences* 105:9140–9144.

Pallas, B., Loi, C., Christophe, A., Cournède, P.H., Lecoeur, J. 2011. Comparison of three approaches to model grapevine organogenesis in conditions of fluctuating temperature, solar radiation and soil water content. *Annals of Botany* 107:729–745.

Sadras, V.O., Denison, R.F. 2009. Do plant parts compete for resources? An evolutionary viewpoint. *New Phytologist* 183(3):565–574.

Sauer, N. 2007. Molecular physiology of higher plant sucrose transporters. *FEBS Letters* 581:2309–2317.

Thompson, M.V. 2006. Phloem: the long and the short of it. *Trends in Plant Science* 11:26–32.

Thompson, M. V., & Holbrook, N. M. 2003a. Application of a single-solute non-steady-state phloem model to the study of long-distance assimilate transport. *Journal of Theoretical Biology* 220(4):419–455.

Turgeon, R., Ayre, B.G. 2005. Pathways and mechanisms of phloem loading. *Vascular Transport in Plants*, eds. N.M. Holbrook, M.A. Zwieniecki. Elsevier/Academic, Oxford, pp. 45–67.

4 Crop Load, Assimilate Partitioning, Translocation and Distribution

Suman Bodh and Praveen Verma

4.1 INTRODUCTION

Crop load signifies the volume of fruits that comes out per unit of a tree and is commonly used as an indicator of orchard productivity. The term "yield efficiency" is commonly used when expressing crop load in relation to metrics such as fruit yield per total leaf area, trunk cross-sectional area (TCA), or tree light capture (Janick, 2010). Crop load is a measurable factor that is commonly characterized by the number of fruits present on a given tree. The assessment of fruit yield is often referred to as the number of fruits to trunk cross-sectional area ratio, which is abbreviated as fruit/TCSA (Whiting et al., 2005; Rana et al., 2023). Nevertheless, it is imperative to bear in mind certain considerations while employing efficiency terminologies. Initially, it may be advisable to consider an alternative term such as crop mass or density, given that efficiency is inherently dimensionless. It has been commonly noted that in the case of sizable trees on sturdy rootstocks, a high quantity of fruit produced per tree does not necessarily correspond to optimal cropping efficiency, as measured by the number of fruits per square centimeter of tree canopy area. It is expected that in high-density orchards, the maximum canopy volume and tree light interception will be achieved a few years after planting. However, the total cross-sectional area (TCA) of the tree will continue to increase over its entire lifespan. The efficiency of yield increases until the tree canopies occupy the designated area entirely; however, it subsequently decreases when measured per unit of TCA. The considerable variation in trunk size with respect to tree age and the potential inadequacy of TCA as a denominator in this context prompt inquiries into the efficacy of utilizing such ratios, as posited by Janick (2010) and Pawar and Rana (2019).

4.2 CROP LOAD MANAGEMENT

The implementation of various techniques pertaining to orchard management has been observed to exert an influence on both crop yield and quality (Rana et al., 2023). It is recognized that a significant portion of these practices are essential for modern fruit cultivation. The scope of our discourse is confined to the variables that affect the yield of arboraceous crops in orchard parcels that are in good condition and adequately managed. Pollination and fertilization are the key determinants of both fruit set and crop load. Studies have shown that inadequate irrigation levels can significantly affect the soil water availability, subsequently influencing the water status of plants and resulting in deviations in fruit size and yield (Erf and Proctor, 1987; Rana et al., 2023). Inadequate nutrient availability may lead to similar outcomes. According to Mika's research in 1986, the act of pruning has been found to have a negative impact on tree growth, resulting in a reduction of both cropping and potential fruit-bearing sites. The degree of this decrease is contingent upon the magnitude of the trimming. This discourse analyzes the effects of rootstock and flower thinning on crop load (Tables 4.1 and 4.2).

 DOI:10.1201/9781003354055-4

TABLE 4.1
Different between Active and Passive Transport

Active Transport	Passive Transport
Movement *against* electrochemical gradient	Movement *down* the electrochemical gradient
From a more negative electrochemical potential to a more positive electrochemical potential	From a more positive electrochemical potential to a more negative electrochemical potential

TABLE 4.2
Relationship between Number of Leaves and Fruiting in Some Fruit Crops

Crop	Species	No. of Leaves Per Fruit
Pineapple	*Ananas comosus*	6–8
Grapevine	*Vitis vinifera*	7–8
Satsuma mandarin	*Citrus unshiu*	25
Pineapple guava	*Feijoa sellowiana*	7

4.2.1 Ways of Crop Load Management

For tree fruits, orchardists use several kinds of produce load control techniques. To get the required number of fruits per tree, growers primarily employ pruning, chemical thinning, and hand thinning.

Pruning: The primary intent of pruning is to decrease the dimensions of the trees and promote the even distribution of light throughout the tree canopy. However, it also has the capacity to regulate the number of floral buds (Costa et al., 2019) and serves as the initial step in any decreasing strategy (Jones et al., 1998). Pruning to reduce the number of floral buds has multiple advantages. Firstly, it can be carried out as part of the regular winter pruning process. Secondly, it results in a reduction of competition for assimilates among flowers and fruitlets, thereby enhancing the distribution of assimilates. Lastly, it is a sustainable approach to reducing crop load. According to Musacchi (2021), pruning aims to sustain an equilibrium between vegetative and reproductive growth.

Flower and fruit thinning: The development of fruit trees demonstrating an excessive amount of crop yield is commonly recognized and requires the application of efficient techniques for managing fruitfulness. The utilization of said techniques presents numerous benefits, such as the mitigation of crop output, resulting in an augmentation of the mean mass of products and an enhancement of commercial value (Palmer et al., 1997). Furthermore, the utilization of these techniques has the potential to surmount the hindrance of flower bud induction, leading to enhanced return bloom and uniform yearly yields, as stated by Tromp (2000). A decrease in the number of fruits per tree results in an increase in the leaf area per fruit, which in turn leads to a greater supply of assimilates to the remaining fruits. According to existing literature, fruit thinning can lead to an increase in fruit size, albeit with a relatively smaller impact compared to the reduction in fruit yield (Wünsche and Lakso, 2000).

Hand thinning: The development of fruit trees demonstrating an excessive amount of crop yield is commonly recognized and requires the application of efficient techniques for managing fruitfulness. The utilization of said techniques presents numerous benefits, such as the mitigation of crop output, resulting in an augmentation of the mean mass of products and an enhancement of commercial value (Palmer et al., 1997). Furthermore, the utilization of these techniques has the potential to surmount the hindrance of flower bud induction, leading to enhanced return bloom and uniform yearly yields, as stated by Tromp (2000). A decrease in the number of fruits per tree results in an increase in the size of the leaves per fruit, which in turn leads to an increased amount of assimilation to the remaining fruits. According to existing literature, fruit thinning can lead to a rise in fruit size, albeit with a relatively smaller impact compared to the reduction in fruit yield (Wünsche and Lakso, 2000).

Chemical blossom thinners: The predominant method of crop load management in different countries, specifically for apples as opposed to pears, entails the application of chemical thinning. The process entails the utilization of plant growth regulators (PBRs) in the bloom and/or post-bloom stages, succeeded by manual thinning. Chemical thinning is a horticultural practice that involves the use of both acidic compounds or synthetic hormonal regulators for development to reduce the number of flowers and/or fruit on a tree. There exist multiple variables that can influence the degree of fatigue and subsequent revitalization of plant life in the springtime. According to Meland (1996), various factors can impact the growth and development of trees, such as the tree's species or cultivar, its overall health, age, stock, vigorousness, blossoming density, pollination, climatic conditions, and the method of application. The thinning response of chemical thinning agents is influenced by multiple interacting factors, which renders the management of crop burden challenging due to the unpredictability of the chemical thinning responses.

4.3 ASSIMILATE PARTITIONING

Leaves in higher plants are responsible for acquiring resources through the process of photosynthesis. This process captures free energy and reduces oxidized nitrogen and carbon into carbohydrates and amino acids, respectively, making leaves the primary site for this assimilation. In higher plants, photosynthesis takes place in the leaves, which are in charge of acquiring resources. As a result, leaves become the main location for this absorption. This process also collects free energy and converts oxidizing nitrogen and carbon into sugars and amino acids, respectively. Up to 80% of the carbon produced during photosynthesis is transferred to the organs that depend on external supplies via the vascular system of the plant. Sucrose, which acts as the main metabolite, is principally responsible for regulating how resources are distributed in plants. This is because, in most plant species, it serves as both the main result of the breakdown of carbon during photosynthesis and the main form of carbon delivered to heterotrophic tissues. Assimilate partitioning, which refers to this systemic distribution of photosynthates, is a critical mechanism for plant development and production. A region where organic materials are synthesized is a source of organic material. For example, the growth of a leaf, root tuber, or tuber, any organ unable to produce enough photosynthates to suit its own needs or one that is not photosynthetic. Different structures, including roots, embryos, fruits, root tubers, and tubers, go through substantial modifications during the developmental process. The most important organ for photosynthesis is the leaf, which is the process through which green plants use chlorophyll in their leaves to catch light energy and use it to create reduced carbon molecules from CO_2 and water. Up to 90% of a plant's dry mass is made up of photosynthates, and photosynthesis produces carbohydrates for growth and energy. Growth and yield depend on a steady stream of nutrients and carbohydrates (Table 4.3).

4.4 TRANSLOCATION AND DISTRIBUTION

4.4.1 Source and Sink Relationship

The movement of organic solute in plant phloem is referred to as a source and sink relationship (Pawar and Rana, 2019). Food transmission can happen in any chosen direction, unlike the unidirectional action of water conduction. The pressure flow/mass flow theory of food/sucrose translocation was put forth by Munch in 1930. The most well-known theory of plant food conduction is this one. This idea states that food is transferred between the source and sink in the direction of the turgor pressure gradient or from high to low turgor pressure.

TABLE 4.3
Process of Loading and Unloading of Photoassimilates

	Phloem Loading
Process	It is an active process that occurs through Sucrose-H symporters
Forces	Provides driving force for translocation.
	Generates osmotic pressure in sieve element-companion cell complex
	Pressure difference results in bulk flow
	Energy used for loading provides a motivating force for translocation.
Movement of molecules	Sucrose moves from the source in sieve tubes towards the sink from high T.P.
Water potential	High water potential to towards the low T.P./Low water potential.
	Phloem Unloading
Process	It is an active process helped by carrier molecules.
Movement of molecules	At the sink, sucrose is unloaded, resulting in a decrease in O.P. it results in the exit of water into the nearby xylem, leading to a decrease in T.P. and water potential of phloem.
Condition at sink	In sink cells, the unloaded sucrose is either changed into starch or consumed to maintain low O.P. and continuous unloading.

4.4.2 Phloem Loading/Sucrose Loading at the Source

The process in controversy is a dynamic one that is facilitated by carrier molecules. As a consequence of phloem influx, the number of sieve cells rises at the source, leading to a rise in osmotic pressure. This, in turn, causes water to flow from the adjacent xylem into sieve cells, resulting in an increase in turgor pressure and water potential. It is imperative to enhance the total pressure at the origin and the filtration channels. Sucrose is transported via sieve tubes from the source to the sink, moving from regions of high turgor pressure and high potential for water to regions of low turgor pressure and low water potential.

4.4.3 Phloem Offloading and Sucrose Unloading at the Sink

This is a process that is facilitated by carrier molecules and involves active transport. The unloading of sucrose at the sink results in a reduction of osmotic pressure (O.P.), the efflux of water into the neighboring xylem, and a decline in total pressure (T.P.) and water potential of phloem. In order to sustain a reduced osmotic potential and uninterrupted offloading, the sucrose that is released in sink cells undergoes either starch conversion or consumption. The process of loading sucrose at the source and subsequently discharging it at the washbasin is ongoing. The maintenance of the turgor pressure differential is crucial for the uninterrupted flow of water from the source to the receptacle. Research conducted by Genard et al. (2007) suggests that the process of phloem conductivity is an active one that necessitates the utilization of metabolic energy within phloem cells.

4.4.4 Passive Transport Versus Active Transport

This pertains to the mechanism of traversing along the electrochemical gradient. The electrochemical potential experiences a shift from a state of elevated positivity to one of reduced negativity. Active transport refers to the movement of ions or molecules in a direction that is contrary to the electrochemical gradient. The phenomenon under consideration involves a change in electrochemical potential, whereby the direction of said potential shifts towards a more positive orientation in relation to a more negative orientation.

4.4.5 Water Absorption Route

The epidermis cortex is a term used to describe the outermost layer of cells in a plant's stem. Water molecules can be transported from the root hairs to the cortex via the apoplastic or symplastic route. The presence of Casparian strips leading to the delays of the apoplast involves the transportation of water through the symplast by means of the passage cells. The term symplast refers to a mode of living that is distinguished by its emphasis on sustainability. The aforementioned statement can be construed as a figurative depiction of the voyage of being. The symplastic pathway is the mode of water transportation across plant cells via plasmodesmata. The apoplast in plants refers to a pathway that is devoid of living components. The apoplast is generated upon hydration through the amalgamation of the cell wall, intercellular space, and xylem cavity.

4.4.6 Assimilate Partitioning

According to Marschner (2012), the process of photosynthesis is essential for the production of assimilates. According to Lakso and Flores (2003), the growth and yield quantities and patterns of plants are determined by the distribution of assimilate. The process of translocation is contingent upon the developmental stage of the plant. The direction and volume of transportation are contingent upon the placement and comparative desirability of the lavatory. According to Friedrich and Fischer's (2000) study, a majority of the solute molecules found in exudate are composed of carbohydrates and exhibit a velocity ranging from 50 to 100 cm/s. Glucose is considered as the predominant form of translocation. According to Teiz and Zeiger (2006). The process of partitioning within a tree is not governed by genetic programming; rather, it is influenced by the interplay of various organs that compete for limited carbohydrates, each with its own relative capacity to do so. The level of rivalry among distinct sinks is contingent upon the metabolic function and proximity to the carbohydrate origin.

4.4.7 Source and Sink Managements

The reduction of fruit load has been found to have a positive impact on the quality of fruit in the present year, as well as on the buildup of reserves that can potentially benefit the development of the tree in future years. However, the occurrence of alternate bearing in fruit crops presents a significant obstacle, which may result in substantial economic losses for cultivators. Alternate bearing is most likely caused by a substantial fruit load. The citation provided is Iglesias et al. (2007). The act of removing fruits in apple trees has been observed to result in a higher degree of leaf area development as compared with trees with fruits that remain intact. Conversely, immature apple trees that bear fruits exhibit a reduction in leaf area (Lenz, 2009). In addition, it was observed that strawberry plants that did not bear fruit had a higher percentage (61.1%) of assimilates in their foliage compared to plants that bore 6 or 12 fruits, which had assimilate percentages of 39.2% and 21.1%, respectively. The impact of defoliation on trees is contingent upon the extent of leaf loss. In cases where defoliation occurs, the leaves that remain experience an increase in photosynthetic rate due to their relatively larger sink capacity.

4.4.8 Girdling

The majority of the techniques used in orchard management have an impact on the carbohydrate economy of trees. However, the objectives of girdling and fruit thinning are achieved by modifying the source-sink relationship. Several research studies have provided evidence that girdling induces the buildup of carbohydrates, particularly glucose, in the tree organs located above the girdle. While the contribution to other hormones and nutritional structures cannot be entirely discounted, compelling evidence suggests that the advantageous outcomes of girdling can be attributed to the augmented accessibility to carbohydrates (Wu et al., 2008). The practice of fruit thinning is a widely utilized agricultural technique that significantly alters the dynamics of source-sink relationships. According to Dejong and Gross (1995), fruit size increases as a result of reduced fruit quantity on the same leaf area after partial fruit removal. The present study aimed to investigate the correlation between fruit development and available leaf area on girdled citrus branches through the manipulation of the number of fruits and leaves. The results indicate that the decline in fruit yield did not exert a noteworthy influence on the photosynthetic capacities of the trees. Thus, it can be deduced that the excess photosynthetic products are probably distributed to alternative storage locations.

4.4.9 Leaf area Index (LAI)

The utilization of the Leaf Area Index (LAI), in combination with the absorption of solar radiation represents a valuable approach for evaluating the productivity of a canopy. Apart from cultural practices, the agroecological conditions and age of the plant can also exert an impact on the growth of the Leaf Area Index (LAI). In contrast to the elevated location, the expedited Leaf Area Index (LAI) growth at an altitude of 2,300 m above sea level facilitated premature and amplified yield generation across the entirety of the cultivation. The leaves in close proximity to the developing fruits exhibit a higher photosynthetic capacity compared to the remaining foliage of the tree.

4.4.10 Training and Pruning

According to Myers (2003), the redistribution of resources, including carbohydrates, water, and growth regulators, via training and pruning techniques, leads to an alteration in the equilibrium between vegetative development and reproductive fruiting. According to Casierra (2007), the act of heavy pruning leads to a decrease in leaf area, which in turn affects the overall process of photosynthesis within the tree. Additionally, this process can also impact the translocation of photosynthesis to various parts of the tree such as fruits and roots. Consequently, the aforementioned phenomenon results in an elevation of the ratio between the root and shoot, thereby promoting the growth of vegetative tissues. According to Serrano et al. (2007), the Psidium guajava plant species exhibits a higher ratio of fruit weight when subjected to minimal and moderate pruning as opposed to intensive pruning. During the reproductive phase, the technique of "fruiting pruning" is utilized to augment fruit output, control the vegetative-reproductive physiological equilibrium, attain a balanced and rational distribution of high-quality production, maintain consistent production over an extended period, and facilitate fruit thinning. The removal of upright water sprouts is a crucial aspect of pruning. These sprouts have been found to direct photosynthates toward the growing shoot tip, thereby compromising reproductive development (Arjona and Santinoni, 2007).

4.5 CONCLUSIONS

Plant source-sink relationships have emerged as one of the most promising research areas in recent years. The topic has important implications for crop management and involves a wide range of physicochemical processes. Several variables, including photosynthesis, the number, and location of competitive sinks, capacity for storage, and vascular transport, influence how nutrients are distributed among competing sinks in plants. The actual regulations controlling assimilation partition at the level of the entire plant appear to be poorly understood, despite the fact that specific plant activities like the process of photosynthesis translocation, and cell development are well recognized. There is an urgent need for additional modeling-based integrative research because many processes are closely interrelated.

REFERENCES

Arjona, C. and Santinoni, L.A. 2007. The fruit trees. Sozzi, G.O. (ed.). *Fruit Trees, Eco-physiology, Cultivation and Utilization.* University of Buenos Aires, Buenos Aires, pp. 243–82.

Casierra-Posada, F., Rodríguez, J. I., andCárdenas-Hernández, J. 2007. The leaf-to-fruit ratio affects the production and quality of the fruit (*Prunus persica* L. Batsch, cv and Rubidoux). *Nal. Agr. Medellin.* 60(1):3657–69.

Costa, G., Botton, A, Vizzotto, G. 2019. Fruit thinning: advances and trends. *Hortic. Rev.* 46:185–226.

Dejong, T. M. and Grossman, Y. L. 1995. Quantifying sink and source limitations on dry matter partitioning to fruit growth in peach trees. *Physiol. Plant.* 75:437–43.

Erf, J. A. and Proctor, J. 1987. Changes in apple leaf water status and vegetative growth as influenced by crop load. *J. Am. Soc. Hortic. Sci.* 112(4):617–20.

Friedrich, G. and Fischer, M. 2000. *Physiological Bases of Fruits.* Ulmer Verlag, Stuttgart, Germany.

Iglesias, D.J., Cercós, M., Colmenero-Flores, J.M., Naranjo, M.A., Ríos, G. andCarrera, E. 2007. Physiology of citrus fruiting. *Braz. J. Plant Physiol.* 19(4):333–62.

Jones, K.M., Bound, S.A., and Miller, P. 1998. *Crop Regulation of Pome Fruit in Australia.* Tasmanian Institute of Research, Hobart; ISBN 1-86295-027-X.

Lakso, A. and Flore, J.A. 2003. Carbohydrate partitioning and plant growth. In: T.A. Baugher and S. Singh (eds), *Concise Encyclopedia of Temperate Tree Fruit.* Food Products Press, New York, pp. 21–30.

Marschner, P. 2012. *Marschner's Mineral Nutrition of Higher Plants.* 3rd ed. Elsevier, Oxford.

Meland, M. and Gjerde, B. 1996. Thinning apples and pears in a Nordic climate. I. The effect of NAA, ethephon and lime sulfur on fruit set, yield and return bloom of four pear cultivars. *Nor. J. Agric.* Sci. 10:437–51.

Musacchi, S. 2021. Physiological basis of pear pruning and light effects on fruit quality. *Acta Hortic.*, 1303:151–62.

Myers, S.C. 2003. Training and pruning principles. In: T.A. Baugher and S. Singha (eds), *Concise Encyclopedia of Temperate tree Fruit.* Food Products Press, New York, pp. 339–45.

Palmer, J.W., Giuliani, R. and Adams, H.M., 1997. Effect of crop load on fruiting and leaf photosynthesis of 'Braeburn'/M. 26 apple trees. *Tree Physiol.* 17(11):741–6.

Pawar, R. andRana, V.S. 2019. Manipulation of source-sink relationship in pertinence to better fruit quality and yield in fruit crops: a review. *Agric. Rev.* 40(3):200–7.

Rana, V.S., Zarea, S.E., Sharma, S., Rana, N., Kumar, V., and Sharma, U. 2023. Differential response of the leaf fruit ratio and girdling on the leaf nutrient concentrations, yield, and quality of nectarine. *J. Plant Growth Regul.* 42:2360–73.

Serrano, L.A.L., Marinho, C.S., Ronchi, C.P., Lima, I.M., Martins, M.V.V., and Tardin, F.D. 2007. Goiabeira 'Paluma' under different systems of cultivation, season and intensities of cutting of fruit. *Fish. Agropec.* 42(6):785–92.

Tromp, J. 2000. Flower-bud formation in pome fruits as affected by fruit thinning. *Plant Growth Regul.* 31:27–34.

Whiting, M.D. and Ophardt, D. 2005. Comparing novel sweet cherry crop load management strategies, *HortScience* 40(5), 1271–75.

Wu, B.H., Huang, H.Q., Fan, P.G., and Li, S.H. 2008. Photosynthetic response to sink-source manipulation in five peach cultivars varying in maturity date. *J. Am. Soc. Hort. Sci.* 133(2):278–83.

Wünsche, J.N. and Lakso, A.N. 2000. Apple tree physiology – implications for orchard and tree management. *Int. Dwarf Fruit Tree Assoc.* 33:82–8.

5 Root and Canopy Regulation

Tanuj Bhardwaj, Vijay Kumar,
Vishal Singh Rana, and Sunny Sharma

5.1 INTRODUCTION

In the plant kingdom, controlling canopy architecture is a key factor in enhancing crop output by boosting the proportion of dry matter allocated to the reproductive phase. The successful enhancement of fruit yield in plants often involves the management of their canopy architecture through strategic positioning and maintenance of their structural components. The most crucial method of managing fruit plants is canopy control. Fruit tree training and trimming are specialized horticulture tasks. The training of pear trees is an essential procedure to ensuring optimal air and light penetration into the foliage, thereby facilitating healthy pigmentation and the generation of high-quality fruit. Young plants must be trained according to a suitable system of training, which requires sufficient attention. The building blocks for plant growth and production are photosynthates. Light, temperature, CO_2 concentration, water, soil fertility, and other environmental as well as plant-related parameters affect net photosynthesis (Proietti et al., 2000). All planting systems aim to manipulate canopy design to increase global light interception because whole-tree photosynthesis is largely light-limited (Wagenmakers et al., 1991).

Fruit productivity and quality in fruit trees are influenced by the light microclimate. The fruit yield of well-nourished and effectively irrigated trees has a positive correlation with the general light interception, as photosynthetic carbon fixation primarily relies on the sunlight assimilated by a tree or an orchard. The performance of a plant is a crucial factor in determining its ecological success, as it is closely linked to its phenotype. If this connection is disrupted, the shape of the plant becomes irrelevant in terms of its ecological and evolutionary significance (Koehl, 1996). The physical framework of a fruit tree, which consists of the stem, branches, shoot, and leaves, is termed the canopy. The quantity and size of leaves have an impact on the canopy's density. Canopy management is the technique of modifying the canopy of trees to increase the yield of quality fruits. Canopy management, especially its components like tree training and pruning, has an impact on the amount of sunlight trees can absorb since the structure of the tree affects how the leaf area is exposed to incoming radiation. Fruit plants grow heavy whenever they are not properly trained and pruned from the beginning. To capture energy from the sun, each plant must establish its perfect canopy spreading at the start of its life cycle. The oversight of canopy architecture is thus one of the most popular methods for efficiently managing big, unmanageable trees to raise their production. It manages flower and fruit ripening rates in addition to trimming and tree training.

5.2 FEATURES OF THE IDEAL CANOPY

The following points are the ideal features of canopy

- Strong, well-spaced secondary and tertiary branches.
- Enough fruiting terminals in the places that are most productive.
- Strong, photosynthetically efficient foliage to make the most of solar energy.
- Enough room inside the canopy for air circulation.
- Support a sufficient amount of shade to shield the fruits from sunburn.

DOI:10.1201/9781003354055-5

- Effective use of light is necessary for the growth and development of flowers and fruits, as it facilitates the production of carbohydrates.
- To provide the fundamental tree form and promote strong, wide crotch angles, which will help the growth of a sturdy tree framework.
- To encourage trees to live longer and bear fruit earlier.

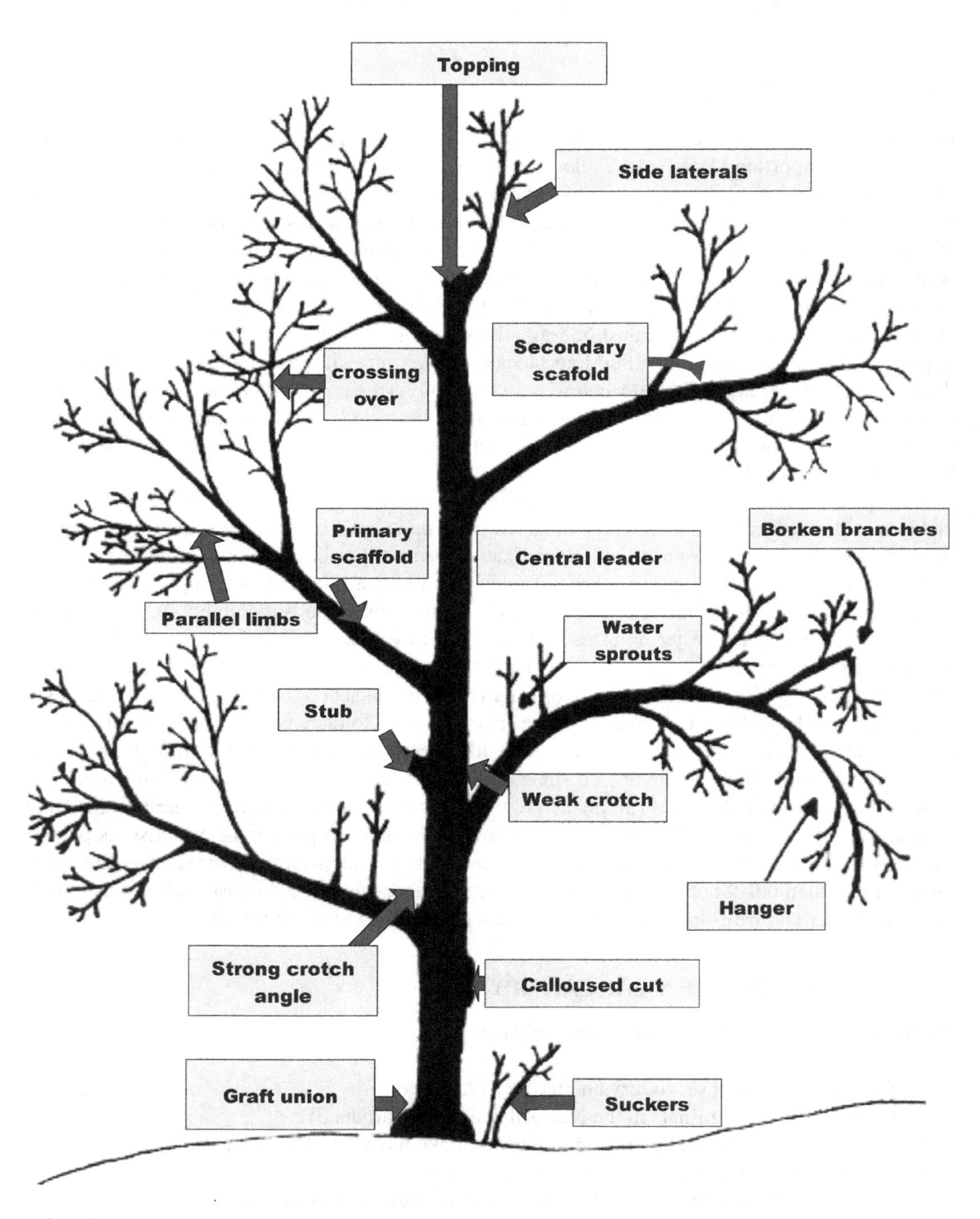

FIGURE 5.1 General aspects of canopy management.

5.3 BASIC PRINCIPLES OF CANOPY MANAGEMENT

The condition of trees and environmental variables, specifically light interception, temperature, humidity, and airflow, have a significant impact on the productivity and quality of fruits. This connection maintains accurate if fundamental characteristics and management factors remain constant. Canopy management is primarily concerned with optimizing tree vigor to enhance both efficiency and quality while mitigating the adverse impact of environmental factors on the trees (Figure 5.1).

The influence of temperature and sunlight exposure on various physiological processes, such as the processes of photosynthesis, bud differentiation, ripening, and fruit quality is of paramount significance. Additionally, increased air movement has been found to mitigate the likelihood of disease occurrence. Excessive humidity is a good environment for disease and pest issues. Comparable to the aforementioned, shade may cause a multitude of detrimental impacts, including but not limited to the promotion of herbaceous traits, the elevation of pH levels, inhibition of inflorescence initiation, reduction of fruit set, and a slowdown of fruit growth and maturation. The following are some fundamental factors in canopy architecture management:

- Create the desired and robust structure and vigor.
- Enhancing the amount of light entering and leaving the canopy
- Preventing the development of a microclimate that is conducive to illness and pest
- Improving fruit quality and production
- Susceptible to automation and practical in cultural operations
- Crop diversification and regulation

5.4 LIGHT CAPTURE AND PLANT STRUCTURE

The rate of canopy photosynthesis is influenced by both the distribution of light within the tree canopy and the metabolic capabilities of the leaves. The goal of canopy management is to change the crown architecture to improve total light harvesting and light harvesting efficiency. The measurable variable that has the most significant impact on the level of radiation assimilated is the assembled surface area of the foliage canopy. Similarly, the dispersion and configuration of foliage in a botanical canopy also have a substantial impact on the quantity of light that can be captured per foliage unit.

The size, the height-to-breadth ratio, and the contour's form can all be used to visualize the shape of the plant crown. The inherent architectural framework of the cultivar varies from varieties, from heavily branched scaffolds to those with low-branched, tip-bearing structures. Since the sun's inclination angle diminishes as one travels from the equator to the pole, crowns with different height-to-width ratios have naturally varied light-interception efficiency. In particular, large solar inclination angles allow light to enter high latitudes, indicating that beam route lengths lengthen with increasing crown flatness. In high-latitude regions, the crowns of trees are characterized by their narrow and vertically elongated shape, which facilitates the interception of direct sunlight across the entire canopy. This is achieved through the maintenance of comparable beam path lengths throughout the canopy. The significance lies in the observation that alterations in the crown configuration can occur at varying height-to-breadth ratios. The optical distance traversed by a narrow ellipsoidal crown significantly rises as the canopy depth increases, particularly for solar inclination angles that are low (Valladares and Niinemets, 2007).

Despite the lower canopy position of thin conical crowns, their branches extend considerably away from the stem. However, the beam path length remains nearly identical, indicating that these crowns can be highly efficient in regions closer to the equator. Typically, there is a positive correlation between the length of the cone and the amount of radiation captured. The optimal absorption of light by horizontal leaves located at the uppermost part of the canopy occurs during noon and summer months when the irradiance exceeds the photosynthetic light saturation point. From the bottom to the top of the canopy, foliage's photosynthetic capacity (Amax) often increases by a factor of 2–4. Therefore, in high light, horizontal leaves' higher light absorption typically results in a small increase in potential carbon acquisition. In general, there is around a meter of effective light penetration into the tree canopy. This fact allows us to divide the canopy of a huge tree into three zones based on the degree of light penetration. Zone one of the trees receive between 60% and 100% of its full sunlight, zone two between 30% and 59%, and zone three less than 30%.

If the level of light exposure falls below 30% of the optimum amount of sunlight, it can hinder the development of flower buds. Additionally, it can reduce the strength of the spurs and result in the production of small and low-quality fruit in that particular area. The phenomenon of fruit spur mortality is expected to result in a decline in fruit yield at this particular site over time. The dimensions of zone three, an unproductive region, are regulated by the height, arrangement, and pruning of trees. This area of insufficient lighting might range from 25% in a large central leader tree to 1.6% in a dwarf central leader tree. One of the factors contributing to the higher productivity of smaller trees as compared to larger ones is the reduced non-productive zone resulting from their smaller size.

5.5 USING AND INTERCEPTING LIGHT

The interception of light by fruit plant canopies is contingent upon the dimensions, alignment, inclination, and surface attributes of the photosynthetic structure. By developing larger leaves, shade plants can increase the amount of light they intercept. The production of large leaves is observed in rainforest understory plants due to their preference for shade. This is attributed to the fact that shade-tolerant plants tend to have larger leaves in comparison to their sun-loving counterparts. Leaf size is capable of variations based on its location within a plant's tree. Smaller leaves tend to develop in areas closer to the top where radiation is at its maximum, while larger leaves tend to develop in areas closer to the interior and base where light levels are lower. Similar to how the angle/orientation of the leaf can alter the light interception. Vertical layouts are known to optimize light interception during early morning and late afternoon when the sun is shining at low angles. However, during solar noon, when radiation levels are at their peak, light interception is reduced. Leaves that are aligned in a horizontal manner have the potential to obstruct light throughout the entire day, with the highest degree of obstruction occurring during midday.

5.6 RELATIONSHIP BETWEEN LIGHT ABSORPTION AND PLANTING DENSITY

The impact of light exposure on fruit set, color, and quality is widely known. It can also cause morphological and physiological changes between leaves that were developed in the light or shadow. Orchard design, or row pattern, is directly related to planting density for any specific training method. However, there is a nonlinear correlation between production and the number of trees per hectare. At increasing planting densities, fruit quality deteriorates, which can also have a detrimental impact on fruit yield. Obstacles in orchard management in double and triple-row designs have an impact on expenses and fruit quality. The total light interception of an orchard can be enhanced by enhancing the density of the canopy, the height of the trees (in relation to the open alley width), or the number of trees (of smaller size) per hectare, without increasing the canopy volume. Despite the potential loss of well-illuminated canopy volume, which is responsible for producing high-quality fruits, increasing the density of the canopy can lead to a greater proportion of total light intercepted (F), even at high LAI values.

Palmer (1980) conducted an estimation of the Leaf Area Index (LAI) for three separate approaches to training, specifically palmitate, slender spindle, and full field. The light interception fractions were found to be 53% for both the palmate and slender spindle and 66% for the entire field. The Leaf Area Index (LAI) values obtained for each of the three systems were 1.6, 1.6, and 1.9, respectively. Especially for trees with a triangular form, growing the height of the trees will boost light absorption. Maintaining a constant leaf area index (LAI) can enhance the spatial arrangement of light throughout the canopy, particularly in instances of elevated LAI, thereby mitigating the density of the canopy. The correlation between the spacing of plants and the height of trees is of utmost importance. Because more light will completely miss the tree in rows that are too far apart, the interception will decrease. On the other side, excessively tight rows will mutually shade one another, lowering the amount of illumination throughout the entire orchard.

5.7 SYSTEM OF INSTRUCTION AND REACTION TO PHOTOSYNTHESIS

Lincoln Training systems such as Canopy, V-trellis, and Y-trellis are commonly employed to enhance the absorption of light by the canopy. However, these systems tend to result in disproportionate growth in height on the viewable sides of scaffolds, with the Canopy being the most culpable, followed by V-trellis and Y-trellis (Palmer and Warrington, 2000). The prevalent tree forms employed for the training of apple trees are globular in shape, which is characteristic of expansive, open-centered trees. The top third of the trees, where the fruit is relatively less accessible, is the most productive region. This form is reminiscent of a Christmas tree with an open framework, either conical or pyramidal.

A substantial amount of the bearing surface is in close proximity to the ground level, and the uppermost part of the tree does not cast a shadow on the lower branches. The canopy's open framework will enable plenty of light to enter. A horizontal canopy provides for effective light penetration throughout the entire canopy since its thickness is kept to a maximum of roughly 1 m. The Y or V-shaped tree form facilitates optimal light penetration, growth control, and production influence, as demonstrated in the Lincoln canopy. Examples of this form are the various systems along with fruit crops, which are shown in Table 5.1 and Figure 5.2.

TABLE 5.1
Different Training System along with Fruit Crops

	Type of Training System
Open center system (Vase open)	Peaches, apricots, and ber
Central leader system:	Apple, pear, pecans, litchi and sapota
Modified leader system:	Apple, pear, cherry, guava, plum (cv. Santa Rosa), apricot, walnut, persimmon, etc.
	Training Systems for Dwarf Trees
Cordon system:	Spur bearing apple and pear, grape vines incapable of standing on their stem, passion fruit, red currant and gooseberries
Dwarf pyramid:	Apples and pears
Vertical axis:	Apple
Tatura trellis:	Peach
Spindle bush:	Apple, pear, plum and cherries
Slender spindle:	Apple
Super spindle:	Apple and pear
Palmette:	Apple, pear, peach, plum and apricot

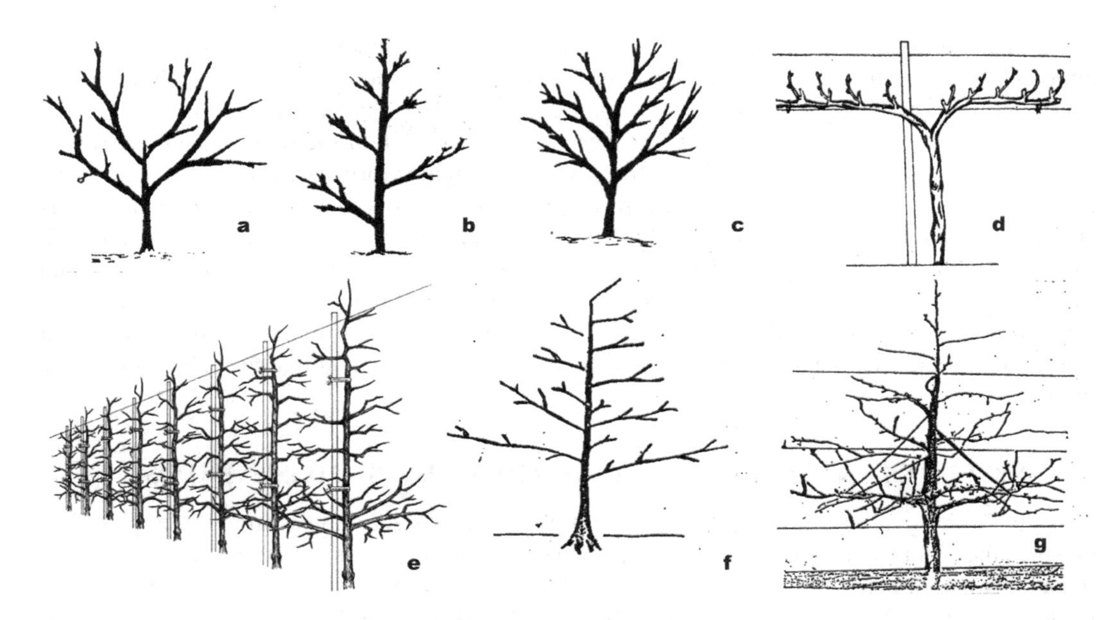

FIGURE 5.2 Different training systems of Fruit Orchards. (a) Open center system; (b) Central leader system; (c) Modified central leader; (d) Espalier system; (e) Vertical axis system; (f) Slender spindle systems; (g) spindle bush.

5.7.1 Open-Centre (Vase) System

Within a year of planting, the main trunk may develop up to 75 cm before being trimmed. Every side branch has a conical form and is directed backward. This system is less effective since it has weak crotches and frail support. More sunlight, though, might be advantageous for the fruits' interior coloring. However, due to the poor foundation and other obvious flaws, this method is not ideal for sub-tropical climates where there is abundant sunshine and strong summer winds. Low-density planting is used (100–200 trees acre).

5.7.2 Central Leader

Choose the right scaffold branches and train them properly. A minimum distance of 45 cm ought to be maintained between the initial scaffold limb and the ground. It is recommended that supplementary support branches be uniformly distributed both along the trunk and from each other, with a minimum distance of 20 cm. It is recommended that the leader be pruned to two-year-old wood once it surpasses the optimal height for harvesting. Remove any branches that cast an excessive amount of shadow. Keep the tree's cone-shaped form. Eliminate non-productive shoots.

5.7.3 Modified Central Leader

This system is a cross between a central leader and an open leader. After permitting the stem to develop for the first two years, it is trained to act as a central leader before being headed back at a height of 75 cm. As with an open center system, lateral branches are permitted to expand and be pruned. With this approach, when the main trunk's length is shortened, the height of the tree decreases. The length and size of the scaffold's branches are encouraged to increase. The trees have sturdy frames and powerful crotches. The technique works best with commercial fruit trees since they are relatively shorter in height, which makes it easier to do tasks like spraying, pruning, and harvesting. It is a planting strategy with a medium density (250–400 trees per acre).

5.7.4 Solaxe Training System

In order to enhance early maturation and diminish excessive growth, arborists cultivate trees in the form of central leaders, with branches trained to a 120° angle through the use of ties. With the exception of the removal of closely spaced juvenile limbs, branch trimming is infrequent. The implementation of restricted branch renewal and the establishment of light tunnels in proximity to the trunk, coupled with the removal of all buds located within a 30–40cm radius of the trunk, are effective measures for enhancing light penetration. Notching of buds or application of Promalin treatments is employed to induce branching in areas where it is deficient. The tree's top is bent to the horizontal as it reaches 3.5–4 m so that the maximum height can be preserved. This is a fairly young system because branch bending and pruning are not used. Little pruning also indicates a generally low incidence of bacterial canker infection. Spur removal regulates the amount of crop and the size of the fruit. The number of spurs that persist on a given branch is influenced by the productivity of the cultivar and the diameter of the branch. Pruning of "Bing" branches is typically carried out by thinning them to a distance of approximately 10cm between spurs, while "Van" branches are pruned to a distance of 20cm. This system is highly intensive and may be quite expensive to build and maintain due to spur removal and branch bending.

5.7.5 Vertical Axis System

In order to enhance the penetration of light into the tree canopy, arborists employ a technique of training and maintaining trees in a conical shape with a prominent central leader. The central leader is directed to exhibit vertical growth until it attains a height of approximately 10 feet. In the initial three-year period following plantation, prior to the establishment of tree canopy closure, minimal pruning interventions are executed utilizing this methodology. After that, 2-year-old or older growth is frequently cut into to rejuvenate lateral branches. A trellis that has at least one wire linking the tops of each conduit or wooden post at each tree supports trees.

5.7.6 Slender Spindle System

The slender spindle system, when employed in the cultivation of apple trees, results in a conical shape that differs from the pyramidal appearance of the central leader or spindle bush trees. By removing any or all of the lateral limbs, one can get this shape. To promote the growth of a new lateral from a single spot on the trunk, a 45° cut is typically performed when amputating limbs. In order to regulate the vitality of trees and facilitate prompt fruit maturation, branches are instructed to grow in a horizontal orientation or below it. The slender spindle system is a suitable method for high-density plantations on dwarf rootstocks, particularly for trees with a height of 2 m or less. The density of trees per hectare may vary between 2,000 and 5,000, contingent upon the utilization of either single- or multi-row bed plantings. The tree in question is a diminutive, cone-shaped specimen with a maximum diameter of 1.5 m and a height of merely 2 m. The cropping process of spindles commences at an early stage with minimal trimming, and the laterals are subsequently bent in a horizontal orientation. The cultivation technique involves planting trees at a high density ranging from 700 to 1,100 trees per acre, or even at an extremely high density of 1,500–2,000 trees per acre using the super spindle method. The trees are grown with a thin cylindrical canopy and dwarfing rootstock, specifically of M9 size, and are supported by structures while their growth is controlled by bending the central leader to regulate their vigor. The term "narrow spindle" refers to a cylindrical object with a small diameter in relation to its length.

5.7.7 Spindle Bush

This training system shares characteristics with the 155 system, namely the use of a central leader training method to create a cone-shaped tree that is supported by a post or wire support system.

The training form is appropriate for medium-to-medium high-density planting with trees ranging in height and spread from 2 to 3 m. This size of tree requires the growth of a series of scaffold limbs that are always present in the lowest third of the tree canopy. Where good growth is attained, early cropping is preferred to regulate tree vigor and keep mature trees within designated tree spacing.

5.7.8 Espalier System

The technique of espalier training is directed towards creating a visually appealing work of creativity that is confined within a specific space. This requires the implementation of rigorous pruning techniques to effectively manage the growth of shoots that emerge from the horizontal limbs. The Kniffen system observed in grape cultivation exhibits similarities to the present system under consideration. The trellis's background, which is often a wall or fence, typically results in the formation of two-dimensional tree shapes. Every design necessitates a support system, particularly during the learning phase. When constructing espaliers, it is important to take into account the species of the plant and the robustness of the design. Peach trees exhibit a wider horizontal limb spacing, ranging from 20 to 24 inches (50.8–61 cm), in comparison to pear trees which typically display a spacing of 12 inches (30.5 cm). While smaller plants require only two or three sets of branches, larger and more vigorous plants may require additional levels. Maintaining flexibility in the limbs during bending is imperative. Maintaining equilibrium requires the preservation of equal length in the opposing limbs situated on both sides of the central axis. As previously mentioned, it is imperative to firmly attach the branches to the structure or other forms of support. The designs of espalier can exhibit either a formal or an informal aesthetic.

5.7.9 Horizontal Espalier

A set of horizontal wire trellises that are fastened to poles, fences, or walls make up the horizontal espalier training system. When beginning growth from a single stem, a seedling plant is pruned directly beneath the fruiting wire. This approach encourages lateral buds to sprout additional branches. Two equally vigorous lateral shoots are chosen, trained, and positioned on stakes that are attached diagonally to the wires. Stakes might not be required. Initially heading the plant directly below each wire allows for the development of horizontal limbs. Trim any growth that is heading in that direction. Equivalent plant vigor is produced by horizontal espaliers. To create the subsequent tier of lateral branches, the entire process is repeated. Usually, the lower scaffold branches lose their vigor at the ends. Following establishment, lateral branches will emerge, primarily on the lower portions of the tree. To create fruiting spurs the next season, these branches are trimmed in the summer. Fruiting spurs should be trimmed off as their population grows over time to prevent overcrowding and decreased plant vigor.

5.7.10 Palmette System

The palmette training framework, a variant of the espalier arrangement, involves the cultivation of plants at a slope of approximately 400°, as opposed to the traditional horizontal branching approach. The candelabra palmette system of training utilizes a lattice framework consisting of horizontal and vertical arms to cultivate aesthetically pleasing and well-proportioned trees. By encouraging earlier bearing and enabling higher planting densities, this strategy increased orchard yield. In a semi-high-density planting technique, an acre might include 400–800 trees.

5.8 PRUNING

Pruning is a horticultural practice that involves the selective removal of vegetative plant parts to regulate production and ensure optimal equilibrium between vegetative and reproductive growth. The act of pruning involves both artistic and scientific elements. The artistic aspect pertains to the precision and accuracy required in executing the pruning cut, while the scientific aspect pertains to the knowledge of the optimal timing and techniques for achieving the most favorable outcomes (Figures 5.1 and 5.3).

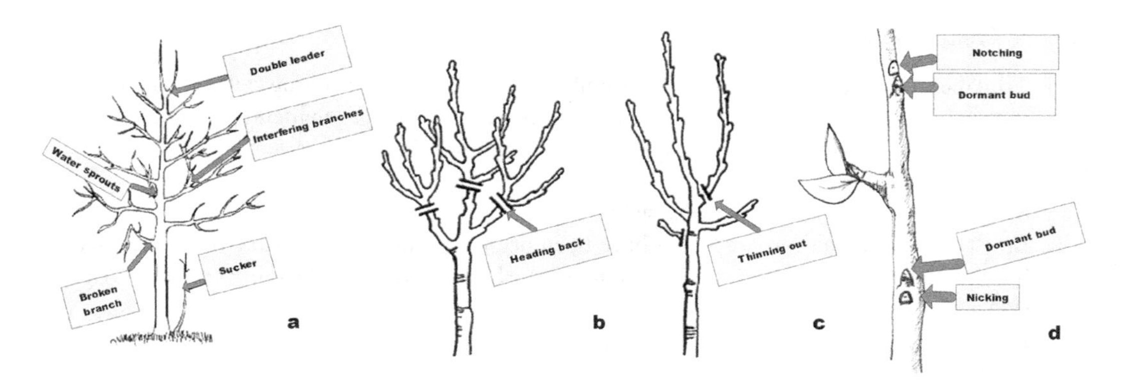

FIGURE 5.3 Different pruning techniques of Fruit orchards. (a) General aspects of crop pruning -production; (b) Heading back; (c) Thinning out; (d) Notching and nicking.

5.8.1 Advantages of Pruning

i. **Pruning is a dwarfing process**: The act of pruning has been observed to enhance the development of vegetative growth in the immediate vicinity of the pruning cut, leading to the perception that pruning exerts a stimulatory effect on growth. The act of pruning is a process that results in dwarfing. Nevertheless, a tree that undergoes annual pruning exhibits a lower weight compared to a tree that does not undergo pruning.

ii. **Pruning reduces yield**: The act of pruning has the potential to reduce yield as it involves the removal of wood that may contain viable fruit and flower buds. Whilst pruning can improve the quality of fruit, pruned trees typically yield less than unpruned trees. The act of pruning has been observed to enhance fruit size by augmenting the leaf area per fruit. Pruning is a crucial horticultural practice that enhances the absorption of light throughout the tree, thereby promoting optimal growth of fruit sugar levels and red coloration.

iii. **Pruning delays fruiting**: The process of pruning has been observed to impede the fruiting of young trees by stimulating the growth of vegetative structures instead of reproductive ones. A tree that does not have pruning is likely to exhibit earlier flowering and fruiting compared to a tree that has undergone pruning. Pruning of juvenile trees is conducted with the aim of promoting the growth of preferred branches and establishing a robust tree framework that can sustain substantial yields as the tree advances in age. As a tree matures, its physiological processes undergo a transition from vegetative development to reproductive growth. In order to attain optimal yearly yields from mature trees, it is imperative to defer the fruiting process until the trees have occupied their allotted space to a significant extent. Pruning is a technique that can be employed to postpone the fruiting of juvenile trees.

5.8.2 Summer Pruning

During the growing season, green shoots are specifically removed during summer pruning. The time of pruning, the degree of pruning, the health of the trees, the location, and the variety all affect how the trees react to summer pruning. In the 1980s, several researchers examined summer pruning, and several general conclusions regarding the process can be drawn. Summer pruning reduces the amount of internal shade, usually enhances the growth of fruit color, and occasionally enhances flowering bud formation. The leaves that create photosynthates (sugars) for the growth of all three sections are removed during summer pruning. Sometimes fruit size and sugar content are decreased by summer pruning.

5.8.3 Objectives of Pruning

i. Trim the tree. It appears that pruning stimulates growth because pruning causes an increase in vegetative growth at the pruning cut. However, a tree subject to annual pruning consistently exhibits a lower weight compared to an unpruned tree.
ii. Control tree shape.
iii. Strengthen the tree's structural integrity.
iv. Boost the fruit's quality. Trees that have been pruned produce fewer fruits than trees that haven't been pruned. By increasing the number of leaves per fruit, pruning enhances fruit size and quality.
v. Increase light permeability.
   - Enhances fruit color.
   - Enhances the commencement of flower buds and blossoms in the next season.
   - Aids in pest management.
vi. Increased light permeability
   - The beginning of a flower bud.
   - Fruit's hue.
   - Pest management.
   - Assist in cultural activities.
   - Maintain a crop that is closely planted.
vii. Elimination of infected wood.

5.9 MANAGEMENT OF THE CANOPY IN DIFFERENT FRUIT CROPS

5.9.1 Apple

To regulate the canopy architecture and capture the greatest amount of solar radiation—which is a limiting element in the mountainous area—training is a potential strategy. Several training schemes are in use as a result of the creation of dwarfing rootstocks and spur cultivars in apples. Open tree canopy and well-exposed fruit-bearing spurs encourage strong fruit production of large, well-colored fruits. By eliminating the central leader at 1 m above ground level, an open center can be created, which is a type of canopy development. The center of the canopy is still open, allowing sufficient sunshine to enter the tree throughout.

A palmette leader is a variation on the central leader tree form designed to increase light penetration, was described by Lakso et al. in 1989. This kind of canopy gives the branches the most sunshine exposure. The most valuable type of canopy management is cordon training, which involves planting trees in rows 60–90cm apart on rootstocks (M-9 and M-26) at a 45° angle to maximize light exposure. A tightly stretched cable supports the trees.

In order to maintain a balance between vegetative growth and reproductive growth, pruning is necessary in the apple industry. Unpruned trees look unkempt and bear meager-quality apples because there is insufficient sunlight interception and dispersion in the inner depth of the canopy. Pruning during this time can boost both flowering and fruit sets because the competition between vegetative and reproductive organs is particularly fierce during flowering, which takes place in the next 3–4 weeks. The recommended procedure involves pruning the lateral branches of the scaffold to a length of 25–30cm, while also eliminating the leader and significant inner branches. The light intensity within the tree was increased by a factor of 2–3, resulting in a doubling of the apple yield. The findings indicate that the act of pruning during the summer season leads to a rise in both fruit and light intensity.

5.9.2 Pear

The strategy is to limit the number of large, permanent openings for light to enter geometrically limited canopies. The thin, constrained planes of foliage, such as A, V, or T-shaped trees, tree walls, and closely spaced hedgerows. To put and maintain the branches in specified places using this method, the canopy is typically severely geometrically restricted, expensive support structures are needed, and extensive labor is required. These various tree types have value because of how they distribute light and because they ultimately improve yield and quality. Spadona pear cultivar on Quince A rootstock with four branches at a 45° angle and numerous secondary branches growing freely worldwide generates an oblique, hedge, or narrow spindle. After 2 years of training the hedge, a high yield was produced, and the uneven palmette thereafter produced yields that were superior to others. Pear cultivar Conference trees were raised using the angled trellis, thin spindle, vertical axe, or Y trellis systems. After 5 years, the Y trellis tree had the most spread, while the vertical and slender spindle trees had the tallest trees. Comparing the Y trolling to other training techniques, the yield was noticeably higher.

To increase the early output, the tree can be planted in single, double, or multiple-row beds at high tree densities due to its narrow, fully dwarfed development. In the pear cultivar Doyenne du Comice, a single-row planting at 2,000 trees/ha intercepted around 10% less light than one at 2,667 trees/ha or a three-row planting at 3,570 trees/ha. In pear growing, the trees on super spindle bushes or slender spindle bushes are not profitable (Teeffelen and Teeffelen, 1993).

5.10 ROOT CANOPY REGULATION

The growth of roots and shoots exhibit interdependence. In broad terms, the shoot system provides the root system with photosynthesis and a specific group of phytohormones. The root system facilitates the provision of structural support, water, nutrients, and a diverse array of hormones to the shoot system in exchange. According to a growing amount of research, roots may be able to sense soil conditions and transfer them to the shoot, which can change stomatal apertures and leaf growth (Masle and Passioura, 1987)

The mechanism used by plants to respond to soil drying has been studied the most. Abscisic acid (ABA) is a hormone that is synthesized by roots in response to soil dehydration. This hormone is then transported to the shoot, where it impedes leaf growth and, in certain instances, induces stomatal closure prior to any alteration in the water and nutrient levels of the leaves (McDonald and Davies, 1996). The availability of nutrients and the pH of the xylem sap have both been shown to influence how leaves respond to ABA.

It is hypothesized that comparable signaling mechanisms are implicated in the reaction of foliage to soil compression, albeit utilizing a distinct array of signaling molecules such as ethylene, nitrate, and ABA. This stands in contrast with the signaling mechanisms delineated in the literature with regard to soil desiccation (Hussain et al., 1999) and Mulholland et al. (1999). The current state of knowledge regarding the regulation of canopies by signaling systems is limited, as noted by Stoll et al. (2000). The modification of signaling has the potential to induce alterations in the response of canopies to varying soil conditions. The integration of modern ideas pertaining to fertilizer and water absorption with the emerging principles of chemical communication is imperative for the formulation of efficacious strategies aimed at augmenting crop management and productivity. The partial root-zone drying irrigation technique was developed by Davies et al. (2000) as a result of their investigation into the response of shoots to soil drying. In certain crops such as grape vines and tomatoes, where robust shoot growth vies with maturing fruits for nutrients, selective irrigation of a singular root section can potentially curtail overall biomass yield without causing a substantial decrease.

5.11 SOIL FACTORS

The soil factor, encompassing drought, availability of nutrients, mechanical impedance arising from compacted layers, water-logging, and root pathogens, plays a crucial role in determining plant growth and development. A decline in biomass production is observed due to a reduction in light interception or radiation use efficiency. This reduction is caused by the impact of the root system and soil conditions on the growth, activity, or lifespan of the canopy.

5.12 MECHANISMS OF ADAPTATION TO COPE WITH CHANGES IN NUTRIENT AVAILABILITY

Plants have developed diverse mechanisms of adaptation to cope with changes in nutrient availability. These mechanisms include morphological changes, such as an expansion in root growth to explore a greater soil volume, or the acidification of the surrounding soil to mobilize additional mineral nutrients. Despite significant progress in this area, and despite the extensive documentation of these adaptations, the precise mechanisms underlying the detection and communication of insufficient mineral nutrition remain inadequately comprehended (Wang and Wu, 2010).

5.12.1 Mineral Nutrients Are Greatly Influenced by Rootstocks

The rootstock, serving as the foundation of the plant's root system, plays an essential part in the absorption of nutrients from minerals in the soil. This process ultimately influences the development of the shoot system and induces alterations in the physiological features of the tree. The nutrient uptake and transport of rootstocks may exhibit selectivity, while the xylem flux may be influenced by the scion, leading to variations in the nutrient concentrations that are delivered to the leaves and fruits.

5.12.2 Approaches for Regulation of Root and Canopy Development under Restricted Soil Conditions

i. The advancement in enhancing crop productivity and excellence is contingent upon the formulation of methodologies to surmount impediments posed by soil and roots on the canopy.
ii. Furthermore, in order to ensure the sustainability of productivity enhancements, it is imperative that practices are designed to optimize the utilization of resources.

5.12.3 Role of Hormone in Root-Shoot Development

Plants are composed of two distinct systems: the root system, which grows below ground and facilitates the absorption of water and nutrients to the soil, and the branching system, which grows above ground and works photosynthetic and reproductive functions. Collaboration among both of these systems is essential throughout the entire life cycle of the plant

i. Shoots are where auxin is largely produced, and it is actively carried to the roots where it encourages root growth.
ii. CKs migrate acropetally to promote shoot development and take part in root-to-shoot signaling.
iii. To encourage plant organogenesis and regeneration in vitro micropropagation, the balance between these two hormones must be adjusted.

REFERENCES

Davies, W.J., Bacon, M.A., Thomson, D.S., Sobeih, W., Rodríguez, L.G. 2000. Regulation of leaf and fruit growth in plants growing in drying soil: exploitation of the plants' chemical signalling system and hydraulic architecture to increase the efficiency of water use in agriculture. *Journal of Experimental Botany* 51:1617–1626.

Hussain, A., Black, C.R., Taylor, I.B., Roberts, J.A. 1999. Soil compaction. A role for ethylene in regulating leaf expansion and shoot growth in tomato. *Plant Physiology* 121:1227–1237.

Koehl, M.A.R. 1996. When does morphology matter? *Annual Review of Ecology and Systematics* 27:501–542.

Masle, J., Passioura, J.B. 1987. The effect of soil strength on the growth of young wheat plants. *Australian Journal of Plant Physiology* 14:643–656.

McDonald, A.J.S., Davies, W.J. 1996. Keeping in touch: responses of the whole plant to deficits in water and nitrogen supply. *Advances in Botanical Research* 22:229–300.

Mulholland, B.J., Black, C.R., Taylor, I.B., Roberts, J.A. 1999. Influence of soil compaction on xylem sap composition in barley (*Hordeum vulgare*L.). *Journal of Plant Physiology* 155:503–508.

Palmer, J.W. 1980. Computed effects of spacing on light interception and distribution within hedgerow trees in relation to productivity. *Acta Horticulturae* 114:80–88.

Palmer, J. W., & Warrington, I. J. 1998, August. Underlying principles of successful apple planting systems. In *XXV International Horticultural Congress, Part 3: Culture Techniques with Special Emphasis on Environmental Implications*, 513 (pp. 357–366).

Stoll, M., Loveys, B., Dry, P. 2000. Hormonal changes induced by partial rootzone drying of irrigated grapevine. *Journal of Experimental Botany* 51:1627–1634.

Valladares, F., Niinemets, U. 2007. The architecture of plant crowns: from design rules to light capture and performance (Eds. F.I. Pugnaire, F. Valladares), *Handbook of Functional Plant Ecology*. Taylor and Francis, New York, pp. 101–149.

Wagenmakers, P.S., Nijsse, F., DeGendt, C.M.E. 1991. Planting systems and light climate. Research Station for Fruit Growing, Wilhelminadorp, The Netherlands. *AnnualReports* 1991:39–40.

Wang, Y., Wu, W.H. 2010. Plant sensing and signaling in response to K^+ deficiency. *Molecular Plant* 3:280–287.

6 Plant Growth Regulators – Structure, Biosynthesis, and Mode of Action

Pooja Sharma, Chunni Lal Sharma,
Vishal Singh Rana, and Sunny Sharma

6.1 INTRODUCTION

For a long time, attempts to boost crop yields were limited to a few key areas: ensuring appropriate fertilizer and water supplies, selecting and breeding more productive plant varieties (hybridization), and safeguarding crops against invasive species (weeds, insects). If we want to achieve the maximum crop output potential, we must first defeat natural growth regulatory mechanisms, the majority of which are controlled by hormones. Chemical messengers are critical for development, stress response, and homeostasis in various routes and gradients (Jaillais and Chory, 2010). Research has demonstrated that reduced concentrations of plant hormones, a group of organic compounds that occur naturally, can impact diverse physiological processes in plants. Aside from stomatal movement, other processes may also be altered, including growth, differentiation, and development. Phytohormones by Went (1926) define a hormone as a chemical that is transported from one portion of an organism to another. As a hormone, it was first used in plant biology, but it was originally developed from the mammalian notion. The word originates from Greek, where it means "to stimulate" or "to set in motion". Natural plant components that can alter physiological processes at concentrations significantly lower than those of minerals or vitamins are the only common properties of these substances. Plant growth regulators, which are manufactured and biotechnologically derived commercial products, are widely recognized for their regulation properties (Andresen and Cedergreen, 2010).

Auxin is a hormone produced by plants that were the first to be identified. It acts as a chemical messenger that is transported to cause a growth reaction at a location that is distant from its site of synthesis. Historically, synthetic plant growth regulators such as ethylene and ethylene-releasing chemicals have held significant importance. Additionally, gibberellin antagonists, such as paclobutrazol, have been utilized as inhibitors. There are already more than 60 commercially marketed plant bio-regulators. Several of these have become quite important in the agricultural sector. Plant hormones are synthesized by independent metabolic processes. The aforementioned molecules play a crucial role in the regulation of plant growth and development through their involvement in the mediation of both stressors and biotic as well as abiotic responses.

There are five distinct categories of plant hormones that have been identified. Auxins are exemplified by native IAA and synthesized 2,4-D. More than 130 distinct structural forms make up the gibberellins, the second largest class of steroid-like chemicals. The cytokinins are all N-6 substituted adenosine compounds of the third category. Abscisic acid, the sole representative of the fourth class, is generated from the same isoprenoid units as gibberellins. Plant hormone ethylene, the simplest of all plant hormones, is a gas and may be easily transferred from the source to the target. Gibberellins and abscisic acids are among the five hormone groups that originate from

 DOI: 10.1201/9781003354055-6

mevalonic acid. Ethylene and indoleacetic acid are formed when methionine and tryptophane are broken down, and the carbon and nitrogen are removed. Among hormones, cytokinins are the most challenging to synthesize. They could be generated by the breakdown of transfer RNA, according to this theory (tRNA).

6.2 AUXIN

Frits Went found the first AUX in 1926. It was a chemical molecule that helped cells grow and gather in the dark zone (Thimann, 1940). Auxins are a class of plant hormones that are synthesized in both roots and shoot apices and subsequently transported from the apex to the elongation zone. According to Kogl et al. (1934), Indole-3-acetic acid (IAA) holds significant physiological relevance and potential for regulating plant growth, making it the most crucial auxin. Various other compounds have been utilized as auxins, such as IBA, IPA, and NAA acid (Enders and Strader, 2015) (see Figure 6.1).

6.2.1 Forms of Auxins

6.2.1.1 Free Auxins

Those auxins are easily dispersed from the plant. These are the auxins that can be easily extracted using a variety of different solvents (*e.g.* diethyl ether at 0°C–5°C).

6.2.1.2 Bound Auxins

After hydrolysis, autolysis, or enzymolysis, these are liberated from plant tissue. Auxins bound to carbohydrates are found in plants.

6.2.2 Auxin Transport

The coleoptile curvature test devised by Went was employed to investigate the action of auxin, which led to the discovery that indole-3-acetic acid (IAA) exhibits polar transport via the apical to the basal pole of detached oat coleoptile segments. The process of polar transfer exhibits unidirectional characteristics. The predominant hypothesis regarding the establishment of an auxin gradient from the apical meristem to the root apical meristem is through polar transport. It is hypothesized that the phloem serves as the primary conduit for the acropetal transport of auxin in the root.

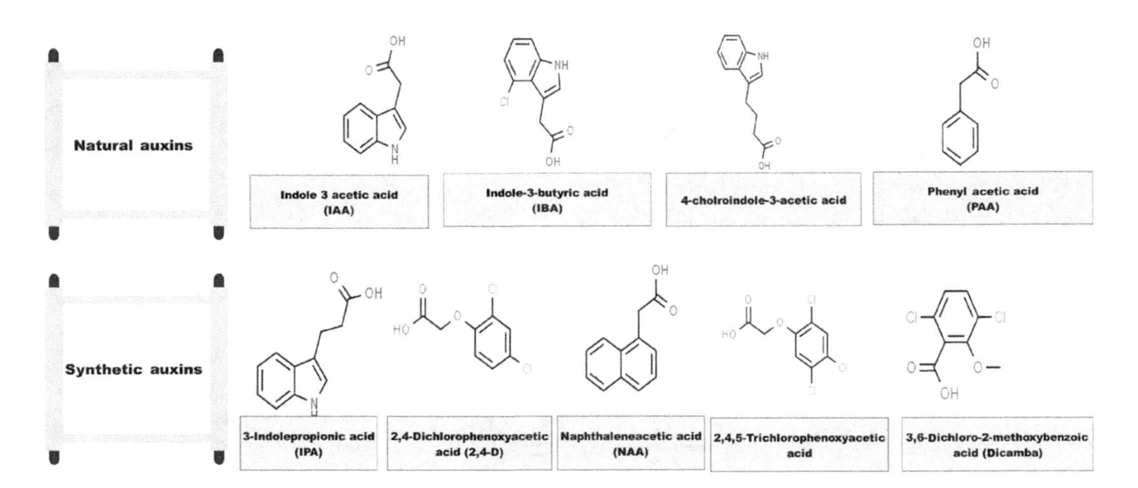

FIGURE 6.1 Some natural and synthetic auxins. (wondershare eddrawmax.)

6.2.3 Biosynthesis

The hormone auxin has been widely recognized as an essential factor for the development and growth of plants in various aspects, as reported by Zhao (2010). The physiological functions of auxin have been deduced through the long-standing observation that plants exhibit responses to exogenous auxin treatments. The aforementioned inquiries have led to an enhanced comprehension of cellular auxin communication and polar transport. The regulating processes of plant growth may be influenced by the quantity of unbound auxin that is present in various plant tissues, cells, and organs. The cellular compartments known as chloroplasts and cytosol are recognized as the primary origins of auxin within the cell. There exist multiple biosynthetic pathways for the production of Indole-3-acetic acid (IAA); however, the most significant pathway is the tryptophan-dependent route that involves the utilization of indole-3-pyruvic acid (IPA) as an intermediary compound. The pathway in controversy commences with the synthesis of tryptophan, followed by the breakdown of an amino acid, which leads to the production of a single ketoacid, namely IPA. The process of decarboxylation results in the formation of Indole-3-acetaldehyde (IAld) and Indole-3-acetamide. Finally, the conversion of these substances into indole-3-acetic acid (IAA) occurs via an enzymatic pathway catalyzed by specific dehydrogenases. Several biosynthetic pathways exhibit similarity to the IPA pathway, albeit with variations in the sequence of enzymatic reactions involved in biosynthesis, as well as the processes of deamination and decarboxylation. Despite the fact that the biologically active form of the hormone is free IAA, the majority of AUXs in crops are covalently bound. Auxins that are "bound" or "conjugated" have been identified in higher-level plants and are believed to lack any discernible hormonal activity. The alternative perspective is depicted in Figure 6.2. Furthermore, biosynthesis may occur through another method that does not require the presence of tryptophan. Indole-3-acetaldehyde is generated as an intermediary in this specific case, and it is

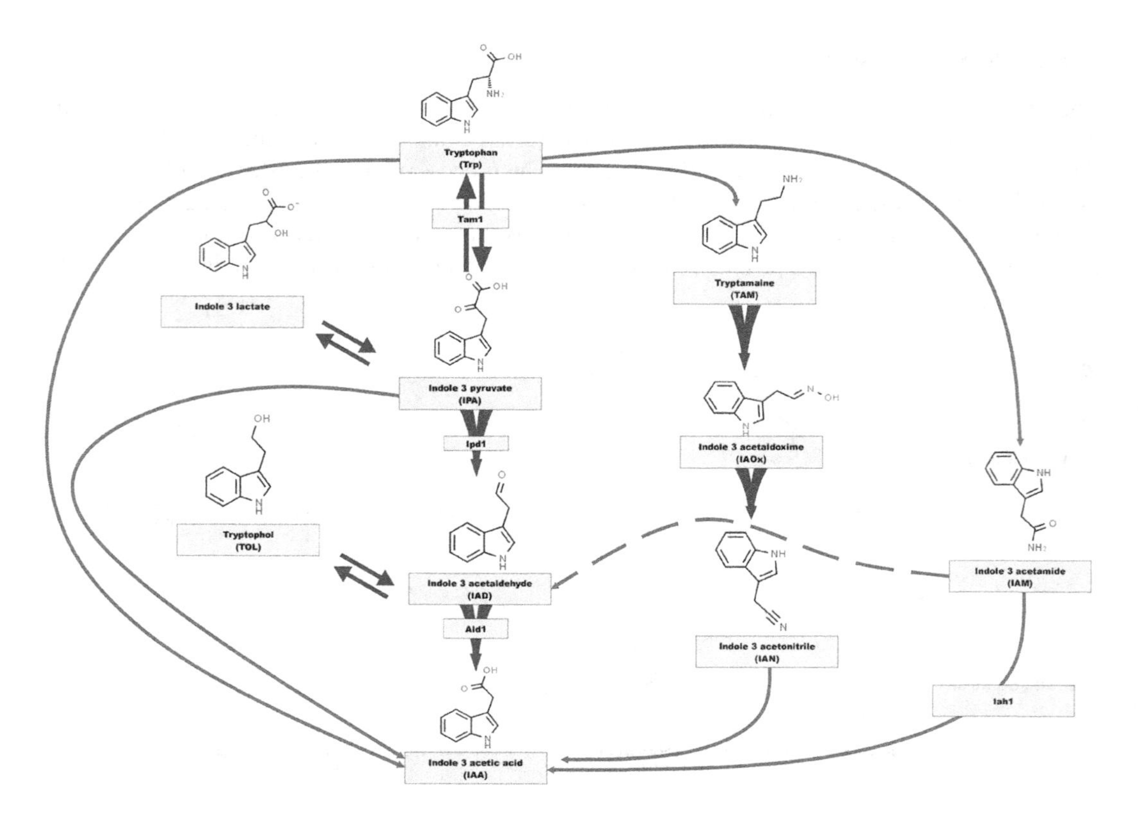

FIGURE 6.2 Auxin biosynthesis pathway. (wondershare eddrawmax.)

the oxidation of indole-3-acetaldehyde that leads to the production of IAA, as stated by Zhao (2012) and Korasick et al. (2013). According to Skoog and Thimann's (1940) findings, the process of auxin synthesis is enzymatic in nature. The conversion of tryptophan into IAA was achieved through a metabolic method built by Wildman, Feri, and Bonner in the year 1944.

6.2.4 Function

Auxin plays various important roles in several plant activities, including:

i. **Cell elongation and cell division**
The interactions of auxin with various binding elements, such as receptors and membranes can result in rapid development as well as a prolonged and sustained response that involves the synthesis of nucleic acids and proteins. Auxin molecules attach to specific locations on the cellular membrane and induce alterations in the conformation of the membrane. This may change membrane permeability, increasing ion flux. Changing the membrane's conformation may release a receptor molecule from the plasma membrane. This receptor may migrate into the target cell's nucleus and boost RNA polymerase activity. This leads to messenger RNA synthesis, which codes for proteins, resulting in auxin-induced growth.

ii. **Phototropism**
Plant shoots exhibit positive phototropism. When a shoot is illuminated from a particular direction, it exhibits phototropism by growing towards the source of light. Initially, the apex of the shoot is employed to ascertain the orientation of the light source. The utilization of blue light is considered the most efficacious method to induce phototropic response in plants. Flavoprotein phototropin is responsible for the absorption of this substance. In plants, auxin transport occurs acropetally (i.e. from the tip down). Cells on the shoot's lateral face are injected with a PIN (Pin-formed) protein, an auxin transporter. The accumulation of auxin in cells located on the shaded side of the plant is attributed to the activity of efflux transporters. The phenomenon of the shoot bending towards the source of light is attributed to the elongation of cells on the shaded side, which are stimulated upon exposure to light.

iii. **Gravitropism**
When a plant senses gravity, it grows. On its side, a plant shoot will grow upwards, yet its roots demonstrate positive gravitropism, which means they grow downwards. By gravity, statoliths (organelles with starch granules) are deposited in root tips when roots are put on their sides. By pumping auxin from the cell's underside, PIN proteins are re-distributed (as they are efflux transporters). As a result, auxin builds up on the root's underside. Root cell elongation is slowed down by this. It is because of this reason that the root grows down as the cells at the top of the root surface elongate.

iv. **Apical dominance**
Apical dominance is a phenomenon commonly observed in higher plants, whereby the development of lateral (axillary) flower buds is inhibited by the growing apical bud. The act of decapitation, or removal of the shoot apex, typically leads to the emergence of one or multiple lateral buds. Shortly following the identification of auxin, it was observed that indole-3-acetic acid (IAA) possessed the ability to act as an alternative for the apical bud in preserving the suppression of lateral buds in bean (Phaseolus vulgaris) flora.

The phenomenon of apical dominance was initially identified by Thimann in 1939. It was suggested that the regulation of apical dominance could be attributed to the influence of auxin, which is synthesized in bud terminals and subsequently transported downwards through the stem. This process impedes the growth of lateral buds. The apical buds of the broad bean plant were excised and substituted with an agar block. As a consequence, there was a swift proliferation of lateral buds. However, upon replacing the bud at the apex with an agar block that contained auxin, it was observed that its lateral buds continued inhibited and exhibited no growth.

v. **Fruit development**

The formation of angiosperm seeds is a consequence of the pollination process that occurs within their flowers. Auxin is released by the seeds as they age, and it is this auxin that aids in the development of the fruit that protects the seeds.

vi. **Abscission**

Leaves and fruits are also affected by auxin. When young leaves and fruits are producing auxin, they remain linked to the stem for as long as they are doing so. The formation of the abscission layer occurs at the basal region of the petiole or fruit stalk in response to a reduction in auxin concentration. When the leaf or fruit's petiole or stalk breaks loose at this point, it falls to the ground.

vii. **Root initiation and development**

Secondary roots begin to form when auxin accumulates in the root's epidermal cells. Many species' adventitious root growth is aided by auxin, which is found in high concentrations of this hormone. In contrast with the stem, a greater concentration of auxin inhibits the elongation of the root, while stimulating the formation of lateral roots. When IAA was mixed into lanolin paste and put on the cut end of a young stem, roots grew quickly and widely. This is a very useful fact that has been used a lot to encourage roots to grow in economically useful plants that are spread by cuttings.

viii. **Auxins as weed killers**

Auxin chemicals are a novel approach to weed control in current crop production, owing to their systemic mobility within the plant and their selective activity against dicotyledonous weeds in cereal crops. Synthetic auxins are commonly employed as herbicides. Dicot weedicides such as 2,4-D and 2,4,5-trichloro phenoxy acetic acid (2,4,5-T) have been documented in the literature (Woodward and Bartel, 2005; Castro et al., 2017).

6.3 GIBBERELLINS

Although gibberellins did not become known to American and British scientists until the 1950s, they had been discovered much earlier by Japanese scientists. Rice farmers in Asia have long known of a disease that makes the rice plants grow tall but eliminates seed production. In Japan, this disease was called the "foolish seedling," or bakanae disease.

Culturing this fungus in the laboratory and analyzing the culture filtrate enabled Japanese scientists in the 1930s to obtain impure crystals of two fungal "compounds" possessing plant growth-promoting activity. One of these, because it was isolated from the fungus Gibberella, was named Gibberellin A. In the 1950s, scientists at Tokyo University separated and characterized three different gibberellins from the gibberellin A sample, and named them gibberellin A1, gibberellin A2, and gibberellin A3. The numbering system for gibberellins used in the past builds on this initial nomenclature of gibberellins A1(GA_1), GA_2, and GA_3 (Gupta and Chakrabarty, 2013). So far, 136GBRs have been identified in higher plants, fungi, and bacteria, of which only a few are bioactive (Figure 6.3).

All the gibberellins are almost similar in structure. They contain a gibbane ring made up of a cyclohexane ring and a 4-lactone ring. They differ in minute details, *viz.*, the number and position of -OH and sometimes $-CH_3$ and -COOH groups at different carbon atoms of the gibbane ring. GA_3 is the most thoroughly studied gibberellin.

6.3.1 Biosynthesis

Gibberellins (GAs) are endogenous plant growth regulators, having tetracyclic, diterpenoid compounds. Biochemical and genetic investigations in higher plants, fungi, and bacteria identified the

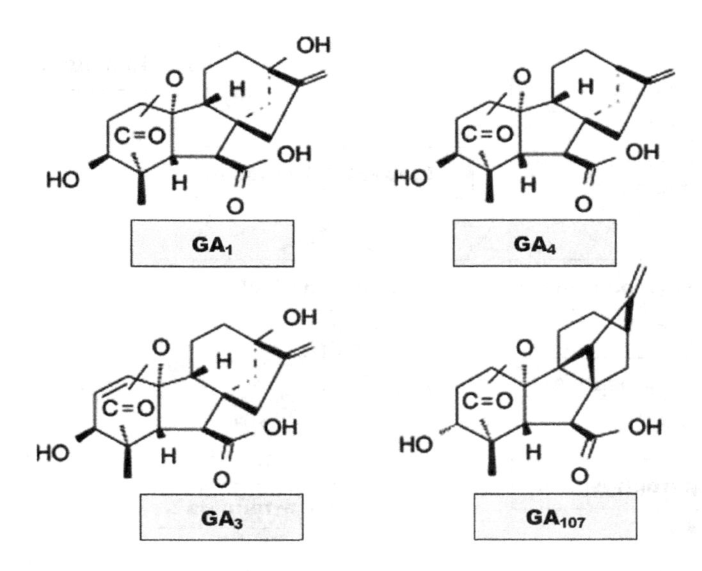

FIGURE 6.3 Some GBRs are biologically active and used as plant growth regulators. (wondershare eddrawmax.)

biosynthetic pathway (Salazar-Cerezo et al., 2018). Dwarf plant bioassay and its quantitative analysis revealed the presence of GA in actively growing tissues, *i.e.* shoot apices, young leaves, and flowers.

Seeds, young leaves, and roots all produce gibberellins in their plastids. The biosynthesis of gibberellins involves the following steps (Figure 6.4):

i. Gibberellins are made by starting with a molecule of acetate. Esterifying acetate with coenzyme (CoA) makes three molecules of acetyl coenzyme A (acetyl Co A), which then go through a series of reactions that make -hydroxyl—methyl glutaryl CoA. (BOG-CoA). Then, BOG-Co A has broken down into two steps that both need NADPH to make mevalonic acid.
ii. Then, mevalonic acid kinase (mevalonate kinase) takes two ATP molecules and phosphorylates mevalonic acid to make mevalonic acid pyrophosphate.
iii. Then, mevalonic acid pyrophosphate is decarboxylated in the presence of ATP to make isopentenyl pyrophosphate (IpPP).
iv. IpPP isomerase changes IpPP into dimethylallyl pyrophosphate (DMAPP), which is another form of IpPP.
v. Then, one molecule of dimethylallyl pyrophosphate accepts one molecule of IpPP, which causes the pyrophosphate to disappear and one molecule of di-isoprenoid alcohol pyrophosphate or geranyl pyrophosphate to form (GPP).
vi. GPP takes in one molecule of IpPP to make farnesol pyrophosphate, which takes in another molecule of IpPP to make geranyl geranyl pyrophosphate (GGPP).
vii. Then, geranyl geranyl pyrophosphate (GGPP) is folded in different ways, and in the presence of ent-copalyl diphosphate synthase, it is changed into a partially cyclized compound called copalyl pyrophosphate (CPP). Then, ent-kaurene synthase changes it into ent-kaurene, a fully cyclic compound.
viii. At C-19, ent-kaurene is broken down into ent-kaurenol, ent-kaurenal, and entkaurenoic acid. This is turned into ent-7-hydroxy kaurenoic acid when it is hydroxylated.
ix. Now, the -ring and the -hydroxylation break down. To change ent-7-hydroxy kaurenoic acid into a 20-carbon GA12-aldehyde, a 6-hydrogen is lost, a 7, 8-bond is moved to the 6,8-position, and a proton is lost from the C-7 that was moved away.
x. Ent-Kaurene Acid Oxidase (KAO) changes GA12-aldehyde into GA12. For C-19 GAs like GA3 to form, there must be a loss of one carbon.

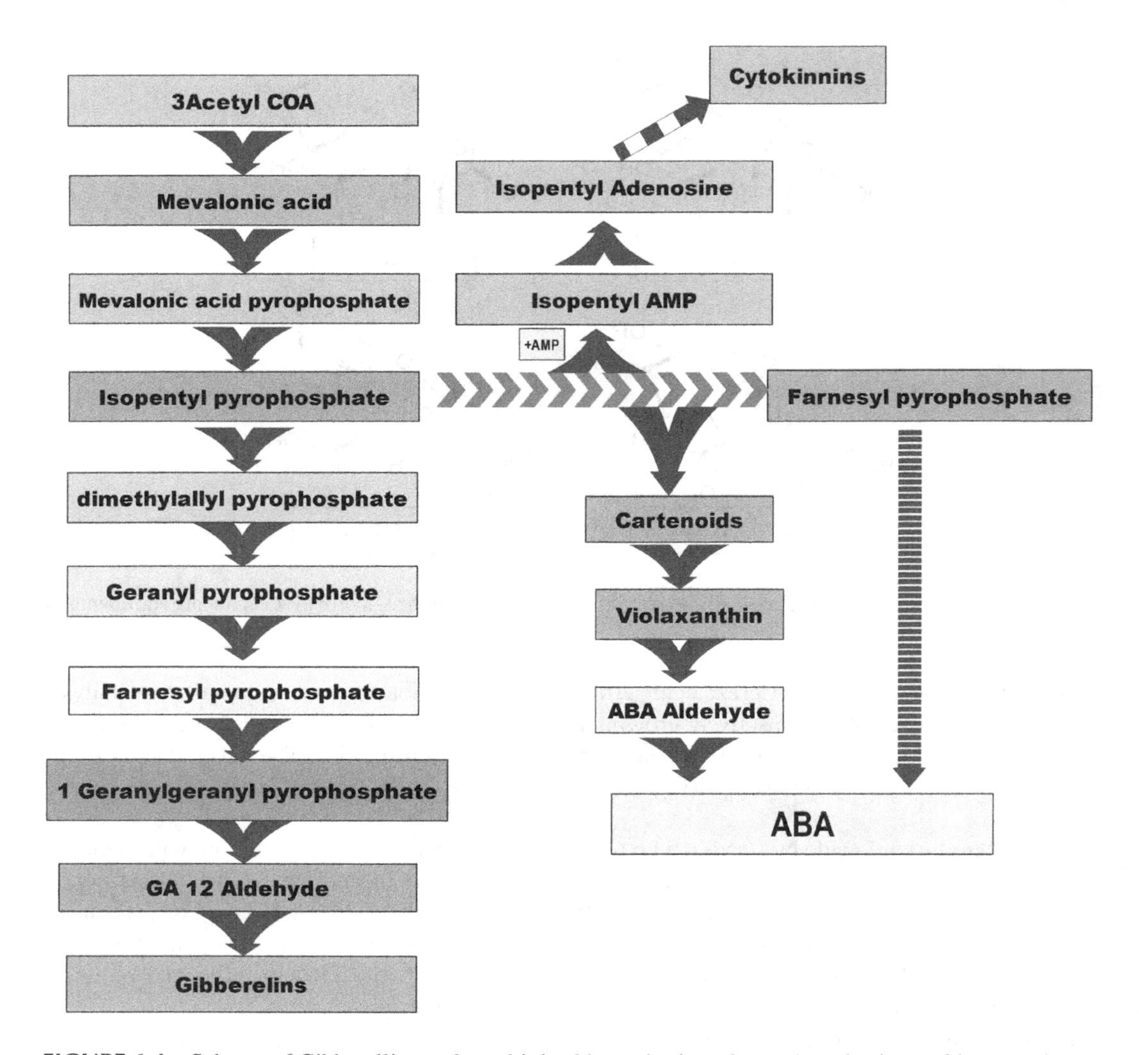

FIGURE 6.4 Scheme of Gibberellins and cytokinins biosynthesis pathway. (wondershare eddrawmax.)

6.3.2 Transport

Gibberellins are transported in the entire conducting system- both in the phloem and xylem. It moves from one part to the other in the phloem similar to the transport of carbohydrates and other substances. GA is translocated in the xylem due to the lateral movement between the two vascular bundles. In general GA movement is non-polar in contrast to polar transport of auxin.

6.3.3 Functions of Gibberellins

i. **Genetic dwarfism**
   Gibberellins cause internode elongation, which overcomes genetic dwarfism in plants like dwarf peas and maize. Dwarf peas have short internodes and enlarged leaves. Gibberellin causes internodes to extend and grow taller. In dwarf plants, (i) the gibberellin gene is lacking or (ii) natural inhibitors are more concentrated. External gibberellin makes up for endogenous gibberellin deficit or overcomes natural inhibitors.

ii. Bolting and flowering

GA acts as a controlling factor in between internode growth- a form of growth called a Rosette and leaf development. The reason that a plant either remains in rosette form or bolt and flowers appears to be related to the amount of native GA present in plants. Native GA is found greater in bolted plants than in non-bolted plants. The influence of GA on bolting includes a 'stimulation of cell division' as well as 'cell elongation'. Bolting is rapid internode growth before reproduction. Such plants need long photoperiods or cold to blossom.

iii. Parthenocarpy

Auxin treatments have little effect on pome or stone fruit. However, GA has caused parthenocarpy in both pome and stone fruit. Parthenocarpy, the development of seedless fruits without pollination and fertilization, can be either spontaneous or artificial.

iv. Light-inhibited stem growth

Light slows stem growth. Dark-grown plants get etiolated and have longer, thinner, pale stems while light-grown plants have shorter, thicker, green stems, indicating that light inhibits stem elongation. Gibberellin additionally darkens the stems of light-grown plants.

v. Mobilization of storage compounds during germination (Breaking of dormancy)

Seed dormancy is known to be broken by gibberellins. When applied to the seeds of Tobacco and Lettuce, they stimulate germination in full darkness. Apple and peach seeds require a period at a low temperature (1°C–7°C) to germinate, but gibberellins can take the place of this requirement. When seed germination is inhibited, gibberellins can be used in conjunction with low temperatures, long days, or red light to get the seeds to germinate. An exogenous supply of gibberellins can also break the hibernation of buds in evergreen and deciduous trees and plants.

vi. *De novo* synthesis of the enzyme-α-amylase

In the aleurone layer surrounding the endosperm of cereal grains, gibberellins play a significant role in promoting the *de novo* (*i.e.*, new) synthesis of α- amylase during germination. To supply energy for the developing embryo, this enzyme hydrolyzes starch, resulting in the formation of simple sugars (Figure 6.5).

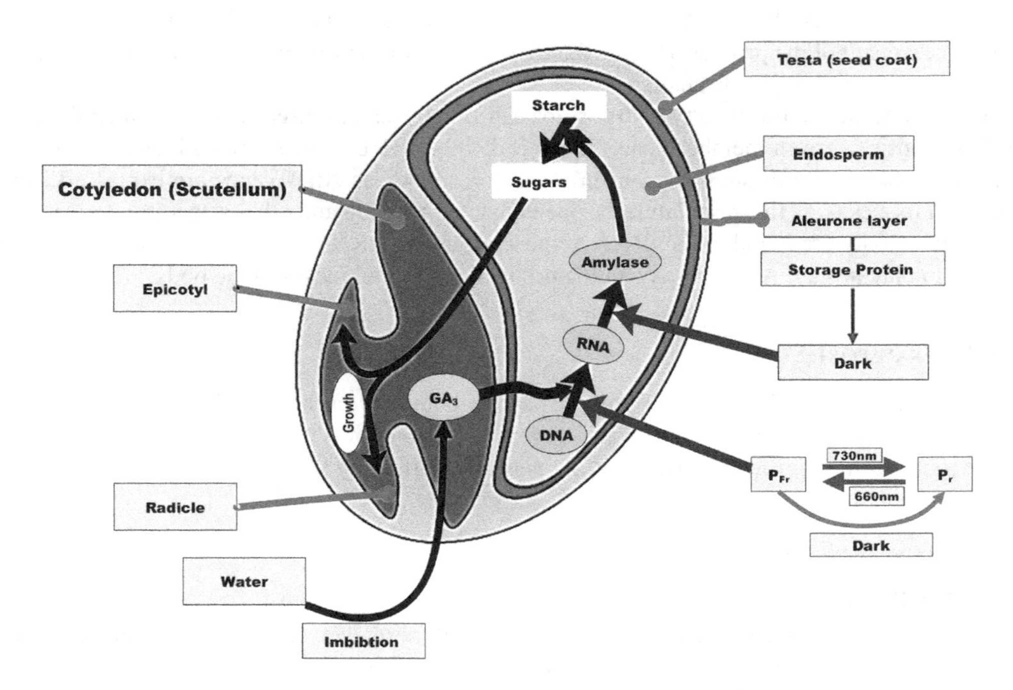

FIGURE 6.5 Induction of amylase enzyme by gibberellin.

6.4 CYTOKININ

Cytokinins are plant-specific chemical messengers (hormones) that play a central role in the regulation of the plant cell cycle and numerous developmental processes. Cytokinins were discovered by F. Skoog, C. Miller, and co-workers during the 1950s as factors that promote cell division (cytokinesis). The first cytokinin discovered was an adenine (aminopurine) derivative named kinetin (6-furfurylaminopurine), which was isolated as a DNA degradation product. The first common natural cytokinin identified was purified from immature maize kernels and named zeatin (chemical name: 6-(4-hydroxy-3-methylbut-2-enylamino) purine; See Figure 6.4).

Haberlandt (1913) found phloem compounds that stimulated cell division. Von Overbeek (1944) found that coconut milk stimulated cell division in tissue-cultured embryos. Skoog in 1955, discovered kinetin, a cell-division stimulant. Kinetin is an artifact, although zeatin and other cytokinins were later identified in various species. They are made in the roots of the plant and then moved to the leaves and stem. There are several naturally occurring cytokinins, and they can be put into two main groups: amino purines and derivatives. In the first one, the N^6 position has a purine ring with a side chain. This is where 6-furfuryl amino-purine (kinetin) fits in. Several urea-based compounds can be used to make kinetin, such as celorophenylurea. It has been shown that benzyl urea blocks the purine that phenyl urea blocks with cytokinins.

6.4.1 Biosynthesis

Cellular cytokinin homeostasis can be influenced by the rate of *de novo* synthesis, metabolic interconversion, and breakdown, as well as transport activities. Cytokinin metabolism primarily involves the conversions of cytokinin bases, ribosides, ribotides, side chain modification, conjugation, and conjugate hydrolysis processes, and the destruction of cytokinins.

The transfer of the isopentenyl moiety from dimethylallyl pyrophosphate (DMAPP) to AMP, ADP, or ATP is the first and rate-limiting step in the biosynthesis of isoprenoid-type cytokinins. DMAPP:AMP/ADP/ATP isopentenyl transferases are involved in the process (IPT). Most known plant IPT enzymes prefer ADP and ATP as substrates, whereas bacterial enzymes prefer AMP. Isopentenyl-AMP, -ADP, and -ATP, which are precursors of physiologically active cytokinins, are formed as a result of the process. (Figure 6.5). Cytokinins of the zeatin type are formed via hydroxylation of the isopentenyl side chain.

Alternately, the adenine moiety's N6 position can already have an already-hydroxylated side chain. In Arabidopsis, a seven-member gene family codes for IPT enzymes (AtIPT1, AtIPT3-AtIPT8). Cellular cytokinin synthesis is evident in all major organs, as AtIPTgenes are expressed in root and shoot tissues (*e.g.*, the vasculature). Isopentenylated adenine and other structural derivatives of cytokinin activity are found in the tRNA of most organisms. tRNAs, on the other hand, are widely thought to play just a tiny role, if any, in the production of cytokineses in the body.

6.4.2 Transport

Cytokinins are present in all plant tissues. They are abundant in the root tip, the shoot apex, and immature seeds. Xylem and phloem transfer cytokinins from roots to shoots. Cytokinins may coordinate root and shoot growth by transmitting nutrient availability information.

6.4.3 Functions of Cytokinin in Plants

i. **Cell division**
   Cytokinins are necessary for the process of cytokinesis; yet, chromosomes can duplicate even without their presence. Even in cells that are intended to be permanent, cytokinins can cause division when auxin is present. It has been discovered that cell division in the

callus, which is an unorganized, undifferentiated irregular mass of dividing cells in tissue culture, requires both hormones. The induction of cell division is the most significant biological effect that kinetin has on plants.

ii. **Cell enlargement**
Cytokinins may cause cell expansion like auxins and gibberellins. Leaves of *Phaseolus vulgaris*, pumpkin cotyledons, tobacco pith culture, tobacco root cortical cells, etc., have enlarged cells.

iii. **Concentration of apical dominance**
External application of cytokinin promotes the growth of lateral buds and hence counteracts the effect of apical dominance.

iv. **Dormancy of seeds**
Like gibberellins, the dormancy of certain light-sensitive seeds, such as lettuce and tobacco can also be broken by kinetin treatment.

v. **Delay of senescence (Richmond-Lang effect)**
The aging process of leaves is typically accompanied by a decrease in chlorophyll content and a speedier decomposition of protein. Therapy with kinetin can delay the onset of senescence by several days. This is accomplished by enhancing RNA synthesis, which is then followed by protein synthesis. During their research on the detached leaves of Xanthium, Richmond and Lang in 1957 discovered that kinetin can delay the onset of senescence by several days.

vi. **Morphogenesis**
It has been shown that high auxin and low kinetin produced only roots, whereas high kinetin and low auxin could promote the formation of shoot buds (Villanueva et al., 2013).

6.5 ETHYLENE

This simple gas is produced in modest quantities by numerous plant tissues and regulates growth and development. They are found in physiologically mature fruits undergoing ripening. Denny and Miller (1935) showed that ethylene could break dormancy, hasten fruit ripening, and induce flowering in pineapple, and was naturally produced by several plant organs (Abeles, 1973). Crocker et al. (1935) suggested that ethylene was an endogenous hormone. The scientific community disagreed that ethylene should be a hormone. A two-carbon molecule floating in the air could not be a hormone. The physiological relevance of ethylene wasn't understood until 1959, when the gas chromatograph was used to measure it (Burg and Thimann, 1959).

Only ethylene is a gaseous plant growth regulator. Volatile gas is found in smoke and industrial emissions. It's produced by practically all parts of higher plants in minute amounts but has noticeable effects. It's a plant metabolite and gaseous hydrocarbon (See Figure 6.6). Ethylene is an inhibitory hormone-like abscisic acid. When pure and unsaturated, it's a colorless, combustible gas with a subtle "sweet and musky" odor. Volatile ethylene gas isn't directly utilized. Commercially, ethephone (2-chloroethyl phosphonic acid) generates ethylene gas after hydrolysis.

6.5.1 Sites of Synthesis

In reaction to physical or chemical stress, the majority of tissues will produce ethylene. Specifically, it is produced in tissues that are going through the process of ripening or senescence.

6.5.2 Transport

Because it is a gas, ethylene moves away from the location where it was synthesized by diffusion (Grierson, 2012; Jiang and Asami, 2018).

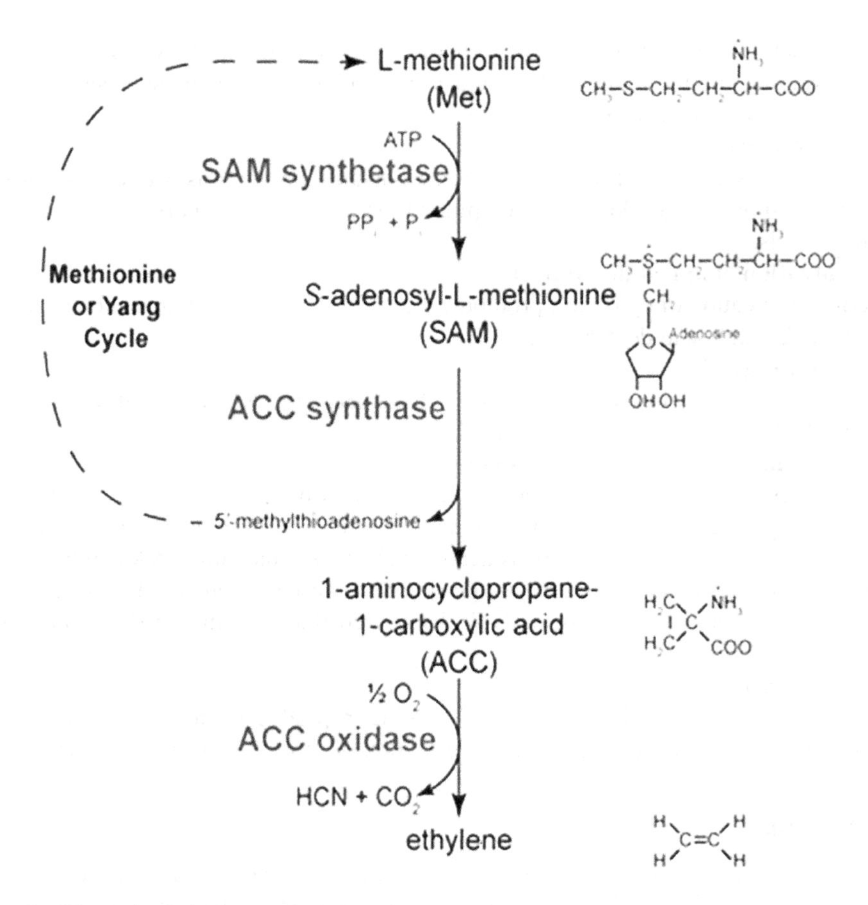

FIGURE 6.6 Biosynthesis pathway of ethylene.

6.5.3 Biosynthesis

Sgang Fa Yang provided the mechanism for the production of ethylene in 1980. This pathway is also known as the Yang Cycle. The following steps make up the biosynthetic route leading from methionine to ethylene. Methionine, an amino acid that contains sulfur, is made active by the presence of ATP, at which point it undergoes conversion to S-adenosyl methionine (SAM), and pyrophosphate is released. SAM synthetase is the enzyme responsible for catalyzing this reaction (ATP-Methionine S-Adenosyl transferase).

The sulfur atom of methionine is bonded to the C-5 position of the ribose moiety in adenosine to form the compound known as S-adenosyl methionine. In several instances, researchers have shown that Selenomethionine (SeMet) is a superior substrate for the ethylene production process compared to methionine. The SeMet molecule has a stronger attraction for the enzyme SAM synthetase than the methionine molecule does.

The enzyme ACC synthase helps turn SAM into 1-aminocyclopropane carboxylic acid (ACC). This is the next step in the biosynthesis of ethylene (ACS). This step is thought to be the slowest step in the process of making ethylene, and ethylene production is affected by anything that changes the activity of ACC synthase. The last step in the biosynthesis of ethylene is the oxidative cleavage of 1-aminocyclopropane carboxylic acid (ACC) by the enzyme ACCoxidase (ACO), which used to be called the ethylene-forming enzyme (EFE). This creates ethylene, carbon dioxide, and hydrogen cyanide (Figure 6.6).

6.5.4 Physiology of Action in Plants

i. **Fruit ripening**
Gaseous ethylene helps fruits mature naturally. Ethylene makes fruit ripen faster, thus, it's dubbed the "ripening hormone." Ethylene speeds up the ripening of fruit by breaking down chlorophyll, cell walls, and carbohydrates (Chen et al., 2018). Both climacteric and non-climacteric fruits react differently to ethylene from the outside, but this is obvious in climacteric fruits. Climacteric fruits like apples, apricots, avocados, bananas, blueberries, figs, mangos, kiwis, tomatoes, and papaya make ethylene and breathe more when mature. This transition is called climacteric, and climacteric fruits show it. In fruits that don't undergo climacteric respiration (cherry, cucumber, grape, lemon, pineapple, strawberry, sweet orange, etc.), ethylene treatment doesn't induce climacteric respiration or additional ethylene to be created, and the rate at which the fruit ripens doesn't vary.

ii. **Flower inhibition and sex expression**
Ethylene stops most plants from blossoming, while mango and pineapple blossoms grow faster. The photoperiod affects how much ethylene inhibits flowers from opening, and it's possible that ethylene injected during the dark period could do so. Ethylene alters unisexual plants' sex displays. Diminishing male flowers, helps plants generate more female flowers, especially cucurbits.

iii. **Epinastic responses**
When roots are saturated or full of water, ethylene produces leaves epinasty. These roots produce anaerobic conditions and make aminocyclopropane-1-carboxylic acid, which is transported by the xylem to the leaf, where it is converted into ethylene in the presence of oxygen. This creates epinasty, in which the upper side of the leaf petiole develops quicker than the lower side, and the leaf slopes downward, which may aid the plant lose water.

iv. **Acceleration of senescence and abscission**
Ethylene accelerates or facilitates plant aging and falling off. Endogenous ethylene increases with age, reducing chlorophyll content in leaves, flowers, and fruits. Ethylene causes leaves, blossoms, and fruits to fall off.

6.6 ABSCISIC ACID

Abscisic acid, sometimes known as ABA, was the final major hormone to be discovered. Abscisin II was found in cotton by a group led by Addicott that was researching hormonal interactions in cotton (Ohkuma et al., 1963), and Dormin was found in woody plants by a group led by Wareing that was researching dormancy in woody perennials. Both studies were conducted independently (Corniforth et al., 1965). At the end of the day, it was discovered that the two chemicals were the same thing, and in 1967, during a symposium on plant hormones held in Ottawa, Canada, they were given the name "abscisic acid."

6.6.1 Transport

Both the roots and the leaves contribute to the export of ABA through their respective xylem and phloem. There is some evidence to suggest that ABA might go from the phloem to the roots of the plant and then make its way back up through the xylem to the shoots.

ABA has at least two biosynthetic pathways: (i) fungi-produced ABA derived directly from farnesyl pyrophosphate, and (ii) plant-produced ABA obtained through an "indirect" process via cleavage of a -carotene precursor via xanthoxin and ABA-aldehyde (C40 pathway). This is then followed by a two-step conversion of the intermediate xanthoxin to ABA via ABA-aldehyde. Finally, ABA is produced.

In the first step, zeaxanthin and antheraxanthin are turned into violaxanthin by a process called epoxidation, this happens in plastids. A zeaxanthin epoxidase (ZEP) is what makes this step happen. After violaxanthin goes through a series of changes to its structure, it becomes a 9-cis-epoxycarotenoid. The 9-cis-epoxycarotenoid dioxygenase (NCED) cuts the major epoxycarotenoid 9-cis-neoxanthin apart when it reacts with oxygen. This makes a C15 intermediate called xanthoxin. The product, xanthoxin, is then sent to the cytosol, where it is changed into ABA in two steps by ABA-aldehyde.

6.6.2 Physiology of Action in Plants

i. **Stomatal regulation**
ABA attaches to stomatal guard cell receptors upon stress. The receptors activate numerous pathways that converge to raise cytosol pH and transport Ca^{2+} from the vacuole to the cytosol. These modifications promote the loss of negatively charged ions (anions), mainly NO and Cl^-, and K^+ from the cell. Loss of these solutes in the cytosol lowers cell osmotic pressure, turgor, and stomata closure. It results in a reduction of osmotically active solutes so that the guard cells become flaccid and stomata get closed.

ii. **Seed and bud dormancy**
In plants, ABA inhibits growth and produces bud dormancy. It inhibits bud growth, seed, and bud dormancy. It inhibits bud and seed growth in temperate plant species, but when it dissipates, growth begins. As ABA levels drop in plants, gibberellin levels rise and growth begins. Without ABA, buds, and seeds would grow in warm winter periods and die when they froze again. ABA fades slowly from tissues and its effects take time to be neutralized by other plant hormones, delaying physiological pathways that protect against premature development. It accumulates in seeds during fruit ripening, limiting germination in the fruit or before winter.

iii. **Seed development and germination**
ABA accumulates in growing seed embryos, either *de novo* or from leaves. It prevents embryonic germination and vivipary. Exogenous ABA suppresses non-dormant seed germination. Once it is eliminated by washing the seeds, germination can occur due to the inhibition of germination enzymes, water intake by developing seeds, etc.

iv. **Senescence and abscission**
Many researchers believe ABA is an endogenous element in leaf senescence and abscission. Exogenous ABA causes leaf yellowing in deciduous and herbaceous plants. Once photosynthetic activity drops below the compensation point, senescing leaves produce more ABA.

v. **Flowering**
In long-day plants, ABA inhibits blooming by counteracting gibberellin's effect on flowering. ABA, on the other hand, promotes flowering in plants that have a short flowering period.

REFERENCES

Abeles, F. 1973. *Ethylene in Plant Biology.* Academic Press, London.

Andresen, M. and Cedergreen, N. 2010. Plant growth is stimulated by tea-seed extract: a new natural growth regulator? *HortScience.* 45:1848–53.

Burg, S.P. and Thimann, K.V. 1959. The physiology of ethylene formation in apples. *Proceedings of National Academy of Sciences.* 45:335–44.

Castro, A., de Souza, C. and Fontanetti, C. 2017. Herbicide 2, 4-D: a review of toxicity on non-target organisms. *Water, Air, and Soil Pollution.* 228:1–12.

Chen, Y., Grimplet, J., David, K. 2018. Ethylene receptors and related proteins in climacteric and non-climacteric fruits. *Plant Science.* 276:63–72.

Crocker, W., Hitchcock, A.E. and Zimmerman, P.W. 1935. *Similarities in the Effects of Ethylene and the Plant Auxins.* Contribution from Boyce Thompson Institute.
Denny, F.E. and Miller, L.P. 1935. *Storage Temperature and Chemical Treatments for Shortening the Rest Period of Small Corms and Cormels of Gladiolus*. Contribution from Boyce Thompson Institute.
Enders, T. and Strader, L. 2015. Auxin activity: past, present, and future. *American Journal of Botany*. 102:180–96.
Grierson, D. 2012. 100 years of ethylene – A personal view. In *The Plant Hormone Ethylene*, ed. Michael T. McManus, Wiley-Blackwell. 44. 1–17.
Gupta, R. and Chakrabarty, S. 2013. Gibberellic acid in plant: still a mystery unresolved. *Plant Signaling and Behavior.* 8:41–5.
Haberlandt, G. 1913. *On the Physiology of Cell Division*. Sitzungsberichte der Koniglich Preussischen Akademie der Wissenschaften. 318–45.
Jaillais, Y. and Chory, J. 2010. Unraveling the paradoxes of plant hormone signaling integration. *Nature Structural and Molecular Biology*. 17:642–45.
Jiang, K. and Asami, T. 2018. Chemical regulators of plant hormones and their applications in basic research and agriculture. *Bioscience, Biotechnology and Biochemistry*. 82:1265–300.
Kogl, F., Haagen-Smit, A. and Erxleben, H. 1934. Uber Ein Neues Auxin ('Hetero-Auxin') Aus Harn. 11. Mitteilung Über Pflanzliche Wachstumsstoffe. *Hoppe-Seyler's Zeitschrift Fur Physiologische Chemie.* 228:90–103.
Korasick, D.A. Enders, T.A. and Strader, L.C. 2013. Auxin biosynthesis and storage forms. *Journal of Experimental Botany*. 64:2541–55.
Ohkuma, K., Lyon, J. Addicott, F. and Smith, O. 1963. Abscisin II, an abscission-accelerating substance from young cotton fruit. *Science*. 142:1592–93.
Overbeek, J. 1944. Growth-regulating substances in plants. *Annual Review of Biochemistry*. 13:631–66.
Salazar-Cerezo, S., Martinez-Montiel, N., Garcia-Sanchez, J., Perez-Terron, R. and Martinez-Contreras, R. 2018. Gibberellin biosynthesis and metabolism: a convergent route for plants, fungi and bacteria. *Microbiological Research*. 208:85–98.
Skoog, F. and Thimann, K.V. 1940. Enzymatic libration of auxin from plant tissues. *Science*. 19:64–92.
Thimann, K. 1939. Auxins and the inhibition of plant growth. *Biological Reviews*. 14:314–37.
Thimann, K. 1940. Growth hormones in plants. *Journal of the Franklin Institute.* 229:337–46.
Villanueva, F., Avila, M., Mansilla, A., Abades, S. and Caceres, J. 2013. Effects of auxins and cytokinins on tissue culture of Ahnfeltia plicata (Hudson) fries, 1836 (Ahnfeltiales Rhodophyta) from Magellan region. *Annals of the institute of Patagonia.* 41:99–111.
Went, F.W. 1926. On growth accelerating substances in the coleoptile of *Avena sativa.* Proceedings of the Section of Sciences, *Koninklijke Akademie van Wetenschappen te Amsterdam*. 30:10–9.
Woodward, A. and Bartel, B. 2005. Auxin: regulation, action and interaction. *Annals of Botany.* 95:707–35.
Zhao, Y. 2010. Auxin biosynthesis and its role in plant development. *Annual Review of Plant Biology*. 61:49–64.

7 Plant Growth Inhibitors and Growth Retardants

Vandana Thakur, Umesh Sharma, and Sunny Sharma

7.1 INTRODUCTION

Growth retardants are a chemical compound that retards the cell division and elongation of plant tissues. It helps to regulate the plant physiology deprived of any determinate effects. Nicotinum derivatives are the first known growth retarded found in 1949. This new class of chemicals, i.e., nicotiniums retard sub apical meristematic activity, without affecting the apical meristem. Some of the reports had confirmed that a compound under the code name Amo-1618, a thymol-based quaternary ammonium complex, has been described to be the most active chemical of the group which retarded the snap beans growth under greenhouse cultivation. Beta-hydroxyethyl hydrazine, encouraged flowering in pineapple plants. In a group of related quaternary compounds, containing a phosphonium cation has growth retarding qualities. Whereas, the foliar application of substituted maleamic and succinamic acids retarded the growth of vine crops, potatoes, legumes, and ornamentals (Table 7.1).

TABLE 7.1
Effect of Growth Regulators in Vegetable and Fruit Crops

Crop	Growth Regulators	Improved Traits	References
Tomato	GA_3 @ 50ppm	Plant height, no. of leaves, no. of clusters per plant, no. of flowers per plant, earliness in flowering & fruit setting	Gawali et al. (2023)
Chilli	GA_3 @ 1 ppm and 10 ppm	Plant height, no. of leaves, and stem	Kumar et al. (2018); Naga et al. (2022)
Capsicum	GA_3 @ 30 ppm	No. of fruits per plant, fruit set, fruit yield per plant, Vitamin C, total soluble solids, and capsaicin	Ahmed et al. (2022)
Cucumber	GA_3 @ 50 & 100 ppm	Enhanced growth and phonological traits, no. of branches, leaf area, earliness in flowering, and vine length	Kumar et al. (2022)
Muskmelon	GA_3 @ 20 ppm and ethereal @ 150 ppm	Increased vine length, earliness in flowering and fruit set, sex ratio	Zankat et al. (2022)
Watermelon	GA_3 @ 40 ppm	Total soluble solids, total sugar, chlorophyll content, fruit yield	Kumar et al. (2022)
Okra	GA_3 & NAA @ 75 ppm	Plant height, leaf area index, no. of nodes and branches per stem, stem diameter, and early flowering	Kumari et al. (2022)
		Fruit Crops	
Wood Apple	GA_3 100–150 ppm	25% germination	Yadav (2016)
Custard apple	GA_3 500 ppm	Improved germination	Yadav (2017)
Mango cv. Amrapali	GA_3 @ 30 ppm & NAA @ 50 ppm	Fruit retention %, fruit weight, fruit yield, fruit width, fruit volume, peel and pulp weight, TSS, total sugar, ascorbic acid and reducing sugar	Parauha and Pandey (2019)

(*Continued*)

DOI: 10.1201/9781003354055-7

TABLE 7.1 (*Continued*)
Effect of Growth Regulators in Vegetable and Fruit Crops

Crop	Growth Regulators	Improved Traits	References
Litchi	NAA @ 2.5 ppm	Improvise the flowering traits	Megu et al. (2021)
Papaya cv. Red Lady	Gibberellic acid and Etheral @ 150 ppm	Total soluble solids, ascorbic acid, total sugar, reducing and non-reducing sugar, sugar acid ratio	Dubey et al. (2020)
Jamun	IBA @ 3,000 ppm (hardwood cutting)	Maximum sprouting on shoots	Pandey et al. (2023)
Sapota cv. Kalipatti	GA_3 @ 100 ppm & CPPU @ 4 & 6 ppm	Number of fruits, fruit yield, reducing sugar, TSS, non-reducing sugar, ascorbic acid, low acidity	Naik et al. (2022)

7.2 ROLE OF GROWTH RETARDANTS

Plant growth retardants play a significant key role in plant growth and inhibition, which are listed as follows (Tables 7.2 and 7.3):

i. Suppress the excessive vegetative growth and encourage the reproductive growth of the plant
ii. Augmented the flowering and fruiting
iii. Control alternate and irregular fruit-bearing
iv. Induce the early maturity in crops
v. Encourages the fruit ripening
vi. Prevent the biosynthesis of growth hormones in plant tissues

Plant growth retardants don't synthesize naturally in plants but are synthetic compounds that act in the retardation of stem elongation, preventing cell division and cell elongation.

Some examples of plant growth retardants include AMO 1618, phosphon D, CCC, MC, and Alar. These growth retardants are described as follows:

AMO-1618 (2-isopropyl-4-dimethylamino-5-methylphenyl-1-piperidinecarboxylate methyl chloride): these are the chemicals that inhibit the biosynthesis and activity of gibberellins in the plant tissues, which cause dwarfism in plants and results in bushy and sturdy growth of the treated crop plants.

TABLE 7.2
Different Growth Retardants Along with their Activities and Different Attributes

Name of the Chemical	Activity	Usages Pattern	Active sites	Application Method(s)	Shelf Life (years)
Ancymidol	2	2	Leaves and roots and	Foliar dip, and drench	3 years
Chlormequat Chloride	1	1	Leaves, roots	Foliar	3 years
Daminozide	1	1	Leaves	Foliar	2 years
Ethephon	2	1	Leaves	Foliar	Indefinite
Flurprimidol	3	3	Stems and roots	Foliar, dip, and drench	2 years
Paclobutrazol	3	3	Stems and roots	Foliar, dip, drench	4 years
Uniconazole	3	3	Stems and roots	Foliar, dip, drench	2 years

Source: Barrett (2001).
* 1 = Low; 2 = Medium, 3 = High.

TABLE 7.3
Functions of Various Growth Retardants in Fruit Crops

Name of the Chemical	Working Against
AMO-1618	Retards synthesis of gibberellic acid (GA)
Phosphon D	Gibberellin biosynthesis inhibitor
	Promotes photosynthetic electron transport by up to 2-fold
Cycocel	Reduce stem-elongating and apical-dominant plant hormones.
	Reduces plant height, strengthens stems, and increases tiller productivity
Mepiquat chloride	Balances crop nutrition and reproduction by reducing gibberellic acid biosynthesis.
	Plant geometry is often managed with DPC

Phosphon D (tributyl-2, 4-dichloro benzyl phosphonium chloride): is identified as a suppressor of gibberellin biosynthesis and augments the transport of photosynthetic electron. Phosphon D inhibits mostly the synthesis of proteins involved in growth.

CCC (Chlormequat chloride): is one of the growth retardants that inhibits the action of gibberellic acid. Chlormequat chloride has been known as an antimetabolite than anti-auxins or anti-gibberellins. This chemical is involved in the metabolism modification of the plant by performing various functions within the plant, such as suppression of internode elongation, altering the development rates, and also interrupting the apical growth of the plant. The effects of growth inhibitors can be altered by exogenous application of gibberellin, but in some of the cases, gibberellin also reverses the effect of it, by decreasing the survival percentage rate of plants, particularly in under water stress conditions.

MC (Mepiquat chloride): this chemical is a composition of ammonium salts having an equimolar quantity of mepiquat cations and chloride anions. It performs various functions within the plant, as it suppresses the biosynthesis of gibberellin in plant tissues, which ultimately reduced the cell division and elongation that results in a reduction in the vegetative growth of plants (Wang et al., 2014).

Alar (Daminozide): Alar is mostly utilized in fruit orchards to make fruits more attractive and to diminish pre-harvest losses. Apart from this, it has also been used to decrease the shoot length to achieve the desired plant height without altering the progressive patterns. These significant effects of alar can be proficiently achieved by lowering the cell division and elongation rate (Tedila, 2022).

7.3 CLASSIFICATION OF PLANT GROWTH RETARDANTS

Plant growth retardants are a commercially important group of plant bioregulators or plant growth regulators. However, they play an insignificant role compared to fungicides, herbicides, and insecticides, corresponding to only a small percentage of the global market of crop-protection synthetic substances. As soon as growth retardants are employed at a suitable concentration, plant structure is altered characteristically. The impact on plant morphology is due to growth retardants acting against auxin and gibberellins, which have a key role in shoot elongation. There are two main groups in which the existing growth retardants can be classified:

7.3.1 Ethylene-Releasing Compounds: Such as Ethephon

Ethylene-releasing chemicals act as growth retardants by inducing shorter and thicker stems due to the effects of ethylene. Ethephon is primarily a growth suppressant to prevent lodging in cereal and grain crops. Ethephon is a plant growth regulator commonly used for encouraging, abscission, flower induction, and fruit ripening in crops. Apart from this, there are various applications of this growth regulator as a growth retardant, where it helps in height control by inhibiting stem elongation, flower, and bud abortion, etc. All this results from the ethylene which is generated after the application of ethephon to the fruit crops. Ethephon does not inhibit GA synthesis, but rather, the ethylene generated from ethephon causes plants to recognize how their cells develop and, as a result, they elongate less when exposed to ethylene. Another important use of ethephon is to keep plants from flowering and is useful in crops that may flower early.

7.3.2 Inhibitors of GA Biosynthesis

Synthetic compound that prevents gibberellins from promoting plant growth are called anti-gibberellin growth retardants or inhibitors of GA biosynthesis. The biosynthesis of gibberellins was shown to involve mevalonic acid as a primary precursor. A key intermediate in the route of GA1 is ent-kaurene, followed after several intermediates by gibberellin GA12-7-aldehyde, from which different metabolic pathways breach off. Among the more important plant growth regulators, applications have been those involving interference in gibberellins biosynthesis. Numerous plant growth retardants are presently accessible for limiting growth by inhibiting or blocking gibberellin biosynthesis. The process of describing the nature of these substances and their mode of action is as follows:

7.3.2.1 Onium Compounds

Compounds containing a positively charged group, including ammonium, phosphonium, and sulphonium inhibit the biosynthesis of gibberellins before ent-kaurene. Some of the most well-known compounds are chlormequat chloride and mepiquat chloride. Other growth retardants include AMO-1618 and Piproctanyl bromide which possess a quaternary ammonium function. These substances block the activity of CPP-synthase in the gibberellin-producing fungus Gibberella fujikuroi as well as in cell-free preparations of this fungus in higher plants. These substances also hinder the function of ent-Kaurene synthase, yet to a lesser extent. Comparing the cyclic transformation of geranylgeranyl diphsophate into cystein-rich polycomb-like protein, catalyzed by CPP-synthase, to the reaction that produces cycloartenol in higher plants throughout sterol biosynthesis. Reduced amount of gibberellin through growth retardant treatment leads to varying declines in shoot length.

7.3.2.1.1 Mode of Action

These substances function as growth inhibitors by inhibiting specific enzymes crucial in the initial stages of GA metabolism. Preventing the cyclization of geranylgeranyl pyrophosphate to copalyl pyrophosphate results in the suppression of gibberellin production.

7.3.2.1.2 Effects

Compounds from the Onium family boost photosynthesis, improve resistance to drought, and lead to the buildup of solutes like amino acids and sugars, helping plants preserve turgor even when leaf water levels drop. Plants that received onium compounds displayed shorter internodes and thicker, greener leaves compared to those that did not receive treatment. A decrease in leaf area caused by onium compounds leads to a decrease in the transpiration surface, resulting in reduced water loss (Davis and Curry, 1991). Plants treated with Onium have shown increased tolerance to abiotic stress like salt and temperature, as well as biotic stresses, including diseases, insects, and nematodes.

7.3.2.2 Chlormequat

Chlormequat is the "most important inhibitor of gibberellins biosynthesis" and is also called cycocel (CCC). It is an organic compound utilized as a plant growth regulator which is characteristically sold as water and ethanol-soluble substances named chloride salt and chlormequat chloride. It is a quaternary ammonium salt and an alkylating agent.

Mechanism of action: The mechanism of action of the plant growth retardants, including chlormequat, has been related to inhibition of synthesis or action of gibberellin.

7.3.2.2.1 Effect of Chlormequat on Fruit Crops

The growth retardant effects obtained with diaminozide are also similar in many respects to those found with chlormequat.

Inhibition of growth: Cycocel @ 2,000 ppm significantly retarded shoot elongation in young as well as mature Langra and young Baramasia mango. Application of cycocel @ 1,000 ppm resulted in 34% inhibition of leaf area in strawberry; minimum increment in percent increase of shoot length 23.62% and 28.35% at 30 days after spray and 60 days after spray, respectively in 'Sardar' guava (Jain and Dashora, 2007). In Nagpur mandarin, training and pruning followed by foliar application of CCC at the rate of 3,000 ppm for 2 years leads to a reduction in volume, canopy spread, and a maximum flower and fruit percentage. Higher concentration foliar application of cycocel at the rate of 500 ppm, reduced the surplus vigor of Tas-A-Ganesh on Dogridge rootstock.

Increase in flowering and fruit quality: Application of CCC @ 1,000 ppm increased the no. of fruits per shoot and percent fruit set in guava cv. "Allahabad safeda" (Lal et al., 2013); reduced shoot growth, increased fruit yield by 15%, and improved the fruit quality in Thompson seedless grape (Shikhamany and Reddy, 1990). In 'Sardar' guava, aspray of CCC @ 250 &500 ppm improved the fruit set and also enhanced the weight and quality of the fruit; improved anthesis & fruit ripening by about 10 days in red raspberries cv. 'Autumn Bliss'; increased flowering, fruit yield, and quality in guava. Spraying grape plants with Cycocel @ 1,500 ppm shows increased bunch size and yield. The foliar spray of 100 and 200 ppm cycocel in grape augmented percent bud eruption to a prior date than control and GA3 treated plants.

Increase in yield: The application of CCC in different concentrations results in increased yield in fruit crops. Plants treated with CCC @ 1,000 ppm increase the number of fruits per shoot at harvest in guava cv. "Allahabad Safeda" (Lal et al., 2013); increased percentage fruit set and no. of fruits per tree in Kinnow mandarin increased number of hands, number of fingers, finger size and thereby bunch weight, yielding a maximum of fruits in banana cv. 'Grand Naine'. The application of CCC @ 400 ppm at the FBD stage resulted in an increased no. of fruits as well as yield/tree in sapota cv. 'Cricket Ball' (Aggarwal and Dikshit, 2008). Application of 3,000 ppm CCC was found to be extremely effective in increasing the yield/vine in grapes when applied at 15 leaf stage (Shikhamany and Reddy, 1989); increased yield in mango by improving the no. of fruits per tree and fruit weight was also observed and more yield in 'Assam' lemon. An increased number of fruiting buds resulted in grapes when applied with 1,500 ppm CCC.

7.3.2.3 Pyrimidines

Pyrimidine growth retardants include various chemical compounds such as Ancymidol (known commercially as A-REST), Flurprimidoltetcyclacis, uniconazole-P, and inabenfide.

Mode of action: Pyrimidine growth retardants mainly hinder the activity of cytochrome P-450, an enzyme that regulates the conversion of kaurene to kaurenoic acid by oxidation. Nevertheless, the primary impact of pyrimidine growth retardants seems to be caused by the suppression of gibberellin production. Pyrimidines disrupt the production of sterols and abscisic acid. These growth retardants function as inhibitors of monooxygenases that catalyze the oxidative reactions involved in the conversion of ent-kaurene to ent-kaurenoic acid.

Effects:

- An insignificant reduction in photosynthesis and a reduction in water use.
- Protect plants from abiotic stresses due to increment in antioxidant content/activity of treated plants.

7.3.2.4 Triazoles

These are an infinitely dynamic group of plant growth retarding chemicals which includes paclobutrazol, uniconazole, triapenthenol, and BAS 111. Among this category, paclobutrazol and uniconazole-P are quite active and have been applied to fruit plants.

Mode of action: The microsomal oxidation of kaurenol, kaurene, and kaurenal is inhibited by triazole compounds. This oxidation is mediated by kaurene oxidase, which is a cytochrome P-450 oxidase. The outcome of this inhibition is a decrease in the progression of plant development.

Triazole compounds block gibberellin biosynthesis. In addition to this, it also inhibits sterol biosynthesis, reduces ABA, ethylene, and indole-3-acetic acid, and upsurges cytokinin as well as chlorophyll content is detected in triazole-treated crop plants.

7.3.2.5 Paclobutrazol

Paclobutrazol exhibits a remarkable level of activity and persistence in comparison to other growth retardants. Paclobutrazol has been referred to as PP-333 or cultar, which is used as a growth retardant. Various methods can be used to apply it to fruit trees, including trunk injection, truck soil-line pour, and foliar sprays, and soil drench. When this growth retardant is applied to the soil around the base of trees, it controls the shoot growth for several seasons without affecting fruit size. It is absorbed through leaves, stems, and roots and is translocated through the xylem. However, the yield can be increased in some cases because of the increased amount of sunlight reaching the fruiting spurs. By minimizing excessive terminal growth of the shoots, the efficiency of trees is enhanced. One possible explanation is that plants often produce an excess of leaves, surpassing the amount required for optimal photosynthesis in the shaded leaves beneath them.

Mode of action: Paclobutrazol results in the inhibition of the gibberellin biosynthesis known to be synthesized following the isoprenoid pathway. Gibberellins stimulate cell elongation and their inhibition results in the inhibition of cell elongation, while cell division still occurs. As a consequence, the shoots exhibit a compact arrangement of leaves and internodes, occupying a reduced length. The isoprenoid pathway not only synthesizes gibberellins but also plays a role in the synthesis of other significant endogenous hormones like abscisic acid and cytokinins.

7.3.2.5.1 Effect of Paclobutrazol on Fruit Crops

Paclobutrazol (PBZ), a gibberellins inhibitor, has proven to be a valuable solution for addressing issues related to flowering and tree vigor in various tropical and subtropical fruits such as litchi, avocado, mango, citrus, and more. Various effect of paclobutrazol on fruit crops is described as follows:

Growth increment: The use of paclobutrazol at a concentration of 150 ppm as a foliar spray and 6 g per tree as a soil drench resulted in a decrease in vegetative vigor, including reductions in shoot length, shoot thickness, and internode numbers in the 'Sultania' fig cultivar. The use of cultar (25% paclobutrazol) effectively suppressed the yearly growth of shoots and enhanced the photosynthetic activity in cherries. When paclobutrazol was applied at a rate of 10 g per tree, there were noticeable decreases in tree height, tree volume, and mean shoot length in mango. Applying paclobutrazol (PP333) as a foliar spray on avocado plants led to a reduction in their height.

°Increase in flowering: Paclobutrazol is widely recognized as a highly effective chemical growth regulator that is utilized to stimulate flower production in mango trees (Nartvaranant et al., 2000). Applying different concentrations of paclobutrazol to mango trees in July and August can result in abundant flowering of the Alphonso variety. The paclobutrazol can be applied either through foliar spraying or by adding it to the soil. Foliar application of paclobutrazol @ 500 ppm in October resulted in the earliest full bloom and maximum percent of flowered shoots (36.20%) in mango cv. Chausa (Sharma et al., 2011); induces early flowering and increases the number of flowers/shoots (7.77/shoot) in Sardar guava (Jain and Dashora, 2007). Paclobutrazol @ 2.5–15 g a.i./tree significantly increases the percentage of reproductive shoots in mango and increases the flowering period. In addition to early flower induction, Cultar is highly effective in inducing off-season flowering in mangoes, as demonstrated in studies conducted by Nafees et al. (2010) and Burondkar et al. (2013). It has also shown promising results in regulating flower production, improving yield, and enhancing the quality of various perennial fruit crops. The effectiveness of PBZ in promoting flowering in Citrus sp. is dependent on the crop load. Trees with heavy fruit load showed scarce flowering, while trees with medium to low fruit load experienced an increase in sprouted buds and floral shoots, along with a reduction in vegetative growth (Mitra, 2017). This technique has been successfully applied to decrease the size of the canopy and enhance the number of flowers in various fruit trees such as peach, plum, almond, grapes, and mango (Nartvaranant et al., 2000). Additionally, it has been found to promote the growth of flowers in the branches of mango trees.

Increase in fruiting: The application of paclobutrazol @ 500 ppm in 'Sardar' guava increased the fruit set, and fruit retention with early harvesting (Jain and Dashora, 2007). In 'Rose Scented' litchi application of cultar reduced shoot growth, resulted in profuse flowering, higher sex ratio, increased fruit set, yield, and TSS content of litchi fruits. Applying cultar at a rate of 5 g a.i/tree to the soil of Langra mango cultivars resulted in enhanced fruit set and retention during the off-year. The application of paclobutrazol (PP333) through foliar sprays resulted in a shorter spring flush length, leading to a more compact growth in both avocado cultivars. Additionally, the dwarfing effect was found to be more pronounced with the use of paclobutrazol. The application of paclobutrazol spray resulted in a significant improvement in berry set, bunch size, yield, and overall quality of grapes, particularly in terms of T.S.S. (Total Soluble Solids) and acidity.

Others: Paclobutrazol application results in increased chlorophyll content by blocking GA production. Paclobutrazol-treated plants also show a high root-to-shoot ratio because of a reduction in shoot growth, induced dwarfing, tolerance to environmental stress, etc. Paclobutrazol also has fungicidal properties as it inhibits steroid production in fungi through the terpenoid pathway which results in plant resistance to fungal disease, etc.

7.4 GROWTH RETARDING COMPOUNDS NOT INHIBITING GIBBERELLIN BIOSYNTHESIS

7.4.1 Morphactins

Morphactins are a class of growth retarding compounds that include fluorine, fluorine-9-carboxylic acid, and chlorflurenol. These growth retardants received the name Morphactinas they are morphologically active substances (ability to affect plant morphogenesis). Inhibiting plant growth is a common effect of morphactins. From a chemical perspective, morphactins are derived from fluorene compounds that are initially inactive. However, their activity is enhanced by the addition of a -COOH group in the 9th position. Most morphactins are created in a laboratory and have a wide range of impacts on the growth and development of plants. There are various morphactins, such as Flurenol, Chloroflurenol, Methyl benzilate, Methyl chloroflurenol, Methyl dichloroflurenol, and more.

Mode of action: The effects of morphactine are normally irreversible, and they obstruct the mitotic activity of meristematic tissues. This leads to a shift in the alignment of the mitotic spindle, disrupting the usual polarity observed in plants. This disruption is a result of a widespread halt in auxin transport, induced by the morphactine.

7.4.1.1 Effect of Morphaction on Fruit Crops

Morphoactins have a very interesting effect. They exhibit both synergistic and antagonistic effects, depending on the relative concentrations, especially when other natural hormones are present. Morphactins typically have adverse impacts on plant morphogenesis. They prevent seed germination, sprouting, seedling growth, internode lengthening, etc. They depolarize cell division in many cases, which most likely results in distorted morphogenesis. Tillering will be abundant since morphactins are particularly effective at stimulating lateral bud growth.

Interestingly, certain morphactins can induce flowering in specific short-day plants. Morphactines not only prevent elongation but also end apical dominance, which radically changes the structure of the plants and gives them a bushy appearance. In addition, the growth of carpels and stamina are affected, along with geotropism and phototropism.

Dikegulac: The primary response is the retardation of apical dominance leading to lateral bud break.

7.4.2 Maleic Hydrazide

It is a growth retardant that prevents cell proliferation by lowering nucleic acid biosynthesis in shoot & root meristems, which interferes with uracil production. There are effects of Maleic Hydrazide on fruit crops.

Reduction in growth: Maleic Hydrazide inhibits plant growth in all fruit crops. Shoot length reduction was observed when Maleic Hydrazide was sprayed @ 500 ppm at the 5-leaf stage followed by 1,000 ppm at the leaf stage &1,500 ppm at the 15-leaf stage in Thompson Seedless grape. MH caused foliar phytotoxicity & harm to the apical meristem in apples, resulting in the inhibition of terminal growth and stimulation of lateral shoot growth (Cline and Bakker, 2017).

Induction of femaleness: MH induces femaleness in fruit crops, especially papaya. Application of MH @ 600 ppm in CO-2 papaya resulted in an increased proportion of female flowers (59.33%) as compared to male flowers (40.66%) (Pusdekar and Pusdekar, 2009).

Increase in flowering: Application of maleic hydrazide @ 600 ppm results in an increased number of flowers per tree and also induces early flowering (138.33) in CO-2 papaya (Pusdekar and Pusdekar, 2009).

7.5 EFFECT OF GROWTH RETARDING CHEMICALS ON PLANTS

1. **Cell division and enlargement**: Schreff confirmed that parenchymatous cortical cells in the first internode of Amo-1618 treated bean plants were 69% shorter than similar cells in untreated plants. Increased stem diameter of the treated bean plants was also found due to the stimulation of cell production in the cambium accompanied by a delay in cell differentiation and an increase in cell volume of the parenchymatous cortical cells.
2. **Shoot elongation**: The application of wide concentrations of Amo-1618 retards stem elongation as the inhibition of cell division and elongation of the subapical meristem.
3. **Root development**: Growth retardants significantly reduced or delayed root formation.
4. **Flower initiation**: Using growth retardants significantly sped up the flowering period for several woody plants. Rapid flower development has been observed on CCC-treated plants grown during natural days and shoot growth stopped. CCC increased the number of flowers but delayed the flowering of plants, depending upon CCC's concentration and length of cool storage.
5. **Sex expression**: The compounds 2-bromo-ethyl trimethyl ammonium bromide (BCB) and alkyl trimethyl ammonium bromide (AMAB) inhibited staminate flower production, greatly increased pistillate flower formation and occasional lypromoted the formation of hermaphroditic flowers. AMAB caused the greatest stimulation of pistillate flower formation under high light intensities.

   Maturation and crop yield: The flower and fruit yields on various horticultural crops showed a decline as the dosage of CCC and AMAB increased.

   Fresh and dry weights: Application of growth retardants results in less weight than those grown in untreated soil. The foliage of plants treated with growth retardants exhibited a significantly deeper shade of green compared to the foliage of untreated plants.
6. **Increased resistance to chemical and physical changes**: The application of growth retarding chemicals has been found to enhance the frost resistance of plants. The use of growth retardants demonstrated a significant enhancement in tolerance to elevated salt levels and fluctuations in soil pH.

   Metabolic changes: The growth of citrus seedlings was not impacted by Amo-1618; however, it resulted in an upsurge in peroxidase activity.

7.6 PLANT GROWTH INHIBITORS

Plant growth inhibitors are synthetic compounds utilized to regulate plant shoot length in a controlled manner, despite affecting patterns of development or causing harm to plants. This is achieved mainly by inhibiting cell elongation, as well as by reducing cell division.

Some natural growth inhibitors like Lunularic acid, Vanillic acid, Ferulic acid, Phaseic acid, Jasmonic acid, Violaxanthin, Coumarin, Quercetin (flavonol), and Genistein are examples of naturally occurring growth inhibitors (Figure 7.1).

7.6.1 Major Naturally Occurring Growth Inhibitors

7.6.1.1 Jasmonic Acid

Jasmonates are a subset of molecules that are closely associated with linolenic acid. Jasmonates are cyclopentanone derivative products that are produced from fatty acids. This group includes jasmonic acid and its esters, like Methyl jasmonates. Jasmonates and their derivatives play a crucial role in governing a wide range of plant functions, including growth, photosynthesis, and reproductive development. The discovery of jasmonate's molecular structure and name was made through the extraction of Methyl Jasmonate from Jasmine oil produced by *Jasminum grandiflorum.* Jasmonic acid is a particularly prevalent member of this particular group. These organisms have been discovered in a wide range of plant species, spanning various families such as ferns, mosses, and fungi. This suggests their presence is widespread across the plant kingdom. Limited knowledge exists regarding jasmonate biosynthesis, but indications point to the stem apex, young leaves, root tips, and immature fruits as having the most significant levels of jasmonates. Jasmonates are synthesized in different plant tissues and are transported via the phloem to induce gene expression in distant plant regions. The concentration of jasmonates can differ depending on the specific tissue and cell type, the stage of development, and the reaction to different environmental factors. Jasmonates are present in significant amounts in the flowers and pericarp tissues that produce reproductive structures, in addition to in the chloroplasts of illuminated plants. Furthermore, the concentration increases when plants are injured. Jasmonic acid and its methyl esters are found in all plants.

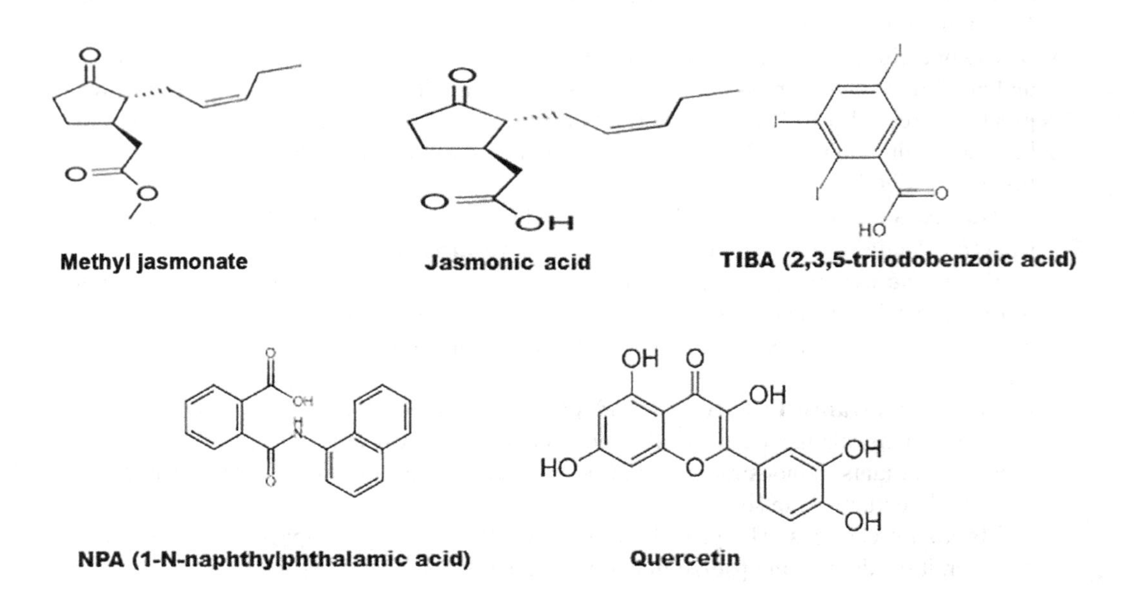

FIGURE 7.1 Structure of different plant growth retardants.

7.6.1.1.1 Biosynthetic Pathway

As previously indicated, linolenic acid serves as the precursor. It is later liberated from the membrane-bound lipids and transformed into jasmonic acid via the so-called octadecanoid route. Chloroplast and peroxisome are the two organelles where biosynthesis largely takes place. Allene oxide synthase (AOS), lipoxygenase (LOX), allene oxide cyclase (AOC), and other important enzymes are involved. Linolenic acid is converted into an intermediate, 12-oxo-phytodienoic acid (OPDA), which is then cycled and delivered to peroxisomes via the ABC transporter, where it is processed by the enzymes of the -oxidation pathway. The end product, jasmonic acid, is produced after three cycles of the - oxidation pathway and is subsequently released into the cytosol (Wasternack et al., 2002).

7.6.1.1.2 Roles of Jasmonates

Jasmonic acid has been demonstrated to limit callus growth, pollen germination, root growth, chlorophyll production, and seed germination, in addition to its promoting effects.

Affects seed germination and growth: JA and MeJA facilitate the germination of dormant seeds while suppressing the germination of non-dormant seeds. The germination of Quercus robur's refractory seeds is hindered by the presence of JA, MeJA, ABA, and ethylene. The production of jasmonate may not be directly involved in regulating germination. Instead, it could be a consequence of membrane damage, as indicated by the observed correlation between the increase in jasmonate levels and lipid peroxidation. Jasmonate enhanced latent embryo germination and elevated alkaline lipase activity in apples. The ABA-inhibited germination mutants jin4 and jarl, which are jasmonate-insensitive, exhibit enhanced sensitivity. As a result, JA may promote seed germination by reducing ABA sensitivity. As an alternative, seed germination could be prevented by JA-mediated growth suppression.

Root growth inhibition: Root development is significantly inhibited by JA by a mechanism unrelated to ethylene. Additionally, JA prevents coleoptile lengthening induced by IAA, probably by preventing the amalgamation of glucose into cell wall polysaccharides. Additionally, ethylene or IAA is not directly engaged in tendril coiling's differential development, which is activated by JA.

Reduction in heat stress: Reports on grapes suggest that jasmonic acid may have the potential to mitigate the adverse impacts of heat stress. The expression of antioxidant enzymes SOD, CAT, and POD is increased in grape seedlings exposed to 50 mM JA, which helps mitigate the detrimental impact of high-temperature stress (42°C).

Insect and disease resistance: Insect and disease resistance in plants is significantly influenced by JA. First, JA builds up in damaged plants and plants or cell cultures treated with elicitors of pathogen defense. Subsequently, JA stimulates the production of protease inhibitor genes, which shield plants from insect harm. Additionally, JA stimulates the production of genes that code for antifungal proteins such as osmotin. Analysis of plants with altered JA levels provides a third type of proof for the role of JA in pest resistance. For instance, applying JA to potatoes boosts their resistance to *Phytophthora infestans*. VspB, which encodes a soybean vegetative storage protein, and Pin2, which encodes a tomato protease inhibitor implicated in plant defense, both have strikingly comparable expression regulatory mechanisms. Both genes have strong expression in the reproductive tissues, flowers, and apical parts of vegetative plants. Both genes are activated when a wound is created and JA is applied. The expression of genes induced by JA is enhanced by sugars and suppressed by phosphate and auxin. Furthermore, the application of JA/wounding treatment to illuminated plants leads to a notable increase in the expression of these genes. The proteins produced by these genes are targeted for vacuoles and often accumulate in large amounts within plant cells.

Tendril coiling and touch: The response to JA, which does not directly involve ethylene or IAA, is the differential growth involved in tendril coiling.

7.6.2 Synthetic Growth Inhibitors

7.6.2.1 Anti Auxins

The synthetic chemicals NPA (1-Nnaphthylphthalamic acid) and TIBA, can function as auxin transport inhibitors (ATIs) (2, 3, 5-triiodobenzoic acid). These inhibitors prevent auxin efflux, which stops polar transport. Quercetin (flavonol) and genistein are auxin transport inhibitors that are found in nature. The auxin transport inhibitors NPA (1-N-naphthylphthalamic acid), TIBA (2,3,5-triiodobenzoic acid), and 1-NOA are not found in plants (1-naphthoxyacetic acid).

7.6.2.1.1 Mode of Action

By adding NPA or TIBA to either the donor or the receiver block in an auxin transport experiment, this phenomenon can be proven. While neither substance affects auxin uptake from the donor block, it does impede auxin efflux into the receiver block. Some ATIs, like TIBA, which are transported polarically and have modest auxin activity, may hinder polar transport in part by competing with auxin for the binding site on the efflux carrier (Michniewicz et al., 2007). Others, like NPA, are hypothesized to obstruct auxin transport by binding to proteins involved in a complex with the efflux carrier even if they are not transported polarly. These NPA-binding proteins are also present in the basal ends of conducting cells, which is congruent with where PIN proteins are located.

A new class of ATIs that blocks the AUX1 uptake carrier has recently been discovered (Parry et al., 2001). For instance, when applied to Arabidopsis plants, 1-naphthoxyacetic acid (1-NOA) prevents auxin uptake into cells and results in root gravitropism that resembles that of the aux1 mutant. None of the other AUX1-specific inhibitors, including 1-NOA, can impede polar auxin transport, much like the aux1 mutant.

7.7 UTILITY OF GROWTH HORMONES IN E CROPS

7.7.1 Seed Germination

The seed germination process can be enhanced by pre-soaking the seed with growth regulators. For example, the application of NAA and IAA at 20 ppm in okra can improve the seed germination rate, while in tomatoes, it has been reported with 2,4 D and GA3 at 0.5 mg/L, respectively. In cucurbit, pre-seed soaking in ethephon at 480 mg/L for 24 hours enhanced the germination rate in bottle gourd, muskmelon, watermelon, and squash melon (Table 7.4).

TABLE 7.4
Use of Growth Regulators/Retardants in Fruit Crops

Fruit Crop	Dose Concentration	Time of Application	Remarks
Pears	Ethephon (250 ppm) + Daminozide (550 ppm)	5 weeks after full bloom	• Increase flowering without reducing fruitset
	Daminozide (1,000 ppm)		• Stimulate flowering flower initiation
Pear cv. Comice	5,000 ppm (CCC)		• Stimulate flower initiation
Cherry	PBZ application as soil drench (0.1 g/tree)		• Increased flowering and productivity
Apricot	PBZ at 1,000 ppm	3 weeks after bloom	• Increased the number of flower buds and fruit yield
Apple	Kinetin (12.5 ppm)		• Advanced the time of flower opening

7.7.2 Sprouting Inhibition

Applying MH at 2,500 ppm 15 days before harvesting stops onion sprouting in storage. The potato tuber's dormancy is broken by soaking it in thiourea at 1% concentration and IAA at 250–1,000 ppm, which also prolongs dormancy.

7.7.3 Flowering

GA3 applied at 50 mg/L can initiate flowering in non-flowering varieties of potato. Additionally, maleic hydrazide helps delay flowering in okra. In contrast, gibberellic acid stimulates early flowering in lettuce.

7.7.4 Stimulating Fruit Set

Inadequate fruit set is a significant issue for *Solanum* crops. Applying 4-CPA, 2,4-D@2–5 ppm, or PCPA 50–100 ppm to tomatoes can improve fruit set and ripeness.

7.7.5 Seed Dormancy

Ethylene chlorohydrin, GA, and thiourea are used to breakdown the potato tuber seed dormancy. These chemicals can be applied in the form of vapor treatment of ethylene chlorohydrin followed by dipping in thiourea and GA at the rates of 1L/20q, 1% (1 hour), and 1 mg/L for 2 seconds, respectively. However, at high temperatures, gibberellic acid treatment has been reported to be effective in breaking seed dormancy in lettuce.

7.7.6 Sex Expression

The application of growth regulators in sex expression therapy has been seen to alter the sexual characteristics of cucurbits, okra, and pepper plants. When applied at the 2–4 leaf stage in cucurbits, GA 3 (10–25 ppm), IAA (100 ppm), and NAA (100 ppm) have been seen to enhance the production of female flowers. GA 3, silver nitrate, and silver thiosulphate, when sprayed on cucurbits at the 2–4 leaf stage with concentrations of 1,500–2,000 ppm, 300–400 ppm, and 300–400 ppm respectively, stimulate the formation of male flowers.

7.7.7 Parthenocarpy

Auxin stimulated the development of seedless fruits in cucumbers and watermelon. The use of PCPA at concentrations of 50–100 ppm induced parthenocarpy in tomato and brinjal. Additionally, the use of 2,4-D at a concentration of 0.25% in lanolin paste, applied to the cut end of styles or as foliar sprays to newly opened flower clusters, has been observed to promote parthenocarpy.

7.7.8 Gametocides

Plant growth regulators have gametocidal effects, resulting in male sterility, which may be used for the creation of F1 hybrid seeds. The concentration of MH in okra, peppers, and tomatoes ranges from 100 to 500 mg/L. GA3 is used in onion. 2,3-dichloro-isobutyrate is applied at a concentration of 0.2%–0.8% in okra, muskmelon, onion, root crops, spinach, and tomato. TIBA is used in cucumber, okra, onion, and tomato. Gibberellic acid (GA) at a concentration of 100 mg/L may be used to induce male sterility in pepper plants.

7.7.9 Hybrid Seed Production

Ethephon has been used to induce feminization in several cucurbit species. An effective F1 hybrid in butternut squash has been developed by using a female line created with the application of 10 weekly sprays of ethephon. Plant growth regulators have also been used for the preservation of gynoecious lines. GA3 sprays have been used in cucumber to stimulate the development of male flowers in gynoecious lines. Studies have shown that a concentration of 500 mg/L of silver nitrate is as efficient as GA3 in stimulating the development of male flowers on gynoecious lines of cucumber. However, when it comes to muskmelon, it was shown that applying silver thiosulphate by foliar sprays at a concentration of 400 mg/L was the most effective method for stimulating the growth of male flowers on gynoecious lines.

7.7.10 Fruit Ripening

Ethephon, a chemical that releases ethylene, has been shown to stimulate the ripening process in tomatoes and peppers. Applying ethephon at a concentration of 1,000 mg/L at the turning stage of the earliest fruits resulted in the acceleration of fruit ripening, leading to a 30%–35% increase in the yield of early fruits. It has been observed that immersing mature green tomatoes in a solution of ethephon at concentrations ranging from 500 to 2,000 mg/L may stimulate the ripening process.

Fruit yield enhancer: It has been observed that soaking tomato seeds in NOA at concentrations of 25–50 mg/L, GA at concentrations of 5–20 mg/L, and CIPA at concentrations of 10–20 mg/L, together with 2,4-D at a concentration of 0.5 mg/L or thiourea at a concentration of 10–1 M, may enhance fruit output. Soaking brinjal seedling roots in NAA at a concentration of 0.2 mg/L and ascorbic acid at a concentration of 250 mg/L has been shown to result in increased fruit output.

REFERENCES

Aggarwal, S., and Dikshit, S. N.. (2008). Studies on the effect of plant growth regulators on growth and yield of sapota (*Achras sapota* L.) cv. Cricket Ball. *Indian Journal of Agricultural Research* 42:207–211.

Ahmed, I. H. M., Ali, E. F., Gad, A. A., Bardisi, A., El-Tahan, A. M., Abd Esadek, O. A., El-Saadony, M. T., and Gendy, A. S. (2022). Impact of plant growth regulators spray on fruit quantity and quality of pepper (*Capsicum annuum* L.) cultivars grown under plastic tunnels. *Saudi Journal of Biological Sciences* 29(4):2291–2298.

Barrett, J. 2001. Mechanisms of Action. In M. L. Gaston, P. S. Konjoian, L. A. Kunkle, and M. F. Wilt (eds.). *Tips on Regulating Growth of Floriculture Crops*. O. F. A. Services, Inc., Columbus, OH.

Burondkar, M. M., Rajan, S., Upreti, K. K., Reddy, Y. T. N., Singh, V. K., and Sabale, S. N. (2013). Advancing Alphonso mango harvest season in lateritic rocky soils of Konkan region through manipulation in time of paclobutrazol application. *Journal of Applied Horticulture* 15:178–182.

Cline, J. A. and Bakker, C. J.. (2017). Prohexadione-calcium, ethephon, tranexamic-ethyl, and maleic hydrazide reduce extension shoot growth of apple. *Canadian Journal of Plant Science* 97:457–465.

Gawali. S. S., Patil, R. A., and Gaikwad, S. B. (2023). Standardization of different plant growth regulators on growth parameters of cherry tomato (*Solanum lycopersicum* var. cerasiforme) under protected condition. *The Pharma Innovation Journal* 12(1):506–510.

Jain, M. C., and Dashora, L. K.. 2007. Growth, flowering, fruiting and yield of guava (*Psidium guajava* L.) cv. Sardar as influenced by various plant growth regulators. *International Journal of Agricultural Sciences* 3:4–7.

Kumar, P. S., Rao, M. C. S., Tamang, A., and Kumar, U. S. (2022). Effect of plant growth regulators on growth, yield and quality attributes of watermelon (*Citrullus lanatus* Thunb.), *Crop Research* 57 (6):375–379.

Kumar, A., Veersain, Dishri, M., Gangwar, V., Singh, N., Kumar, D., and Bajaj, S. (2022). Effect of different plant growth regulators on growth and phenological parameters of cucumber (*Cucumis sativus* L.) cv. Punjab Naveen. *International Journal of Environment and Climate Change* 2282–2288.

Kumari, S., Meena, M. L., Moond, S. K., Mandeewal, R. L., Kalirawna, A., and Kalirawna, S. (2022). Effect of plant growth regulators on vegetative growth and flowering of okra [*Abelmoschus esculentus* (L.) Moench.]. *International Journal of Environment and Climate Change* 12(3):67–72

Lal, N., Das, R. P., and Verma, L. R. (2013). Effect of plant growth regulators on flowering and fruit growth of guava (*Psidium guajava* L.) cv. 'Allahabad Safeda'. *The Asian Journal of Horticulture* 8:54–56.

Megu, O., Hazarika, B. N., Wangchu, L., Sarma, P., Singh, A. K., Debnath, P., and Angami, T. (2021). Effect of plant growth regulators and micronutrients on vegetative growth, flowering and yield attributes of litchi (*Litchi chinensis* Sonn.). *International Journal of Plant & Soil Science* 33(18):176–181.

Michniewicz, M., Brewer, P. B., and Friml, J. (2007). Polar auxin transport and asymmetric auxin distribution. *The Arabidopsis Book/American Society of Plant Biologists*, Arabidopsis Book 5:e0108, 1–28.

Mitra, S. K. (2017). Paclobutrazol in flowering of some tropical and subtropical fruits. In *XIII International Symposium on Plant Bioregulators in Fruit Production* 1206, Japan, 27–34.

Nafees, M., Faqeer, M., Ahmad, S., Khan, M. A., Jamil, M., and Aslam, M. N. (2010). Paclobutrazol soil drench suppresses vegetative growth, reduces malformation, and increases production in mango. *International Journal of Fruit Science* 10:431–440.

Naga, B. L., Deepanshu, Singh, D., andBahadur, V. (2022). Effect of plant growth regulators on growth, yield and quality of chilli (*Capsicum annuum* L.). *The Pharma Innovation Journal* 11(10):227–233.

Naik, K. M., Bhosale, A. M., and Dapurkar, N. D. (2022). Effect of different plant growth regulators and bio-mix on yield and physio-chemical parameters of sapota fruits cv. Kalipatti. *The Pharma Innovation Journal* 11(12):4561–4563.

Pandey, J., Pandey, S. K., Rana, G. K., Deshmukh, K. K., Singh, N. K., and Pandey, A. (2023). Impact of growth regulator and cutting time on growth of jamun [*Syzygium cuminii*, (L.) Skeels] hardwood stem cuttings. *The Pharma Innovation Journal* 12(1):1685–1687.

Sharma, S. K., Chauhan, J. K., and Kaith, N. S. (2011). Induction of flowering and fruiting in unproductive 'Chausa' mango orchards in Himachal Pradesh. *International Journal of Farm Sciences* 1:23–29.

Tedila, K. (2022). Effect of alar, cycocel and bonzi on growth performance of ivy geranium (*Pelargonium peltatum*). *Advanced Crop Science and Technology* 10:506.

Wang, L., Mu, C., Du, M., Chen, Y., Tian, X., Zhang, M., and Li, Z. (2014). The effect of mepiquat chloride on elongation of cotton (*Gossypium hirsutum* L.) internode is associated with low concentration of gibberellic acid. *Plant Science* 225:15–23.

Zankat, S. B., Leua, D. H. N., and Hathi, H. S. (2022). Effect of time of spray and plant growth regulators on growth and flowering of muskmelon (*Cucumis melo* L.). *The Pharma Innovation Journal* 11(9):3031–3035.

8 Absorption, Translocation, and Degradation of Phytohormones

Pramod Kumar, Pratibha Chib, and Sandhya Thakur

8.1 INTRODUCTION

Plants are living entities that exhibit dynamic behaviors by responding to and adjusting to their environment. Plants need to regulate their growth and development in order to adjust to different external stimuli and a dynamic environment. Phytohormones play a significant role in the adaptation of plants to various environments. They regulate growth, development, nutrient distribution, and source/sink transitions in plants (Fahad et al., 2014). Plant hormones, also known as phytohormones, are chemical compounds synthesized by plants that can exert their effects locally or at distant sites inside the plant itself after being transported. Plant hormones, as a whole, regulate all aspects of plants's growth and advancement (Peleg and Blumwald, 2011). Plant hormones serve as chemical messengers that, at very low concentrations regulate a wide range of physiological and biochemical processes of higher plants. Auxins, abscisic acid, cytokinins, gibberellins, and ethylene are the five families of phytohormones that are often found in plants, along with their precursors and synthetic analogs. In addition to these, other classes of plant hormones are also present including brassinosteroids, jasmonates, salicylic acid, strigolactone, etc. These growth substances work together to support every element of plant life, from the development of patterns to reactions to biotic and abiotic stress. Charles Darwin first proposed that specific chemical compounds might be able to spur crop growth in the 19th century (Darwin and Darwin, 1881). Since then, a lot of research has been done on the stimulatory effects of phytohormones on the quantity and quality of fruit crops. The classes of natural growth hormones not only differ chemically, but they also have unique effects on how plants develop. We separate them based on differences in chemical makeup as well as based on their physiological responses. However, many plant growth hormone classes frequently have similar developmental effects. Growth hormones of various types frequently interact with one another, resulting in a net effect that may be less than, larger than, or entirely different from the effects of either hormone alone.

Hormone homeostasis is regulated by hormone synthesis, metabolism, transport, perception, and signal transduction, which collectively control several plant functions. One distinguishing feature of all plant hormones is their capacity to induce reactions in various regions of the plant by being produced in specific tissues and subsequently delivered there in very small quantities. Auxin's local and long-distance transport plays a vital role in various areas of plant growth and development, unlike the transportation of volatile chemicals such as ethylene and methyl jasmonate, which is primarily important for plant defense (Santner et al., 2009). Hormones can be perceived either near or distant from their point of production. Hormones can be regulated in their distribution by active transporters, leading to a range of reactions (Park et al., 2017). For development to occur, growth hormones must not only be present but also be available at the proper time, place, and concentration. This indicates that growth hormone synthesis is closely regulated, the inactivation mechanisms govern the concentration of growth hormones in tissues, and the patterns of plant hormone translocation within plants are crucial.

 DOI: 10.1201/9781003354055-8

8.2 ABSORPTION

The potential effects of plant growth regulators depend on their method of application owing to differences in their mode of absorption by the plant because some of them are absorbed only through root, leaves, or stem, whereas some are absorbed through all listed organs, providing a benefit to apply in any way. Often, growth agents can penetrate deeper into the lower leaf surface than the upper leaf surface. As the leaf ages, there is a continuous accumulation of waxes in the cuticle layer on the upper surface of the leaf which widens this difference (Norris and Bukovac, 1969). Further, the uptake of plant hormones from the lower surface of the leaf was found to be significantly enhanced in the presence of light. In a study, Greene and Bukovac (1971) observed that with the increase in temperature penetration of NAAm (naphthalene-acetamide) into pear leaf disks was increased.

Ethylene is a simple, gaseous plant hormone that is essential for regulating growth and development in plants. Although ethylene gas can be supplied exogenously, however, it might not always be practical to employ in a lab setting. A common substitute for ethylene that has been used is the organophosphorus compound ethephon, though many other chemicals can also be used in its place. It is sprayed or misted over plants, where it passes via the cuticles and stomata to the apoplasts, where it decomposes at pH 5.0 and higher to produce ethylene, chloride, and phosphate (Bhadoria et al., 2018). Ethephon, on the other hand, is a non-gaseous "predrug" of ethylene that can be used much more successfully. It progressively decomposes into ethylene together with phosphoric and hydrochloric acid after being taken by the plant and then being present in a somewhat acidic environment.

8.3 TRANSLOCATION OF PHYTOHORMONES

8.3.1 Auxins

Auxin is translocated through two different pathways in higher plants. The first one involves a quick, nondirectional passive transport through the phloem from source tissues (synthesis site) to sink tissues (Vanneste and Friml, 2009). Around 20% of the auxin present in the acidic apoplastic compartment is neutral and undissociated, allowing it to passively diffuse across the plasmic membrane (Kramer and Bennett, 2006). However, in the second pathway, auxin transporter proteins facilitate the immediate transportation of auxin through the plasma membrane, resulting in a slow, directed cell-to-cell transport referred to as polar auxin transport (PAT) (Grones and Friml, 2015). According to multiple researchers, PAT involves at least three main groups of auxin transporters: AUXIN1-RESISTANT1 (AUX1)/LIKE AUX1 (referred to as AUX1/LAX), PIN-FORMED proteins (referred to as PINs), and ABCB members belonging to the B subfamily of ATP binding cassette (ABC) transporters (Geisler et al., 2016). The ABCB19/MDR1/PGP19 protein was first identified as a result of its interaction with the anion channel blocker NPPB (5-nitro-2-(3-phenylpropylamino)-benzoic acid) during a screening process. On the other hand, the aux1 and pin1 mutants were initially detected through genetic screenings aimed at identifying resistance to the plant hormone auxin (2,4-D) and abnormalities in flower development, respectively (Park et al., 2017). Additionally, numerous abcb mutants were discovered by traditional genetic screening methods.

Solid experimental evidence demonstrating the major members of all three families to be genuine auxin transporters has been produced, and their substrate specificities have been investigated by several researchers (Geisler et al., 2014). IAA is classified as a weak acid with a pKa value of 4.75. As a result, while it is within an acidic apoplastic compartment, a portion of IAA exists in a lipophilic state while being protonated. Furthermore, research has shown that the process of cellular uptake through the plasma membrane depends on the AUX1/LAX proteins, which function as high-affinity auxin-proton symporters (Swarup et al., 2008). AUX1 participates in both the upward and downward movement of auxin. Nitrate Transporter 1.1 (NRT1.1) has been identified as a mechanism for auxin uptake. It is believed that in conditions of limited nitrogen availability, NRT1.1 functions as both an auxin transporter and a nitrate transceptor (Krouk et al., 2010). PINs are auxin transporters with a permease-like structure, which were given their name due to the pin-formed phenotype observed in the pin1 mutant with a loss-of-function mutation. PIN transporters can be classified as either long PINs or short PINs based on the extent of the hydrophilic loop between the two transmembrane domains. Arabidopsis has two types of PIN proteins: short PINs (PIN5, 6, and 8), which are found in endomembrane structures, and long PINs (PIN1, 2, 3, and 7), which function as exporters in the plasma membrane (Geisler et al., 2014). Unlike PINs and AUX1/LAX proteins, a certain group of ABCBs acts as primary active auxin pumps capable of transporting auxin against significant auxin gradients., PAT is now associated with the ABCB isoforms ABCB1, 4, 14, 15, 19, and 21. However, only the transport actions of ABCB1, 4, 19, and 21 have been thoroughly confirmed. It is worth mentioning that ABCB1 and ABCB19 have been shown to function as dedicated auxin transporters, while Arabidopsis ABCB4 and 21 and Oryzasativa ABCB14 have been shown to have the ability to both import and export IAA, depending on the circumstances (Kamimoto et al., 2012).

8.3.2 Gibberellins

While gibberellic acid (GA) can migrate both upwards (acropetal) and downwards (basipetal) within the plant, research suggests that the acropetal movement is more efficient (Binenbaum et al., 2018). Out of the 100+ GAs, only a few are confirmed to have biological activity and are acknowledged by the permeable receptor protein Gibberellin Insensitive Dwarf 1 (GID1) and the negative regulator DELLA proteins (Gupta and Chakrabarty, 2013). Multiple members of the NPF (nitrate transporter 1/peptide transporter family) possess the ability to transport GA (gibberellic acid). The identification of NPF3.1 as a putative GA transporter was achieved by the examination of mutant plants that displayed abnormal root accumulation of GA-Fls (GA molecules tagged with fluorescein). The transport activity of GA via NPF and sugar transporter SWEET family (AtSWEET13 and AtSWEET14) was tested using Xenopus oocytes. The study by Tal et al. (2016) found that NPF3.1 imports GA4, and to a lesser amount, GA1 and GA3, in a manner that is reliant on pH. This finding aligns with the known dependence of the NPF transporter family upon the proton motive force. Nevertheless, it was discovered that NPF2.10 plays a role in the uptake of GA3, but not in the movement of GA1 and GA4.

8.3.3 Cytokinins

Roots and shoots generate distinct types of cytokinins. Roots produce tZ-type cytokinins (trans-zeatin), while shoots produce iP-type cytokinins (isopentenyl adenine). These cytokinins migrate in opposite directions through the plant's vascular system. tZ-typecytokinins go upwards (acropetally) through the xylem, while iP-type cytokinins move downwards (basipetally) through the phloem (Kudo et al., 2010). Based on a recent study conducted by Ko et al. in 2014, it was found that CK is transferred from the root to the shoot through an ABC transporter belonging to the G family. AtABCG14, which is primarily expressed in the root, was mentioned in the article as being integrally involved in the CK transport to the shoot by enabling the loading of CKs into the xylem sap, which in turn facilitates all subsequent translocations. It was also proposed that the PUP (purine permease) family and ENT (equilibrative nucleoside transporter) serve as cytokinin transporters (Hirose et al., 2008; Zurcher et al., 2016). The uptake transport activity of different members of the ENT family, such as rice ENT2 and Arabidopsis ENT3, ENT6, ENT7, and ENT8, was shown when they were expressed in yeast, utilizing radioactive iP-riboside (iPR) and tZ-riboside (tZR). Nevertheless, the sensitivity of iPR and tZR was only marginally diminished in ent3 and ent8 mutant plants, as reported by Sun et al. (2005). Consequently, the specific roles of these mutants in cytokinin transport in plants remain uncertain. PUP14 was recently characterized as a plasma membrane transporter responsible for the uptake of cytokinins. This was achieved by utilizing a synthetically produced fluorescent cytokinin reporter. Furthermore, it was found that PUP14 plays a critical role in the proper growth of the embryo in Arabidopsis, as reported by Zurcher et al. in 2016. While the uptake transport activity of other PUP members, namely PUP1 and PUP2, has been investigated in yeast (Burkle et al., 2003), there is currently no genetic or biochemical evidence supporting their activities in plants.

8.3.4 Ethylene

Due to its high diffusibility through membranes, ethylene can freely pass between cells without requiring transporter proteins (Davies, 2010). 1-aminocyclopropane-1-carboxylic acid (ACC), which is the direct precursor of ethylene, has been identified as a non-gaseous mobile signal capable of long-distance movement across the xylem and phloem (Van and Van, 2014). ACC-resistant2 (ARE2) was identified due to the mutant plant's lack of sensitivity to externally administered ACC, although its reactions to ethylene remained unaltered. LYSINE HISTIDINE TRANSPORTER1 (LHT1), an element of the amino acid transporter relatives, is encoded by ARE2. The study conducted by Shin et al. (2015) discovered that the protein LHT1, which is located in the plasma membrane, has the ability to absorb ACC. This was determined using transport experiments using 14C-ACC in protoplasts obtained from lht1 mutant plants. Therefore, LHT1 serves as the basis for uncovering the ACC transport mechanism and regulating ethylene responses. N-malonylACC (MACC), a compound derived from the ACC conjugate, is synthesized, in the cytosol and transferred into the vacuole for preservation using ATP as an energy source. The molecular identification of the MACC transporters is still to be confirmed (Pech et al., 2010).

8.3.5 Abscisic Acid

Due to its weak acid nature with a pKa value of 4.7, ABA has the ability to pass through the plasma membrane via the apoplastic pathway (Zhang et al., 2021). During drought stress, the concentration of the easily diffusible form of ABA drops as the pH in the space outside the cells increases. This suggests that ABA transporters are necessary to move ABA from where it is produced (vascular tissues) to where it is needed (the guard cells). ABCG25/WBC25 and ABCG40/PDR12, which belong to the G subfamily of ABC transporters, are recognized for their ability to transport ABA from vascular tissues to guard cells via the xylem across the plasma membrane. This transport process induces stomatal closure in response to drought stress. ABCG30 and ABCG31 were found to transport ABA from the endosperm of seed coats to the embryo, hence contributing to the maintenance of seed dormancy. ABCG25 and ABCG31 facilitate the transportation of ABA from the endosperm, whereas ABCG40 and ABCG30 are responsible for the absorption of ABA into the embryo (Kang et al., 2015). A recent finding revealed that the ABA importer MtABCG20 plays a significant role in both the germination of Medicagotruncatula and the formation of lateral roots (Pawela et al., 2019). ABCG22, DTX 50, belonging to the MATE (Multidrug and toxic compound extrusion) type transporter family, together with NPF 4.6, are other transporters that demonstrate ABA translocation properties. The vacuole and endoplasmic reticulum (ER) store abscisic acid (ABA) in its inactive storage state, specifically ABA-glucose ester (ABA-GE). UDP-glucosyltransferases, also known as ABA-glucosyltransferases, facilitate the synthesis of ABA-GE, the inactive form of ABA, by combining ABA with glucose. During periods of stress, the compound ABA-GE undergoes decomposition, resulting in the release of ABA into the cytosol. Both AtABCC2 and an unidentified proton antiporter have been discovered to facilitate the absorption of ABA-GE into vacuoles. The transporters responsible for transporting ABA released by glucosidases from the ER and vacuole, as well as the transporters that sequester ABA-GE into the ER, have not been identified yet (Nambara and Marion-Poll, 2005). ABCG17 and ABCG18 transport ABA into the plasma membrane. ABCG17 and ABCG18 have a prominent expression in shoot mesophyll and stem cortical cells, where they

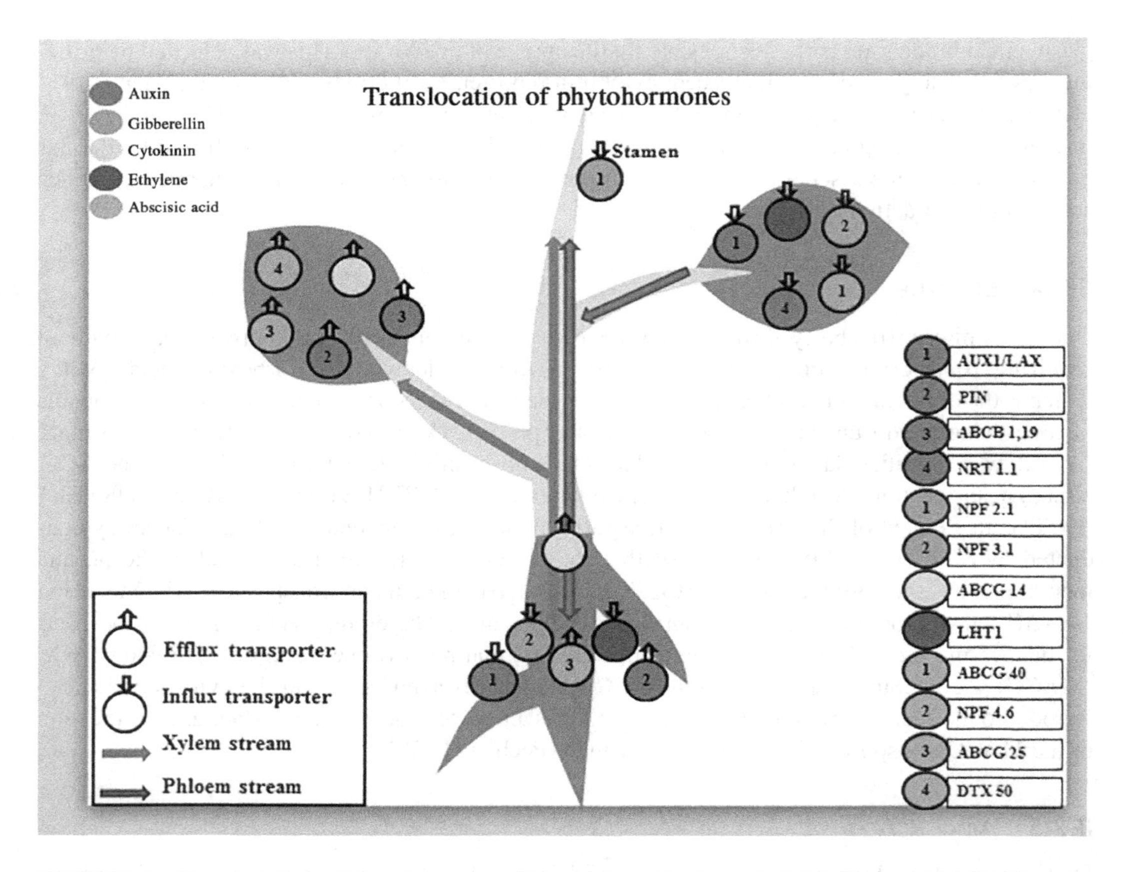

FIGURE 8.1 Translocations of phytohormones. (Edraw Max – Geisleret al., 2016; Kang et al., 2015; Koet al., 2014; Lacombe and Achard, 2016; Shin et al., 2015; Swarup and Peret, 2012; Tal et al., 2016.)

facilitate the import of ABA and the formation of ABA-GE. These ABA-GE sinks prevent the accumulation of active ABA in guard cells and limit the long-distance transportation of ABA to areas where lateral root development occurs. Under abiotic stress circumstances, inhibiting the transcription of ABCG17 and ABCG18 is more beneficial for active ABA to travel, aggregate, and respond, as demonstrated by Zhang et al. (2021) (Figure 8.1).

8.4 DEGRADATION

8.4.1 Auxins

IAA is the predominant endogenous auxin. IAA is believed to exist in its active state as Free IAA. Throughout their growth and development, plants not only produce auxins but also decrease the level of IAA through conjugation, primarily with amino acids and sugars, as well as through an oxidative degradation process that is not well understood (Woodward and Bartel, 2005; Normanly, 2010; Rosquete et al., 2012). The carboxyl group of IAA can rapidly react with amino acids, sugars, and other small molecules, leading to the conversion of IAA into forms that are believed to be inactive. The IAA conjugates may serve as the initial step in the complete degradation of IAA. The genes responsible for the process of combining indole-3-acetic acid (IAA) with an amino acid to produce amides, as well as the genes involved in the breakdown of IAA conjugates, have been identified mainly in Arabidopsis. The mentioned individuals are part of the GH3 family of amido synthases and various amido hydrolases, which are known to be induced by auxin.

This information is supported by the studies conducted by Hagen et al. in 1991, Liu et al. in 1994, and Staswick et al. in 2005. IAA conjugates can be classified as either reversible or irreversible preservation compounds. Although certain IAA-amino acid conjugates can be hydrolyzed to release free IAA, there are some conjugates that appear to be resistant to hydrolysis in living organisms (LeClere et al., 2002). The latter class of IAA conjugates has the ability to deactivate IAA. When IAA-Asp is formed, it does not undergo hydrolysis, causing the conjugated IAA to remain permanently inactive. IAA-Asp is also known to be susceptible to oxidative degradation. The oxidative degradation of IAA can begin either with the decarboxylation of the side chain or with the oxidation of the indole ring. However, there is currently limited knowledge regarding the process of oxidative degradation. According to specific reports, peroxidase might have a function in the process of oxidative decarboxylation of IAA, as stated by Normanly in 2010. The primary degradation products of indole-3-acetic acid (IAA) in plants are 2-oxoindole-3-acetic acid (oxIAA) and oxIAAglucose (oxIAA-Glc) (Kai et al., 2007; Novák et al., 2012). However, the specific genes responsible for the breakdown of IAA have not yet been discovered. In addition, various other metabolites of indole-3-acetic acid (IAA) have been discovered in Arabidopsis. The primary oxidative product is OxIAA-Glc. These include N-(6-hydroxyindol-3-ylacetyl)-phenylalanine (6-OH-IAA-Phe), N-(6-hydroxyindol-3-ylacetyl)-valine (6-OH-IAA-Val), and 1-O-(2-oxoindol-3 -ylacetyl)-beta-d-glucopyranose (Pencik et al., 2013).

8.4.2 Gibberellins

Plants need to accurately regulate their GA concentration in order to promptly adapt to environmental changes. Disabling or breaking down GAs provides a means to regulate homeostasis and allows for a rapid reduction in GA concentration when needed, similar to other plant hormones (Hedden and Thomas, 2012). The capacity of GA 2-oxidases (GA2oxs) to facilitate the 2β-hydroxylation reaction, which transforms active GAs (and/or their precursors) into non-functional forms, has been widely recognized for a considerable period of time. Based on its roles, this enzyme can be divided into two significant groups. The first group operates on C19-GAs, which consists of bioactive compounds and their non-3β hydroxylated precursors. The second group acts on C20-GAs (Schomburg et al., 2003; Thomas et al., 1999). Recently, two new mechanisms to deactivate GAs have been found: methylation of GAs in Arabidopsis and epoxidation of non-13-hydroxylated GAs in rice (Zhu et al., 2006; Varbanova et al., 2007). The EUI (ELONGATED UPPERMOST INTERNODE) gene in rice was found to encode a cytochrome P450 mono-oxygenase (CYP714D1) by in vitro tests. This enzyme is responsible for catalyzing the 16α,17-epoxidation of non-13-hydroxylated GAs, as reported by Zhu et al. in 2006. The over-expression of the EUI gene in rice resulted in significant dwarfism and a reduction in GA4 content in the upper internodes. This provides evidence that epoxidation is responsible for deactivating GA, indicating that epoxidation may be a commonly occurring mechanism for GA deactivation. GAMT1 and GAMT2, which belong to the SABATH group of methyl transferases, are enzymes identified in Arabidopsis that catalyze the methylation of the 6-carboxy group of C19-GAs. This mechanism inhibits the biological activity of GA, leading to dwarfism in Arabidopsis (Brian et al., 1967). The GAMT genes regulate the GA content in developing seeds, so their roles may be restricted to seed development and early germination.

8.4.3 Cytokinins

The degradation of cytokinin is essential for controlling the accumulation of cytokininin plants, and the reaction is brought by enzyme known as cytokinin oxidase/dehydrogenase (CKX, EC 1.5.99.12). Paces et al. (1971) was the first to demonstrate the oxidative cleavage of cytokinins in a tobacco culture. Cytokinin oxidase deactivates cytokinins by breaking down their N6 side chain, resulting in the production of adenine or a related derivative for N9-substituted cytokinins, and an aldehyde from the side chain (Brownlee et al., 1975). The reaction initiates through the dehydrogenation of cytokinin, resulting in the formation of an imine intermediate. Previously, it was widely accepted that the activity of CKX, an enzyme called cytokinin oxidase, depended on the presence of molecular oxygen. However, when CKX was extracted from wheat, it was found to be a flavoprotein containing FAD, which was capable of breaking down cytokinin even in the absence of oxygen. Consequently, the enzyme was reclassified as cytokinin dehydrogenase, also known as CKX, with the enzyme commission number EC 1.5.99.12. The enzyme CKX breaks down isoprenoid cytokinins that have an unsaturated isoprenoid side chain in both base and nucleoside forms. However, the nucleotide derivatives are not affected by CKX and remain intact. The degradation of isopentenyladenine (iP) by cytokinin oxidase/dehydrogenase (CKX) leads to the production of adenine and 3-methyl-2-butenal. The enhanced resistance to CKX is a result of lowering the double bond. Therefore, the presence of the delta-2 double bond in the isoprenoid side chains of tZ and iP is critical for substrate identification. Therefore, CKX is ineffective in breaking down DZ-type cytokinins. Side chain changes, such as O-glucosylation, also contribute to the resistance of CKX. All known plant CKX enzymes possess a covalently attached FAD molecule. The interaction between FAD molecules and proteins promotes both redox catalysis and the oxidizing ability of flavin (Galuszka et al., 2008).

8.4.4 Ethylene

Although ethylene can be generated by nearly all components of higher plants, the rate of production is contingent upon the specific tissue type and developmental stage. Scientists have conducted research on the breakdown of ethylene by introducing 14C2H4 into plant tissues. The metabolic breakdown products have been identified as carbon dioxide, ethylene oxide, ethylene glycol, and the glucose derivative of ethylene glycol. Nevertheless, the process of ethylene breakdown does not have a significant impact on the regulation of ethylene levels in plants (Raskin and Beyer, 1989). Ethylene is capable of forming conjugates. The conversion of ACC to ethylene in plant tissue exhibits variability. ACC can undergo conversion into N-malonylACC, a conjugated form that exhibits resistance to degradation and accumulates within the tissue. Another derivative of ACC, known as 1-(γ-L-glutamylamino) cyclopropane-1-carboxylic acid (GACC), has also been found. Just like how auxin and cytokinin are combined, the combination of ACC may play a vital role in controlling ethylene synthesis (Tiaz and Zeiger, 2002).

8.4.5 Abscisic Acid

Plants quickly obtain ABA when they encounter different environmental conditions. It is crucial to finely adjust the levels of free ABA in plants to ensure optimal growth and development in various environmental conditions. ABA is deactivated in plants through many pathways, such as hydroxylation of ABA at the C-7′, C-8′, or C-9′ locations. Hydroxylation at the C-8′ position is the most common form of modification in plant species. The enzyme CYP707As, which belongs to the cytochrome P-450 type mono-oxygenases, is responsible for facilitating the hydroxylation reaction at the C-8′ position of ABA (Kushiro et al., 2004). The 8′-hydroxy-ABA undergoes spontaneous isomerization to form phaseic acid (PA), which is unstable. Therefore, the generated PA undergoes additional metabolism to yield either dihydophaseic acid (DAP) or epi-DAP. This process is facilitated by an unidentified reductase enzyme, which is currently not functioning. The ABA 8′-hydroxylation pathway, mediated by CYP707A, requires both NADPH and P450 reductase (Kushiro et al., 2004; Saito et al., 2004). In Arabidopsis, the excessive expression of CYP707A effectively reduces the natural levels of ABA, while Arabidopsis mutants with several CYP707A genes acquire a significant quantity of ABA (Okamoto et al., 2006, 2010, 2011). Reports indicate that the enzyme 8'-hydroxylase ABA was detected in the microsomal fraction of maize cells maintained in suspension. Additionally, it was observed that the fusion protein CYP707A-green fluorescent protein (GFP) is specifically located in the endoplasmic reticulum (ER). According to Krochko et al. (1998) and Saika et al. (2007), it is anticipated that the ABA 8′-hydroxylation reaction will take place in the endoplasmic reticulum (ER).

In addition to the ABA 8′-hydroxylation pathway, hydroxylation of ABA at positions C-7′ and C-9′ occurs, resulting in the formation of 7′-hydroxy-ABA and 9′-hydroxy-ABA, respectively. The hydroxylated 9′-hydroxy-ABA undergoes instability and is converted into neophaseic acid (NeoPA), which is present in a wide variety of plant species (Zhou et al., 2004).

REFERENCES

Bhadoria P, Nagar M, Bharihoke V and Bhadoria AS. 2018. Ethephon, an organophosphorous, a fruit and vegetable ripener: has potential hepatotoxic effects? *J. Fam. Med. Prim. Care* 7:179.

Binenbaum J, Weinstain R. and Shani E. 2018. Gibberellin localization and transport in plants. *Trends Plant* Sci. 410–21.

Brian PW, Grove JF and Mulholland TPC. 1967. Relationships between structure and growth-promoting activity of gibberellins and some allied compounds in 4 test systems. *Phytochemistry* 6:1475–1499.

Burkle L, Cedzich A, Dopke C, Stransky H, Okumoto S and Gillissen B. 2003. Transport of cytokinins mediated by purine transporters of the PUP family expressed in phloem, hydathodes, and pollen of *Arabidopsis*. *Plant J.* 34:13–26.

Darwin C. and Darwin F. *The Power of Movement in Plants*. New York: D. Appleton and Company, 1881; p. 592.

Davies PJ. 2010. The plant hormones: their nature, occurrence, and functions. In: Davies PJ, editor, *The Plant Hormones: Biosynthesis, Signal Transduction, Action!* 3rd ed. Dordrecht: Springer Netherlands, pp. 1–15.

Fahad S, Hussain S, Matloob A, Khan FA, Khaliq A, Saud S, Hassan S, Shan D, Khan F, Ullah N, Faiq M, Khan MR, Tareen AK, Khan A, Ullah A, Ullah N, Huang J. 2014. Phytohormones and plant responses to salinity stress: a review. *Plant Growth Regul.* 75:391–404.

Geisler M, Bailly A and Ivanchenko M. 2016. Master and servant: regulation of auxin transporters by FKBPs and cyclophilins. *Plant Sci.* 245:1–10.

Geisler M, Wang B. and Zhu J. 2014. Auxintransport during root gravitropism: transporters and techniques. *Plant Biol. (Stuttg)* 16:50.

Greene DW and Bukovac MJ. 1971. Factors influencing the penetration of naphthaleneacetamide into leaves of pear. *J. Am. Soc. Hort. Sci.* 96:240–246.

Grones P. and Friml J. 2015. Auxintransporters and binding proteins at a glance. *J. Cell Sci.* 128:1–7.

Gupta R. and Chakrabarty SK. 2013. Gibberellicacid in plant: still a mystery unresolved. *Plant Signal Behav.* 8.

Hagen G, Martin G, Li Y. and Guilfoyle TJ. 1991. Auxin-inducedexpression of the soybean GH3 promoter in transgenic tobacco plants. *Plant Mol. Biol.* 17:567–579.

Hedden P. and Thomas SG. 2012. Gibberellin biosynthesis and its regulation. *Biochem. J.* 444:11–25.

Hirose N, Takei K, Kuroha T, Kamada-Nobusada T, Hayashi H. and Sakakibara H. 2008. Regulation of cytokinin biosynthesis, compartmentalization and translocation. *J Exp. Bot.* 59:75–83.

Kai K, Horita J, Wakasa K. and Miyagawa H. 2007. Threeoxidative metabolites of indole-3-acetic acid from Arabidopsis thaliana. *Phytochemistry* 68:1651–1663.

Kamimoto Y, Terasaka K, Hamamoto M, Takanashi K, Fukuda S. and Shitan N. 2012. *Arabidopsis* ABCB21 is a facultative auxin importer/exporter regulated by cytoplasmic auxin concentration. *Plant Cell Physiol.* 53:2090–2100.

Kang J, Yim S, Choi H, Kim A, Lee KP and Lopez-Molina L. 2015. Abscisic acid transporters cooperate to control seed germination. *Nat. Commun.* 6:8113.

Ko D, Kang J, Kiba T, Park J, Kojima M, Do J, Kim KY, Kwon M, Endler A, Song WY, Martinoia E, Sakakibara H. and Lee Y. 2014. Arabidopsis ABCG14 is essential for the root-to-shoot translocation of cytokinin. *PNAS* 111**:** 7150–7155.

Kramer EM and Bennett MJ. 2006. Auxintransport: a field in flux. *Trends Plant Sci.* 11:382–386.

Krochko JE, Abrams GD, Loewen MK, Abrams SR. and Cutler AJ. 1998. (+)-abscisic acid 8′-hydroxylase is a cytochrome P450 monooxygenase. *Plant Physiol.* 118:849–860.

Krouk G, Lacombe B, Bielach A, Perrine-Walker F, Malinska K. and Mounier E. 2010. Nitrate-regulated auxin transport by NRT1. 1 defines a mechanism for nutrient sensing in plants. *Dev. Cell* 18:927–937.

Kudo T, Kiba T and Sakakibara H. 2010. Metabolismand long-distance translocation of cytokinins. *J. Integr. Plant Biol.* 52:53–60.

Kushiro T, Okamoto M, Nakabayashi K, Yamagishi K,Kitamura S, Asami T, Hirai N, Koshiba T, Kamiya Y and Nambara E. 2004,The *Arabidopsis* cytochrome P450 CYP707A encodes ABA 8′-hydroxylases: key enzymes in ABA catabolism. *EMBO J.* 23:1647–1656.

Lacombe B. and Achard P. 2016. Long-distance transport of phytohormones through the plant vascular system. *Curr. Opin. Plant Biol.* 34:1–8.

LeClere S, Tellez R, Rampey RA, Matsuda SP and Bartel B. 2002. Characterizationof a family of IAA-amino acid conjugate hydrolases from Arabidopsis. *J. Biol. Chem.* 277:20446–20452.

Liu ZB, Ulmasov T, Shi X, Hagen G and Guilfoyle TJ. 1994. Soybean GH3 promoter contains multiple auxin-inducible elements. *Plant Cell* 6:645–657.

Nambara E and Marion-Poll A. 2005. Abscisicacid biosynthesis and catabolism. *Annu. Rev. Plant Biol.* 56:165.

Normanly J. 2010. Approachingcellular and molecular resolution of auxin biosynthesis and metabolism. *Cold Spring Harb. Perspect. Biol.* 2:a001594.

Norris RE and Bukovac MJ. 1969. Some physical-kinetic considerations in penetration of naphthaleneacetic acid through isolated pear leaf cuticle. *Physiol. Plant.* 22:701–712.

Novák O, Hényková E, Sairanen I, Kowalczyk M, Pospíšil T and Ljung K. 2012. Tissue-specific profiling of the Arabidopsis thaliana auxinmetabolome. *Plant J.* 72:523–536.

Okamoto M, Kushiro T, Jikumaru Y, Abrams SR, Kamiya Y, Seki M and Nambara E. 2011. ABA9'-hydroxylation is catalyzed by CYP707A in Arabidopsis. *Phytochemistry* 72:717–722.

Okamoto M, Kuwahara A, Seo M, Kushiro T, Asami T, Hirai N, Kamiya Y, Koshiba T and Nambara E. 2006. CYP707A1and CYP707A2, which encode abscisic acid 8′-hydroxylases, are indispensable for proper control of seed dormancy and germination in *Arabidopsis. Plant Physiol.* 141:97–107.

Okamoto M, Tatematsu K, Matsui A, Morosawa T, Ishida J, Tanaka M, Endo TA, Mochizuki Y, Toyoda T, Kamiya Y, Shinozaki K, Nambara E. and Seki M. 2010. Genome-wide analysis of endogenous abscisic acid-mediated transcription in dry and imbibed seeds of Arabidopsis using tilingarrays. *Plant J.* 62:39–51.

Paces V, Werstiuk E and Hall RH. 1971. Conversion of N6-(D2-isopentenyl)adenosine to adenosine by enzyme activity in tobacco tissue. *Plant Physiol.* 48,775–778.

Park J, Lee Y, Martinoia E and Geisler M. 2017. Plant hormone transporters: what we know and what we would like to know. *BMC Biol.* 15:1–15.

Pawela A, Banasiak J, Biała W, Martinoia E. and Jasinski M. 2019. MtABCG20 is an ABA exporter influencing root morphology and seed germination of *Medicagotruncatula. Plant J.* 98:511–523.

Pech J, Latche A and Bouzayen M. 2010. Ethylene biosynthesis. In: Davies PJ, editor, *The Plant Hormones: Biosynthesis, Signal Transduction, Action!* 3rd ed. Dordrecht: Springer Netherlands, pp. 115–136.

Peleg Z and Blumwald E. 2011. Hormonebalance and abiotic stress tolerance in crop plants. *Curr. Opin. Plant Biol.* 14:290–295.

Pencik A, Simonovik B, Petersson SV, Henykova E, Simon S, Greenham K, Zhang Y, Kowalczyk M, Estelle M, Zazimalova E, Novak O, Sandberg G and Ljung K. 2013. Regulation of auxin homeostasis and gradients in Arabidopsis roots through the formation of the indole-3-acetic acid catabolite 2-oxindole-3-acetic acid. *Plant Cell* 25:3858–3870.

Raskin I and Beyer EMJr. 1989. Role of ethylene metabolism in Amaranthusretroflexus. *Plant Physiol.* 90:1–5.

Rosquete MR, Barbez E and Kleine-Vehn J. 2012. Cellular auxin homeostasis: gatekeeping is housekeeping. *Mol. Plant.* 5:772–786.

Saika H, Okamoto M, Miyoshi K, Kushiro T, Shinoda S, Jikumaru Y, Fujimoto M, Arikawa T, Takahashi H, Ando M, Arimura S, Miyao A, Hirochika H, Kamiya Y, Tsutsumi N, Nambara E. and Nakazono M. 2007. Ethylene promotes submergence-induced expression of OsABA8ox1, a gene that encodes ABA 8′-hydroxylase in rice. *Plant Cell Physiol.* 48:287–298.

Saito S, Hirai N, Matsumoto C, Ohigashi H, Ohta D, Sakata K, Mizutani M. 2004. Arabidopsis CYP707As encode (+)-abscisic acid 8′-hydroxylase, a key enzyme in the oxidative catabolism of abscisic acid. *Plant Physiol.* 134:1439–1449.

Santner A, Calderon-Villalobos LIA and Estelle M. 2009. Planthormones are versatile chemical regulators of plant growth. *Nat. Chem. Biol.* 5:301–307.

Schomburg FM, Bizzell CM, Lee DJ, Zeevaart JAD and Amasino RM. 2003. Overexpression of a novel class of gibberellin 2-oxidases decreases gibberellin levels and creates dwarf plants. *Plant Cell* 15:151–163.

Shin K, Lee S, Song WY, Lee RA, Lee I and Ha K. 2015. Genetic identification of ACC-RESISTANT2 reveals involvement of LYSINE HISTIDINE TRANSPORTER1 in the uptake of 1-aminocyclopropane-1-carboxylic acid in Arabidopsis thaliana. *Plant Cell Physiol.* 56:572–582.

Staswick PE, Serban B, Rowe M, Tiryaki I, Maldonado MT, Maldonado MC and Suza W. 2005. Characterizationof an Arabidopsis enzyme family that conjugates amino acids to indole-3-acetic acid. *Plant Cell* 17:616–627.

Sun J, Hirose N, Wang X, Wen P, Xue L. and Sakakibara H. 2005. *ArabidopsisSOI33/AtENT8* gene encodes a putative equilibrative nucleoside transporter that is involved in cytokinin transport in planta. *J. Integr. Plant Biol.* 47:588–603.

Swarup K, Benkova E, Swarup R, Casimiro I, Peret B and Yang Y. 2008. The auxin influx carrier LAX3 promotes lateral root emergence. *Nat. Cell Biol.* 10:946–954.

Swarup R. and Peret B. 2012. AUX/LAXfamily of auxin influx carriers—an overview. *Front. Plant Sci.* 3:225.

Tal I, Zhang Y, Jorgensen ME, Pisanty O, Barbosa ICR and Zourelidou M. 2016. The Arabidopsis NPF3 protein is a GA transporter. *Nat. Commun.* 7:11486.

Thomas SG, Phillips AL and Hedden P. 1999. Molecular cloning and functional expression of gibberellin 2-oxidases, multifunctional enzymes involved in gibberellin deactivation. *Proc. Natl. Acad. Sci. U. S. A.* 96:4698–4703.

Vanneste S and Friml J. 2009. Auxin: a trigger for change in plant development. *Cell* 136:1005–1016.

Woodward AW and Bartel B. 2005. Auxin:regulation, action, and interaction. *Ann. Bot.* 95:707–735.

Zhang Y, Kilambi HV, Liu J, Bar H, Lazary S, Egbaria A, Ripper D, Charrier L, Belew ZM, Wulff N and Damodaran S. 2021. ABA homeostasis and long-distance translocation are redundantly regulated by ABCG ABA importers. *Sci. Adv.* 43(7): 1–17.

Zhou R, Cutler AJ, Ambrose SJ, Galka MM, Nelson KM, Squires TM, Loewen MK, Jadhav AS, Ross ARS, Taylor DC and Abrams SR. 2004. A new abscisic acid catabolic pathway. *Plant Physiol.* 134:361–369.

Zhu YY, Nomura T, Xu YH, Zhang YY, Peng Y, Mao BZ, Hanada A, Zhou HC, Wang RX and Li PJ. 2006. Elongated uppermost internode encodes a cytochrome P450 monooxygenase that epoxidizes gibberellins in a novel deactivation reaction in rice. *Plant Cell* 18:442–456.

Zurcher E, Liu J, Donato M, Geisler M and Muller B. 2016. Plant development regulated by cytokinin sinks. *Science* 353:1027–1030.

9 Growth Manipulation through Canopy Architecture

Debashish Hota and Subhash Chander

9.1 INTRODUCTION

A fruit tree is composed of its canopy, which includes the stem, branches, shoots, and leaves. The density of the canopy is influenced by elements such as the number and size of the leaves. Canopy management involves altering the structure of tree canopies to maximize the production of high-quality fruits. Canopy management, namely, tree training and trimming practices, significantly impacts how trees absorb sunlight. These methods are essential for assessing the extent to which the leaf area of the plant is vulnerable to incoming radiation. If plants are not properly trained and pruned from the beginning, fruit plants grow tall and enormous. Each plant must develop its ideal canopy spread at the beginning of its life cycle to harvest the sunlight. Therefore, one of the most common approaches for effectively managing large, unmanageable trees to increase their productivity is the management of canopy architecture. In addition to trimming and tree training, it also controls how quickly flowers and fruits ripen.

Canopy management includes modifying the canopies of trees to maximize the production of high-quality fruits and improve the structure of the trees to capture the most sunlight and enhance tree productivity. Improvement of light interception and light penetration in the orchard is one of the major objectives of fruit tree architecture manipulation. Photosynthesis relies on light as its primary source of energy, and this process significantly impacts flowering, fruit set, quality, and the development of fruit color. The training and pruning of canopy management procedures have a crucial role in determining the structure of the canopy and the arrangement of plants in the orchard. As a result, they have a significant impact on the interception and distribution of light inside the canopy. Principles of training involve the formation of a strong main framework of scaffold branches. All the branches must be well-spaced and balanced in all directions. The choice of a suitable training technique for fruit trees can vary based on the particular geographical area. The primary objective of selecting a training system and designing an orchard is to attain optimal light interception, often around 60%–70%, in order to maximize yields. This entails creating sparse canopies with a height of around 70–90 cm and a reduced number of leaves. This arrangement facilitates the even dispersion of light, resulting in enhanced fruit quality. Furthermore, it is imperative to achieve a harmonious equilibrium between optimizing the potential fruit production and guaranteeing exceptional fruit quality (Anthony et al., 2021).

In fruit trees, besides the conventional training system, many advanced training systems have been developed and recommended. The selection of a suitable training system for fruit trees depends upon many factors, such as the growing environment, technological measures, machinery, and economy. Fruit yield and fruit quality of fruit trees are correlated with the intercepted light. Fruit trees planted at high density efficiently intercept solar radiation of available Photo Synthetically Active Radiation (PAR) and its distribution in the canopy (Hampson et al., 2002). Nowadays, all the training methods have certain merits and demerits. Many of the latest training systems have replaced the traditional system of training due to certain advantages. Higher vegetative growth, in general correlated with poor fruit yield and quality. In the solaxe system of training, bending of all the branches, including the top portion is followed to check the vigorous growth for induction of flowering.

Pruning is an old-age practice used to maintain the desired size of tree. Pruning needs a thorough understanding of the physiology and development of the tree species. In different fruit crops, pruning severity and methodology depend upon tree habit and its purpose of cultivation.

DOI: 10.1201/9781003354055-9

Generally, two kinds of pruning systems are in practice, i.e. summer pruning and winter pruning. Summer pruning is practiced to maintain the size of fruit trees to the form that we need, while winter or dormant pruning promotes excessive growth for building of tree framework and renewal shoots. Therefore, harvesting adequate fruit yield with good quality fruits depends on the ideal canopy size, which can utilize the maximum amount of sunlight. An ideal fruit tree canopy enables a grower to obtain high and regular fruit yields besides improved farm management practices, leading to higher productivity.

9.2 CANOPY AND MANAGEMENT

The canopy of a fruit tree encompasses its physical structure, including the branches (old mature stem), stem (mature shoot but younger than a branch), shoots (young stem of less than one year having leaves), and leaves. Leaf area and leaf number constitute a major part of canopy density. The overall canopy architecture is governed by canopy and canopy density, which is influenced by a few factors, such as the physical composition of stems, branches, and shoots with respect to their number, area, orientation, and angle, etc. The configuration of the canopy is largely dependent on the orchard production system, which encompasses four variables: variety, rootstock, spacing, and training system.

The fundamental idea behind the three-dimensional growth of a canopy of a perennial fruit tree is to maximize the utilization of the land (spacing and density) and climatic elements. One of the most crucial management techniques for fruit plants is canopy architecture. The orchard's commercial life and early flowering behavior are completely dependent on canopy architecture. The alteration of tree and vine canopies to enhance fruit yield and quality is known as canopy management. Another way to explain it is that fruit tree canopy management entails the development and maintenance of the framework in relation to the size and shape, branch placement, and light absorption for the highest levels of productivity and quality A variety of approaches are used in canopy management to control the position and quantity of leaves, shoots, and fruits in space, which in great part affects the geometry of the plant, including the spatial distribution of leaf area and leaf orientation. A link of the relationship between the physiology of vegetative growth and maximum qualitative production is referred to as canopy management. The primary goal of canopy management is to optimize the distribution of carbon to fruit-bearing parts of the tree while minimizing any negative impact on the growth and development of other parts of the tree. The influence of temperature, light, humidity, and tree vitality on fruit yield and quality can be controlled by training, pruning, and applying growth inhibitors to modify the tree canopy, hence optimizing utilization and harvest outcomes.

A robust scaffold system plays a crucial role in enabling fruit trees to bear substantial crop loads of high-quality fruits without experiencing significant limb breakage. Certain trees, such as peaches and grapes, require annual pruning to consistently bear fruit. Regular pruning is essential to stimulate sufficient new growth, which ensures optimal fruit-bearing areas across the entire tree. The photosynthetic activity of the trees gets adversely affected, leading to the poor yields of low-quality fruits. Orchards become economically unproductive. To keep the fruit trees within the area provided for their spread for the whole of their life, canopy management is of prime importance.

9.3 PRINCIPLES OF CANOPY ARCHITECTURE

Canopy management of fruit trees affects fruit yield and quality by influencing light interception and distribution within the canopy. The canopy geometry of trees should be in such a way that it intercepts maximum light by training, pruning, branch and tree orientation, using chemicals, planting space, and design. The principles of canopy management involve: (i) Optimizing light utilization (ii) Preventing the development of disease-friendly microclimates (iii) Enhancing convenience in cultural practices (iv) Maximizing productivity and fruit quality (v) Achieving cost-effective canopy architecture. Fruit trees, which are planted at high density or old plantations generally, achieve less amount of total light interception. Poor light penetration and interception not only cause a heavy reduction in yield by reducing flower intensity, poor photosynthetic activity, and poor fruit set, but also, decrease the orchard longevity in the long run by the incidence of disease and pest. The quantity and quality of photon flux are highly essential inside the tree canopy to stimulate plant developmental and growth activity along with qualitative yield attributes. Conversion of photon energy to carbon allocation in growing fruits has been found more in smaller trees compared to large ones.

9.3.1 Light Interception

Light is the source of energy for the photosynthesis process in plants, determining flowering, fruit set, fruit quality, and the productivity of fruit trees. Light interception is the amount of light intercepted by the tree canopy, a fundamental requirement for crop growth, biomass production, and plant growth modelling. It improves both the physical and the biochemical quality of fruits. Canopy shape, tree orientation, canopy leaf area index, plant density, and foliage density in the canopy all impact light interception. Light interception can be increased by raising the density of foliage in the canopy, increasing tree height, and adjusting the number of trees per hectare. As the outer portion of the canopy shades the inner canopy, light intensity decreases within the tree canopy. The amount of light intercepted by the canopy depends on tree density, canopy shape, tree orientation, and leaf area index within the canopy. Light interception varies during a season, solar angle, and with distance from the center of the tree or row, time of the day, and leaf surface area development. Improvement of light penetration in fruit trees is one of the major objectives of fruit tree architecture manipulation. Both light interception and light penetration are the major factors under high-density planting to yield good quality fruits. In high-density tree planting, light serves as the sole limiting factor. It acts as the energy source for photosynthesis, which in turn affects flowering and fruit set and enhances fruit quality and color development. Specifically, the photo-synthetically active part of the spectrum (400–700 nm), known as Photo-synthetically Active Radiation (PAR), directly drives photosynthesis. Orchards can employ reflective films and colored nets to achieve the desired quality of light. Therefore, achieving a satisfactory fruit yield with high-quality fruits relies on favorable light conditions, which can be enhanced through the establishment of an appropriate tree canopy.

9.3.2 Light Distribution

Optimizing light interception and distribution plays a crucial role in achieving abundant yields of high-quality fruits. Light interception is influenced by factors such as orchard density, canopy architecture, alley width, row orientation, size, and leaf area index. On the other hand, adjustments to light distribution can be made through practices such as dormant pruning with minimal heading cuts, summer pruning (branch thinning), and branch bending. The distribution of light within the tree canopy depends on various interconnected factors, including the inherent architectural pattern of the cultivar, the planting system (including planting distance and row orientation), as well as the training of the tree canopy through pruning and bending techniques.

9.3.3 Photosynthesis

Light is essential for photosynthesis and affects flowering, fruit set, color development, and fruit quality. Light interception (LI) serves as the driving force for photosynthesis and directly influences the productivity of fruit trees. Consequently, improving light conditions within tree canopies has always been a fundamental goal in fruit tree cultivation. LI represents a valuable source of energy for tree growth and development, but it is specifically the photo-synthetically active part of the spectrum (400–700 nm) known as Photo-synthetically Active Radiation (PAR) that directly fuels photosynthesis. The quantity of PAR available determines the energy and carbon supply essential for sustained tree and fruit development. Shade diminishes PAR, thereby reducing local photosynthetic activity and canopy temperature and altering the wavelength distribution of transmitted light. Since whole-tree photosynthesis is primarily limited by light availability, manipulating canopy architecture to optimize light interception becomes a key objective across all planting systems (Lakso and Corelli-Grappadelli, 1992).

9.3.4 Fruit Quality

The design and implementation of orchard systems are crucial for improving productivity and fruit quality in the orchard. An optimal canopy design is essential for maximizing light interception, ensuring uniform light distribution, and ultimately generating high yields of top-quality fruits. By altering the structure of the canopy, the illumination dynamics within the tree can be enhanced to optimize the quality of the fruit. The fruit's quality is affected by the dispersion of light, although the yield of the fruit is strongly correlated with light interception.

In recent decades, multiple training schemes have been devised for fruit trees. For peach trees, conventional multi-leader 3D systems such as the open vase and delayed vassette can contribute to a greater fruit output per tree since they have a bigger canopy volume. Nevertheless, these systems often suffer from insufficient light penetration in the lower and inner regions of the canopy, leading to reduced productivity per unit of land. Furthermore, these canopies frequently intercept a greater quantity of light in the upper and outer areas of the tree. In the case of peach trees, it is especially accurate to say that around 80% of the leaves are found in the upper 40% of the tree in these three-dimensional systems. The presence of shade within the inner and lower regions of the canopy, particularly in cultivars with strong growth, can result in decreased tree productivity, lower crop yields, and compromised fruit quality, including reduced color intensity and lower levels of soluble solids content (SSC). Studies have demonstrated that open vase systems in peaches capture less light and provide lower yields in comparison to higher-density 2D cordon and KAC-V systems. In their study, Nuzzo et al. (2003) found that the Y-system, which utilizes a particular arrangement of the canopy, demonstrated superior levels of leaf area index, light interception, and yields when compared to the open vase system in peach cultivation. Uniform distribution of light throughout the canopy results in consistent fruit quality. It is worth noting that fruit quality is mostly determined by the environmental conditions that the fruit experiences during its growth, rather than its specific location within the canopy (Anthony and Minas, 2021).

9.3.5 Flower Bud Differentiation/Flowering and Fruit Set

Light plays a vital role in both flower induction and fruit development by facilitating carbohydrate synthesis. While increased assimilates in the shoots are necessary for flowering in fruit crops, achieving high yields of quality fruits is closely linked to maximizing light interception and distribution within the tree canopy. Effective canopy management strategies prioritize reducing excessive shading within the canopy and enhancing air circulation in the fruiting zone. These measures help create optimal conditions for fruit production and contribute to the overall quality of the harvest.

9.4 MODIFICATION OF THE TREE CANOPIES

Tree shape and size are the two most important factors in determining the percentage of total foliage that receives adequate irradiance. The volume: surface area ratio plays an important role in the exposure of the foliage to the light. Unmanageable tree shapes reduce production efficiency. The most common tree forms in temperate fruits are discussed.

9.4.1 Globular

The area is covered with a continuous expanse of tall trees with open centres. The most productive area of the tree is located at the outside edge towards the top, where the fruit is least easily reached. A significant portion of the tree remained unproductive.

9.4.2 Conical Shape with Open Framework

The tree's canopy creates a light-efficient structure where the upper part of the tree does not cast excessive shade on the lower branches, and a significant portion of the surface area for growing fruit is near the ground. The open structure enables the infiltration of sunlight into the innermost regions of the tree.

9.4.3 Vertical Tree Wall

This configuration provides a branch-spread only up to 2 m across the row. The light penetrates the canopy to a depth of 1 m. Hence, illumination from both sides provides adequate light through the tree.

9.4.4 Horizontal Canopy

This provides uniform light exposure to the entire bearing surface from the top; hence the effective thickness of the canopy is limited to 1m. This canopy shape has the potential for the mechanical harvesting of fruit.

9.4.5 "Y" or "V" Trellis

Tree shape provides maximum light exposure to the bearing surface, and has adaptability to mechanical harvesting.

9.5 VIGOUR CONTROL THROUGH GENOTYPE

Nature has enormous potential to create variability in fruit species. Many dwarf genotypes are available in fruit species, which can be further used for breeding as well as to reduce the canopy volume, increase precocity, produce quality fruit, and be tolerant to biotic and abiotic stress. The temperate fruit crops have more variability in dwarfing genotypes.

9.5.1 Apple

After the introduction of dwarfing rootstocks in Europe, they multiplied exponentially throughout the world. A lot of dwarfing rootstocks, including M9, M26, and B9 are generally used by orchardists for high-density planting. Semi-vigorous rootstocks like MM.111, MM.106 or M.793 are there as an alternative for HDP of apple orchards. The other promising rootstocks in this situation seem to be P 63, P 65, P 16, B 491, and P 66, which exhibit vigour intermediate between very dwarfing and standard dwarfing. Country-wise preference of rootstocks varies However, the orchards from countries like France, Germany, Italy, UK have a higher ratio (>90%) of dwarfing rootstock and HDP by using this M and MM series of rootstocks (Wang et al., 2019). Various apple rootstocks possessing various rhizospheric activities also change the growth and canopy of apples (Singh et al., 2018). The genes regulating dwarfing in the case of apple is given in Table 9.1.

9.6 TRAINING

Training involves shaping and directing the growth of trees into a desired form. It can be described as a deliberate practice of selectively removing plant parts to maintain the shape, size, and framework

TABLE 9.1
Important Gene Resources in Apple for Exploiting Canopy Architecture

Genotype	Gene	Characteristics	Reference
M9	MdDRO/MdPIN11	Root angle-related gene	An et al. (2017a)
Malus×domestica	MdMIEL1	Lateral root growth	An et al. (2017b)
Apple cv. Royal Gala	MdMYB88/MdMYB124	Root architecture	Geng et al. (2018)
*Malushupehensis*Redh. var. *pingyiensis*	MhGAI1	Repressor in the GA signaling pathway	Wang et al. (2012)
Malusdomestica cv. Hanfu	MdGA20-ox	Biosynthesis of gibberellic acid	Zhao et al. (2016)
M9	MdIPT5b	Isopentenyl transferase	Feng et al. (2017)
M9	MdPIN1b	Auxin efflux carrier	Gan et al. (2018)
M26	MdWRKY9/MdDWF4	Brassinosteroids (BR) synthetase	Zheng et al. (2018)

of the plant. When tree canopies are left unmanaged, it not only reduces fruit productivity but also compromises the quality of the produce.

Proper training of young fruit trees is essential for their appropriate development. It is more effective to guide tree growth through training rather than relying solely on corrective pruning. The primary objectives of fruit tree training include facilitating sunlight and airflow, directing growth for easy and cost-efficient cultural operations, protecting against sunburn and wind damage, and ensuring a balanced distribution of fruit-bearing parts.

9.6.1 Training System

The training system encompasses a combination of tree arrangement, planting density, support systems, and training methods. When it comes to tree crops, there is no universal planting system that suits all situations. The optimal combination for each orchard is determined by various factors, including soil conditions, rootstock, variety, management system, and socio-economic considerations. Tree density and light interception play significant roles in the early production of fruit trees. The primary objective of fruit tree training is to guide tree growth and establish sturdy frameworks that support the production of high-quality fruits. Effective training techniques promote canopy openness, ensuring maximum light interception (LI) and distribution. In mature trees, the shape of the canopy influences light distribution within it, and excessive shading can hinder flower bud development, fruit set, and fruit coloration. The planting area is determined by the spacing between rows and within rows. Choosing the appropriate training system is an integral part of overall orchard management, as it determines the planting distances and light interception and ultimately contributes to the success of the orchard. The choice of training system is highly dependent on planting density and is most important in the case of planting at very high densities. In peach, the traditional Open centre training system resulted in poor yield during its early years of planting, whereas the implementation of higher planting densities and different tree shapes has led to a substantial increase in fruit yield in the early stage of planting.

The conventional systems of training are (i) the central leader system (ii) the open center or vase system (iii) the delayed open center or modified leader system.

9.6.1.1 Central Leader System

A central leader system of training is recommended in areas that are prone to heavy snowfall, leading to breakage of branches. In a comparative study, among all the training systems, the Central Leader, which possessed a leader, formed a higher tree in terms of tree development (Cetinbas, 2021). In the central leader system, central leader branches are permitted to grow independently to have faster and more vigorous growth than the side branches. Several side branches grow at different heights in various directions. It forms a pyramid-shaped canopy. In this system, summer pruning is practiced to increase light interception in the interior of the canopy.

9.6.1.2 Open Centre or Vase System

In an open center system, the main stem is generally permitted to grow only up to a certain height (1.5–1.8 m), and then it is cut for the development of side branches. All side branches are headed back. There are no strong crotches in this system, which provides a weak frame. This system permits full sunlight to reach each branch and it appears like a vase structure. The trees trained to open the centre allow maximum sunlight to reach the branches.

9.6.1.3 Delayed Open Centre or Modified Leader System

This system is an interim solution between the central leader and open-centre training approaches. At first, open-centres are trained using a method similar to the central similarly system, where the main stem is allowed to develop for the first 2 or 3 years before being pruned back to a height of 75 cm. Subsequently, the growth of lateral branches is promoted and subsequently trimmed, adhering to the principles of the open center training system. In this system, emphasis is placed on developing scaffold branches that are larger and have greater length. The branches are strategically distributed, allowing ample light to penetrate the interior of the tree canopy. This training system is well-suited for most commercial fruit trees due to its relatively shorter tree height, which facilitates tasks such as pruning, spraying, harvesting, and other operations.

9.6.1.4 Spindle Bush System

The spindle bush training system is the modification of a dwarf pyramid system or intermediate between a bush form and a vertical cordon of training. It differs from the dwarf pyramid training system as it does not have specific scaffold branches. Trees trained on this system start bearing fruits within 2–3 years of planting. Fruit plants trained on the spindle bush system had significant results for fruit yield and other parameters.

9.6.1.5 Tall Spindle

This system is the integration of the Slender Spindle, the Vertical Axis and the Super Spindle training systems. It involves high plant densities, dwarf rootstocks, highly feathered trees, bending of the feathers below horizontal at planting, minimal pruning, and no permanent scaffold branches. Different fruit crops, responded quite differently when trained on different training systems. Robinson (2011) recorded higher fruit yield from Tall Spindle-trained trees as compared to trees trained on the solaxe system.

9.6.1.6 Slender Spindle or Fussetto

The slender spindle training system is a modification of the spindle bush training system. Compared to the vertical axis system, this system requires more extensive pruning. It is well-suited for a tree density ranging from 2,000 to 5,000 trees per hectare. In the slender spindle system, each tree is supported by an individual post. The fruit trees are trained to form a slender shape with a lower whirl of branches. Leader control is important, and the tree height is restricted to 2–2.5 m in this system. Cetinbas (2021) reported that a slender spindle training system in peaches with a 1 m row distance was most suitable in terms of fruit quality, economic yield, and profitability.

9.6.1.7 Vertical Axis System (VA)

The vertical central leader (axis) training system involves the development of a strong central leader from relatively "weak" fruiting branches that originate around the main leader. Typically, this training system maintains a tree density ranging from 1,000 to 2,500 trees per hectare. The tree forms a conical shape and is restricted up to 3.5–4.0 m. To ensure weak fruiting branches, keeping apical dominance is essential in the VA system, especially during the early stages of development. Trees, generally receive minimal training and pruning, which results in early production. Vertical axis systems (VA) allow trees to grow taller than slender spindle systems. The support system of the VA system is more complex than the slender spindle training system to install as it involves a high level of skill to form the support system to hold the trees and adequate crop load

9.6.1.8 Solaxe System

Solaxe system is a modification of the vertical axis (VA) in which renewal pruning is replaced by bending. In this system, fruiting branches and the central axis are generally bent in a similar pattern as of the free-growing fruiting branches. Under this system, fruiting branches are distributed spirally down the trunk of the tree to improve light penetration and fruiting yields. This system involves bending the top of the tree and the branches to suppress vigorous growth and to enhance flower bud induction.

9.6.1.9 Palmette System

The palmette is a trained form rather like an espalier, except the arms are raised at an angle of about 40°–45° instead of being horizontal. It is a modification of the central leader system. The palmette system is not suitable for very high-density planting because its "squeezes" trees along the row. This type of system is popular because the bending of branches on trellises controls the growth and provides a balance between vegetative growth and fruiting. This system of training has been adopted in many temperate fruit crops. In cherry, Moreno et al. (1998) recorded higher fruit yield efficiency but comparatively poor fruit quality with the palmette system as compared to plants trained over the multiple leader vase system.

9.6.2 Pruning

Pruning involves the elimination of unwanted or excessive branches, shoots, roots, or any other plant elements to allow the remaining portions to grow in a desired manner and attain an optimal canopy structure. Pruning is a traditional technique used in deciduous and temperate fruit crops. It is also employed in subtropical species, including grapes, figs, and phalsa. It serves as a means to dwarf the tree and maintain its desired size. When a branch is removed through pruning, it not only reduces stored carbohydrates but also decreases the potential leaf surface. This practice results in increased fruit size, more nitrogen per growing point, and stimulates new growth near the pruning cut. Pruning serves several purposes, including controlling plant size and shaping the tree form by managing the number, placement, relative size, and angle of branches. It also enhances fruit quality by promoting better light distribution, removes diseased, criss-crossed, dried, and broken branches, eliminates non-productive parts to redirect energy towards fruit-bearing areas, helps establish a proper root-to-shoot ratio, regulates the fruit crop, and reduces the likelihood of insect pests, diseases, and winter injury.

Reasons for pruning of fruit trees at different stages:

i. Pruning of newly planted trees restores the balance between the root and shoot system.
ii. Small young trees are generally pruned to develop a strong tree framework.
iii. Pruning of mature fruit trees is typically carried out to maintain the desired spread, height, and density of the canopy, which facilitates various orchard operations such as chemical spraying, fruit thinning, and harvesting. By managing the canopy structure through pruning, it becomes more convenient and efficient to perform essential tasks within the orchard, ensuring effective management and maintenance of the fruit trees. The prime aim for pruning fully grown trees is to achieve high-quality fruit yield.

Two types of pruning systems are in practice, i.e. summer pruning and dormant pruning. Summer pruning maintains the size of the fruit tree to the form that we need, while winter or dormant pruning promotes extra growth for building framework and renewal shoots. It is necessary for trees trained to a restricted structure, but can also maintain the standard trees from outgrowing their given space. In the summer pruning, water shoots and excess vigorous growth can be removed by leaving only a choice few after the heavy dormant pruning.

9.6.3 Summer Pruning

It used to perform mainly in dwarf fruit trees to control the shape and size of traditionally trained fruit trees. Nowadays, summer pruning is recommended mainly for high-density fruit trees to control tree size and improve fruit yield and quality. Summer pruning has been used as a management method for fruit trees. It was reported to be a good technique for controlling the growth of trees & increasing flowering (Bayazit et al., 2012), enhancing soluble solids concentration (Demirtaş et al., 2010b) for decreasing titratable acidity.

9.6.4 Apical Dominance

Pruning performed on shoots removes apical dominance, releases buds from correlative inhibition, and changes the branching pattern and tree form. In apical dominance, the top or upper bud has an initial advantage over the lateral buds, which starts to develop a little faster and produces more auxin that stimulates cambial activity by enabling the bud to develop better vascular connections with the main axis. Pruning practices encouraged efficient canopy management for maximum utilization of sunlight and break apical dominance, thus allowing lateral bud growth.

9.6.5 Leaf Area and Photosynthesis

Pruning practices influence the leaf area as well as the photosynthetic rate of plants by improving the various physiological parameters in the plant system. Manipulating the apical dominance of guava plants resulted in more intense pruning, which in turn led to an increased number of leaves and a larger leaf area. Enhanced leaf count and expanded leaf surface area enhance the process of photosynthesis, leading to improved fruit size. Adhikari and Kandel (2015) observed similar results in guava, where the highest leaf area was achieved when the branch was clipped to a length of 30 cm. In addition, the intensity of pruning reduced the number of shoots per branch, leading to increased nutrient availability for each individual leaf.

9.6.6 Carbohydrate Reserves

Pruning practices changes' the partitioning of total dry weight in a way that a higher amount of dry weight is translocated to new shoots. Therefore, the anticipated consequence of greater production of new shoots is a reduction in the reserves of nutrients, particularly carbohydrates. Summer and dormant pruning in peaches had different effects on carbohydrate reserve. Higher carbohydrate was recorded in dormant pruned and unpruned peach trees. Also, the time of pruning also influenced the carbohydrate reserve. Peach trees pruned during late summer had higher carbohydrate content than trees pruned in early summer (June or July) (İkinci et al., 2014). Among different times of summer pruning, post-harvest summer pruning resulted in the accumulation of maximum sugar and starch content in apricot (Demirtaş et al., 2010a).

9.6.7 Root Pruning

Root pruning has been employed as a management technique in fruit crops for a considerable time, with the purpose of restraining excessive shoot growth (Geisler and Ferree, 1984). This practice is commonly used to regulate the growth of vigorously growing fruit trees. By pruning structural roots, the efficiency of functional roots is enhanced. Root pruning is extensively applied during the transplantation of saplings to promote better establishment. The timing and selection of the distance from the main stem (trunk) for root pruning are crucial factors that influence not only vigor control but also vegetative growth, flowering, and fruit set. However, successful root pruning necessitates additional experience, horticultural expertise, and a comprehensive understanding of the physiological principles underlying this technique.

9.6.8 Molecular and Physiological Aspect behind Dwarfing through Rootstock

The practice of manipulating the above-ground portion by altering the below-ground portion, the canopy structure, and gene editing is fairly widespread. However, comprehension of the mechanism is still lacking. Many apple rootstocks that can tolerate pests, diseases, poor soils, and harsh climatic conditions have been produced by selective breeding and development programs around the world. The qualities that a rootstock must possess have evolved through time, but the basic goals have remained the same: preventing excessive growth, improving cropping efficiency (yield per tree size), and shortening the period needed for a tree to enter cropping (precocity). It may not always be possible to anticipate rootstock performance in sites with varied climatic or soil conditions using the data gathered from these evaluation trials. Because of this ignorance, breeding programs must always be time- and money-consuming operations. A more predictive method of evaluating rootstocks can be devised with an understanding of the essential physiological mechanisms that determine their benefits, and by identifying molecular markers, breeding programs can be accelerated. In simplest terms, dwarfing rootstocks lessen the scion's dry weight. The molecular approach can be exploited to identify the different genes responsible for inducing dwarfing in fruit crops, which is explained in Table 9.2.

9.6.9 Anatomical Aspect

The bark (periderm, cortex, and phloem tissue) content of the dwarfing rootstocks is higher than that of the wood (xylem tissue). Additionally, a significant portion of the dwarfing rootstock is occupied by functionally dead xylem arteries, axial parenchyma, and ray parenchyma cells. When compared to vigorous rootstock, dwarfing rootstocks for apples contain fewer xylem vessels, active root tips, and fine roots. Because of their anatomical structure, plants are able to absorb fewer nutrients and liquids as they grow and develop, which results in a decreased canopy volume.

9.6.10 Hormonal Aspect

Dwarfing occurs due to the obstruction of auxin flow at the graft union interface, leading to a decrease in xylem production and limited access to water and minerals for the scion. The root system of dwarf rootstocks experiences a decrease in auxin transport, which leads to a reduction

TABLE 9.2
Identification of different genes imposing dwarfing in fruit crops

Sr No	Genotype	Name of the gene	Functions	Reference
1.	M26 apple	WRKY transcription factors	Imposing dwarfism	Zheng et al. (2018)
2.	Dwarfing apple	MdWRKY9	Inhibiting brassinosteroid synthetase MdDWF4 expression	
3.	Transgenic apple	MdNAC1	Regulate endogenous ABA and BR biosynthesis	Jia et al. (2018)
4.	*Malusbaccata*	galactinol synthase 1-like (GolS1-like)	Sugar metabolism and control tree architecture	Chen et al. (2020)
5.	Dwarfing apple rootstock	MdAUX1 and MdLAX2	increasing flavonoids and reduced auxin movement which leads to an imbalance in carbohydrate distribution and reduced cell growth and metabolism	Foster et al. (2017)
6.	M9 Apple	MdFT1/2, MdBFTa/b, MdCO, MdGI, and MdSOC1	Induction of early flowering by suppressing the vegetative growth	Foster et al. (2014)
7.	Citrus	rolABC and GA20-oxidase	Induction of dwarfism	La Malfa et al (2011)

in cytokinin levels. This alteration in root metabolism ultimately affects cytokinin biosynthesis (Hartmann and Kester, 2011). Cytokinin that has been translocated moves upward from the root system to the shoots, causing a decrease in shoot growth. Dwarfing apple rootstocks exhibited elevated levels of ABA, reduced ABA:IAA ratios, and decreased GA in contrast to vigorous ones. Dwarfing rootstocks have a higher proportion of bark to wood (xylem) due to this mechanism. When selecting a dwarfing apple rootstock, it is advisable to use a rootstock with a higher concentration of ABA in the branch bark (Hartmann and Kester, 2011).

9.6.11 Phenolic Compound Metabolism

Not only do the hormones control the dwarfing behavior, but also phenols control the IAA metabolism in the wood and bark skin of dwarfing genotypes and control the dwarfness of scions. Monophenols and polyphenols have diverged roles in IAA metabolism in plants. Some monophenols increase the IAA oxidative decarboxylation, while some polyphenols like ferulic acid, chlorogenic acid, and caffeic acid prevent IAA oxidation (Habibi et al., 2022).

9.6.12 Carbohydrate Partitioning and Nutritional Levels

The rootstock influences the distribution of carbohydrates on both sides of the graft union. Dwarfing apple rootstocks have a greater capacity to allocate carbon to reproductive areas compared to vigorous rootstocks. In contrast to dwarfing stock, robust rootstocks exhibit greater accumulation of dry matter in both the shoot and root systems (Foster et al., 2017). Additionally, the robust rootstock's increased nutrient intake aids in the development of new vegetative growth, which competes with reproductive growth for nutrients (Habibi et al., 2022).

9.7 EXPLOITING PLANT GROWTH REGULATORS TO CONTROL GROWTH

The commercial use of PGRs in manipulating growth and development in agriculture, however, developed slowly in comparison to other chemicals such as fertilizers, insecticides, and pesticides. However, as fruit production becomes more intensive, it permits the use of advanced technologies, including contemporary inputs to enhance production, and the role of PGRs becomes more vital (Darshan et al., 2022). Maintaining the balance between vegetative and reproductive growth is of utmost priority if, in any case, vegetative growth exceeds more, it could affect a variety of fruit production processes, including flower bud development, fruit set, fruit quality, physiological problems, insect control, and fruit life after harvest (Hota et al., 2017) (Table 9.3).

9.7.1 Auxin

9.7.1.1 Auxin-Cytokinin Interactions Control the Shoot Branching

In many plant species, axillary bud development (which leads to further lateral branching) is suppressed, and the shoot apical meristem grows vertically leading to taller plants. Apical dominance is a term for this phenomenon. The axillary buds undergo dormancy release and initiate growth upon decapitation of the stalk apex. The production of auxin by an intact shoot apex hinders the growth of axillary buds, while the production of cytokinin resulting from the removal of the shoot apex stimulates the growth of axillary buds. Auxin suppresses bud expansion and is actively distributed basipetally in the entire shoot. Cytokinins, on the other hand, move acropetally and encourage bud development. Such interactions between auxin and cytokinin regulate the branching of the shoot.

TABLE 9.3
Optimum Dose of PGRs in Fruit Crops

Crop	Dose	Timing	References
Apple	PP_{333} @ 500ppm	As multiple (3–4) foliar sprays with beginning at PF and subsequent spray(s) at 3 weeks intervals	Richardson et al. (1986)
Apple	Prohexadione-Ca 125–250ppm	At, late bloom- early PF, followed by 2nd spray 14–21 days later	Rademacher and Kober (2003)
Pear:	CCC @ 1,000ppm	At Petal Fall (PF) and 2–3 weeks after PF	
Apricot:	Foliar spray of PP_{333} @ 1,000ppm,	3 weeks after full bloom	Kuden et al. (1995)
Peaches	PP_{333} 2 g a.i./tree	Before bud-break to July Elberta peach on seedling r.s. under HDP (2,500 trees/ha)	Negi and Sharma (2009)
Peaches	PP_{333} 2 g a.i./tree	Foliar spray of PP_{333}@ 2,000ppm at bloom.	Negi and Sharma (2009)
Plums	PP_{333} @4 L/ha	In the end of March.	
Cherries	PP_{333} @3 L/tree	Before the end of April.	

9.7.1.2 Cytokinin

Cytokinin generally helps in increasing plant canopy through cytokinesis (Priyadarshi et al., 2017;Priyadarshi et al., 2018;Hota et al., 2020). However, benzyl adenine (BA) treatments decreased terminal shoot length, BA treatment at a concentration of 0.04% decreased the intermodal number and average length of internodes on terminal shoots. Compared to untreated trees, BA-treated plants decreased the plant height by 10%–15%. By increasing the growth of laterals while decreasing the growth of terminal shoots, BA treatments boosted the total growth of the trees, which was distributed between terminal growth and the growth of laterals. The concurrent growth of laterals could slow down the growth of terminal shoots.

9.7.1.3 Ethylene

Ethrel increases floral bud formation and reduces vegetative branch growth in non-bearing apple plants. However, the time of application is very much crucial for achieving the intended result. In bearing trees, ethrel should be sprayed in diluted version 1–2 weeks after full bloom to thin out excessive flowers. However, in a young tree or tree just completing the juvenile stage, the sparing of ethrel should postponed to 3–5 weeks after bloom to protect the plant from excessive flower and fruit thinning and deformed fruit formation. The application of ethrel at this time of the growing season will result in terminal growth suppression after a period of 2–3 weeks of active growth.

9.7.1.4 Promalin

Application of promalin (a mixture of GA_{4+7} and 6-benzyladenine) increased feathering in many sweet cherry cultivars, including Germersdorfióriás', BiggareauBurlat', and 'Linda'. Increased short shoot with reduced apical growth was observed with higher concentration. When promalin was applied during the bud swell and bud burst stage in cherry, it increased the spur production and lateral shoot development with increased crotch angle on the one-year-old primary scaffold branch (Veinbrants andMiller, 1981). Cultivars Cortland and Gloster branched poorly in the nursery whereas cv. Melrose and cv. Jonagold branch well. In nursery pinching or spraying with growth substances before the end of June gave better results than at the beginning of July. The best effect of promalin at a concentration of 500 mg/L on the branching of cv. Cortland and cv. Gloster was obtained. Maximum no. of lateral shoots was obtained when these cultivars grafted either on M26 or MM106 and treated with promalin as compared to control, pinching, and mixture of GA_3 and BA spray.

9.7.1.5 Paclobutrazol [Cultar, Bonzi]

Paclobutrazolis very helpful in high-density planting since it makes the plants smaller by slowing the vertical growth of trees by inhibiting the synthesis of gibberellin. Increased flower bud density, fruit set, fruit size, fruit maturation, and fruit quality can all be achieved by manipulating the vegetative growth of plants. Therefore, by increasing the plant population coupled with management of their canopies, especially with the use of growth retardants, may be emphasized under high-density orcharding concept.

9.7.1.6 Chlormequat [CCC, Cycocel]

A screening procedure for quaternary ammonium compounds for growth retardant action led to the discovery of chlormequat chloride. Since the 1960s, it has been widely used to regulate the development of potted greenhouse crops, including chrysanthemum, poinsettia, geranium, azalea, and hibiscus, as well as to prevent lodging in grain and cereal crops.

9.7.1.7 Ancymidol

The suggested generic name for cu-cyclopropyl-a-(P-methyl-phenyl)-5-pyrimidinemethanol is ancymidol. An inhibitor of gibberellin production is ancymidol. These substances cause internode compression on a variety of plants by partially blocking the synthesis of gibberellin in plants. By using gibberellic acid, the effects of this class of plant growth regulators can be neutralized. Many different plant species and cultivars' growth was found to be effectively slowed down by ancymidol. Both woody and herbaceous species respond well to it. In addition to causing a significant slowdown in plant growth, using higher dosages extended their effectiveness. Treatments with ancymidol may result in fewer nodes. Ancymidol can be used to maintain the prolonged dwarfing of plants that are actively growing.

9.7.1.8 Prohexadione Calcium [Apogee]

When used on apple trees, apogee slows the growth of vegetative shoots. This product inhibits the manufacture of gibberellins, which control cell elongation. The improvement in fruit color, better disease, and insect management, and decreased severity of fire blight shoot infections can all be a result of the reduction in shoot growth (internode length).

9.8 CONCLUSION

A fruit tree's canopy is its physical structure, which includes its stem, branches, shoot, and leaves. The density of the canopy is influenced by the number and size of the leaves. The yield and quality of the fruits have been improved over time by using canopy control in fruiting trees. One of the most common approaches for effectively managing large, unmanageable trees to increase their productivity is the management of canopy architecture. Canopy management is the alteration of tree canopies to maximize their capacity for producing fruits of the highest quality. Canopy management is concerned with the growth and upkeep of the structure of fruit crops in relation to their size and form in order to maximize yield with high-quality fruits. Utilizing the various methods available, such as training, pruning (dormant, summer, and root pruning), branch orientation (bending), scoring, girdling, selection of the right rootstock, use of plant growth regulators, appropriate use of fertilizer, deficit irrigation, and use of genetically modified plants with altered architectural characters would be helpful, canopy management deserves greater attention. Maximizing light interception, enhancing light dispersion within the canopy, and maintaining optimum airflow are the key goals of canopy management. Canopy management increases output, boosts fruit quality, supports cultural traditions, and aids in pest and disease management. In new plantations, early training and pruning are given to establish strong tree frameworks; however in older plantations, the goal of canopy management is to reduce tree height and provide for solar radiation inside the canopy by thinning out surplus biomass.

REFERENCES

Adhikari, S., and T.P. Kandel. 2015. Effect of time and level of pruning on vegetative growth, flowering, yield, and quality of guava. *International Journal of Fruit Science* 15(3):290–301.

An, H., H. Dong, T. Wu, Y. Wang, X. Xu, X. Zhang, and Z. Han. 2017a. Root growth angle: an important trait that influences the deep rooting of apple rootstocks. *Scientia Horticulturae* 216:256–263.

An, J., X. Liu, L. Song, C. You, X. Wang, and Y. Hao. 2017b. Functional characterization of the apple RING E3 ligase MdMIEL1 in transgenic Arabidopsis. *Horticultural Plant Journal* 3:53–59.

Anthony, B.M., and I.S. Minas. 2021. Optimizing peach tree canopy architecture for efficient light use, increased productivity and improved fruit quality. *Agronomy* 11:1961.

Anthony, B.M., J.M. Chaparro, D.G. Sterle, J.E. Prenni, and I.S. Minas. 2021. Metabolic signatures of the true physiological impact of canopy light environment on peach fruit quality. *Environment and Experimental Botany* 191:104630.

Bayazit, S., B. İmrak, and A. Küden. 2012. Effects of tipping applications on yield and fruit quality of some peach and nectarine. *MKU ZiraatFakültesiDergisi* 17:23–30.

Chen, Y., X. An, D. Zhao, E. Li, R. Ma, Z. Li, and C. Cheng. 2020. Transcription profiles reveal sugar and hormone signaling pathways mediating tree branch architecture in apple (*Malusdomestica*Borkh.) grafted on different rootstocks. *PLoS One* 15:1–20, e0236530.

Darshan, D., D. Hota, R. Devi, and J.K. Shukla. 2022. Micronutrients and plant growth regulators affecting the yield and quality of fruit crops: a review. *Emergent Life Science Research* 8(2):92–103.

Demirtaş, M.N., I. Bolat, S. Ercişli, A. İkinci, H. Ölmez, M. Şahin, M. Altındağ, and B. Çelik. 2010a. The effects of different pruning treatments on the growth, fruit quality and yield of 'Hacıhaliloglu' apricot. *Acta Scientiarum Polonorum Hortorum Cultus* 9:183–192.

Demirtaş, M.N., I. Bolat, S. Ercişli, A. İkinci, H. Ölmez, M. Şahin, M. Altındağ, and B. Çelik. 2010b. The effects of different pruning treatments on seasonal variation of carbohydrates in 'Hacihaliloglu' apricot cultivar. *Notulae Botanicae Horti Agrobotanici Cluj-Napoca* 38:223–227.

Feng, Y., X. Zhang, T. Wu, X. Xu, Z. Han, and Y. Wang. 2017. Methylation effect on ipt5b gene expression determines cytokinin biosynthesis in apple rootstock. *Biochemical and Biophysical Research Communications* 482:604–609.

Foster, T.M., A.E. Watson, B.M. Van Hooijdonk, and R.J. Schaffer. 2014. Key flowering genes including FT-like genes are upregulated in the vasculature of apple dwarfing rootstocks. *Tree Genet Genomes* 10:189–202.

Foster, T.M., P.A. McAtee, C.N. Waite, H.L. Boldingh, and T.K. McGhie. 2017. Apple dwarfing rootstocks exhibit an imbalance in carbohydrate allocation and reduced cell growth and metabolism. *Horticultural Research* 4:1–13.

Gan, Z., Y. Wang, T. Wu, X. Xu, X. Zhang, and Z. Han. 2018. MdPIN1b encodes a putative auxin efflux carrier and has different expression patterns in BC and M9 apple rootstocks. *Plant Molecular Biology* 96:353–365.

Geisler, D., and D.C. Ferree. 1984. Responses of plants to root pruning. *Horticultural Reviews* 6:156–188.

Geng, D., P. Chen, X. Shen, Y. Zhang, X. Li, L. Jiang, Y. Xie, C. Niu, J. Zhang, X. Huang, F. Ma, and Q. Guan. 2018. MdMYB88 and Md-MYB124 enhance drought tolerance by modulating root vessels and cell walls in apple. *Plant Physiology* 178:1296–1309.

Habibi, F., T. Liu, K. Folta, and S. Sarkhosh. 2022. Physiological, biochemical, and molecular aspects of grafting in fruit trees. *Horticulture Research* 9:0uhac32.

Hampson, C.R., H.A. Quamme, T. Robert, and R.T. Brownlee. 2002. Canopy growth, yield, and fruit quality of 'Royal Gala' apple trees grown for eight years in five tree training systems. *HortScience* 37 (4):627–631.

Hartmann, H.T., and D.E. Kester. 2011. *Geneve. Hartmann & Kester's Plant Propagation Principles and Practices*, Upper Saddle River (Nueva Jersey, EstadosUnidos). 8th ed. Nueva Jersey, EstadosUnidos, US State: Prentice Hall; p. 915.

Hota, D., A.K. Karna, S.D. Behera, and G. Mishra. 2017. Use of plant growth regulators in canopy management. *Biomolecule Reports – An International e-Newsletter*. Date: 6th November 2017.

Hota, D., A.S. Kalatippi, and J.K. Nayak. 2020. Forchlorfenuron: A new generation growth regulator for fruit crops. *Agriculture & Food: e-Newsletter* 2(7):201–202.

Jia, D., X. Gong, M. Li, C. Li, T. Sun, and F. Ma. 2018. Overexpression of a novel apple NAC transcription factor gene, MdNAC1, confers the dwarf phenotype in transgenic apple (*Malusdomestica*). *Genes (Basel)* 9:229.

Kuden, A., A.B. Kuden and N. Naska. 1995. Physiological effects of foliage applied paclobutrazol on 'Canino' and 'Precoce de Colomer' apricot cultivars. *Acta Horticulturae* 384:419–423.

Lakso, A.N., and L. Corelli-Grappadelli. 1992. Implications of pruning and training practices to carbon partitioning and fruit development in apple. *Acta Horticulturae* 322:231–239.

Moreno, J., F. Toribio, and M.A. Manzano. 1998. Evaluation of palmette, marchand and vase training systems in cherry varieties. *Acta Horticulturae* 46:485–489.

Negi, N.D., and N. Sharma. 2009. Effect of paclobutrazol application and planting systems on growth and production of peach (*Prunuspersica*). *Indian Journal of Agricultural Sciences* 79 (12):1010–1012.

Priyadarshi, V., D. Hota, S.P.S. Solanki, and N. Singh. 2018. Effect of growth regulators and micronutrients on yield attributing character of litchi (*Litchi chinensis* Sonn.) cv. Calcuttia. In: Singh, J., R. Nigam, W. Hasan, A. Kumar, and H. Singh. *Advances in Horticultural Crops*. Germany: Weser Books, pp. 269–277. ISBN: 978-3-96492-079-9.

Priyadarshi, V., K. Mehta, D. Hota, G. Mishra, and A. Jogur. 2017. Effect of growth regulators and micronutrients spray on vegetative growth of litchi (*Litchi chinensis* Sonn.) cv. Calcuttia. *Agriculture Update* 12 (TECHSEAR-3):707–712.

Rademacher, W., and R. Kober. 2003. Efficientuse of prohexadione-Ca in pome fruits. *European Journal of Horticultural Sciences* 68:101–107.

Richardson, P.J., A.D. Webster, and J.D. Quinlan. 1986. The effect of paclobutrazol sprays with or without the addition of surfactantson the shoot growth, yield and fruit quality of the apple cultivarsCox and Suntan. *Journal of Horticultural Sciences* 61:439–446.

Robinson, T. 2011. Advances in apple culture worldwide. *Rev. Bras. Frutic., Jaboticabal* – SP, 37–47.

Singh, N., D.P. Sharma, V. Kumar, and D. Hota. 2018. Impact of different rootstocks and soil agro-techniques on rhizospheric biological activities and growth traits under apple replant sick soil. *Multilogic in Science* 7 (Special issueicaaastsd-2018):88–92.

Veinbrants, N., and P. Miller. 1981. Promalin promotes lateral shoot development of young cherry trees. *Australian Journal of Experimental Agriculture and Animal Husbandry* 21(113)618–622.

Wang, S., Z. Liu, C. Sun, Q. Shi, Y. Yao, C. You, and Y. Hao. 2012. Functional characterization of the apple MhGAI1 gene through ectopic expression and grafting experiments in tomatoes. *Journal of Plant Physiology* 169:303–310.

Wang, Y., W. Li, X.F. Xu, C.P. Qiu, T. Wu, Q.P. Wei, F.W. Ma, and Z.H. Han. 2019. Progress of apple rootstock breeding and its use. *Horticultural Plant Journal* 5:183–91.

Zhao, K., F. Zhang, Y. Yang, Y. Ma, Y. Liu, H. Li, H. Dai, and Z. Zhang. 2016. Modification of plant height via RNAi suppression of MdGA20-ox gene expression in apple. *Journal of American Society of Horticultural Sciences* 141:242–248.

Zheng, X., Y. Zhao, D. Shan, K. Shi, L. Wang, Q. Li, N. Wang, J. Zhou, J. Yao, Y. Xue, S. Fang, J. Chu, Y. Guo, and J. Kong. 2018. Md WRKY 9 overexpression confers intensive dwarfing in the M26 rootstock of apple by directly inhibiting brassinosteroid synthetase MdDWF4 expression. *New Phytology* 217:1086–1098.

10 Growth Regulation Aspects of Propagation

Pramod Verma and Naveen C. Sharma

10.1 INTRODUCTION

Propagation (reproduction) is the main activity of humanity, and it was this realization that sparked the rise of civilization and the beginning of human dominion over the world (Hartmann et al., 2015). Growth hormones have a significant impact on the propagation of fruit crops because they can stimulate certain responses, such as root induction, in addition to being a component of the internal mechanism that regulates plant activity. The three primary methods of fruit crop propagation are sexual (seeds), asexual (layering and cuttings), and micropropagation. These methods are employed in the commercial multiplication of fruit crops, and either natural or synthetic growth hormones are used (Table 10.1).

TABLE 10.1
Important Plant Growth Regulators and Hormones Used in Fruit Crops

Name	Chemical Name	Mol. Weight	Use[a]
	A. Auxins		
IAA[b]	indole-3-acetic acid	175.20	CM;D,S
IBA[b]	indole-3-butyric acid	203.20	CM;D,S
KIBA	indole-3-butyric acid-potassium salt	241.30	CM;D,S
NAA	A-naphthaleneacetic acid	186.20	CM;D,S
2,4-D	2,4-dichloro-phenoxy-acetic acid	221.0	CM;Ap
2,4,5-T	2,4,5-tri-chloro-phenoxy-acetic acid	255.50	Ap
	B. Cytokinins		
BA	6-benzyl-amino-purine	225.30	CM
4-CPPU	N-(2-cholro-4-pyridyl) n'-phenyl urea	247.70	CM
DPU	1,3-diphenyl-urea	212.30	CM
2iP	6(di-methyl-allyl amino) purine	203.20	CM
kinetin		215.20	CM;D,S
TDZ	Thidiazuron	219.20	CM
Zeatin[b]			CM
	C. Gibberellins		
GA_3[b]	Gibberellic acid	346.40	CM;D,S;Ap
KGA_3	Gibberellic acid potassium salt	384.50	CM;D,S;Ap
	D. Inhibitors		
ABA[b]	Abscisic acid	264.30	CM;D,S;Ap

Source: Modified from *Plant Cell Culture* 1993 Catalog. Sigma Chemical Co., St. Louis, MO.
[a] CM = culture medium; [b]Occur Naturally; D,S = dip or soak; Ap = apply to plant.

DOI: 10.1201/9781003354055-10

10.2 PROPAGATION BY SEED

Most fruit crops are recognized to have three distinct maturation stages. The stages consist of Stage I (histodifferentiation - a quick rise in seed size mostly caused by cell division); Stage II (cell expansion - the highest increase in seed size for the accumulation of food reserves), and Stage III (maturation drying - a significant decrease in seed fresh weight due to water loss). Naturally, growth hormone is synthesized by the fruit component and exerts a substantial influence on the metabolism and development of each stage in various ways. These include promoting growth and differentiation of the embryo, facilitating the accumulation of food reserves for storage throughout germination and early seedling growth, and contributing to the development of fruit tissue. Seeds generally have a greater concentration of plant hormones compared to other parts of plants (Bewley and Balck, 1994). IAA (indole-3-acetic acid) exists in both free and conjugated forms in large quantities in growing seeds. In Stages I and II of seed development, free IAA is high; however, mature seeds have less free IAA. In mature seeds and during germination, conjugated IAA is produced in large quantities. Throughout seed development, gibberellins in a variety of forms are prevalent (Stages I and II). At seed maturity, conjugated forms of gibberellins replace active forms, which start to diminish. These conjugated forms, like auxin, are used during germination. Gibberellins might not have a significant impact on seed development. Several cytokinins both free and conjugated, are abundant in growing seeds. The stages of embryogenesis during which cells divide contain the highest concentration of cytokinin (Stage I and early Stage II). Plants use cytokinin to regulate cell division, and it also appears to be significant during Stage I embryos' differentiation period. ABA levels are elevated in growing seeds (Stage II). In Stage II seeds, ABA seems to promote the buildup of storage reserves. Additionally, ABA is a strong germination inhibitor. Most likely, ethylene contributes very little to seed growth.

10.2.1 Hormonal Regulation during Chilling Treatment (Stratification) for Seed Germination

Temperate fruits require specific cold temperature (0°C–10°C) treatment to break seed dormancy and improve seed germination (Ciacka et al., 2019) in comparison to tropical and subtropical fruits. In-depth research has been done on the hormone changes that occur during chilling stratification in apples (Lewak, 2011), peaches, and apricots (El-Yazal et al., 2021). Figure 10.1 depicts the fluctuations in hormones along with additional regulator levels in apple seeds following the cessation of dormancy, specifically under stratified conditions at a temperature of 5°C. (Lewak, 2011). On the 15th day of stratification, there was a notable increase in the release of hydrogen cyanide (HCN), and the levels of all stimulants' levels gradually increased after the disappearance of abscisic acid (ABA) during the initial 20 days of cold therapy. During the ensuing phase, there is a considerable increase in the levels of GA, CK, and JA. The levels of all these hormones reached their highest point during the 30th and 40th days of stratification. During this period, a second, less intense peak of HCN is observed. During the last phase of seeds after ripening, the concentrations of all the examined regulators fell to levels similar to those found in dormant seeds. During the last stage of stratification, only JA exhibited a second, minor peak. Contrary to the previously mentioned hormone level changes, the ethylene emission rate increased gradually rather than in distinct phases throughout stratification. The levels of gibberellic acids (GAs), cytokinins (CKs), and jasmonic acid (JA) begin to rise at the same time as the decrease in abscisic acid (ABA) levels and reach their highest point when ABA is no longer detectable (on day 30). Upon surpassing this barrier, the rate at which embryonic dormancy is eliminated escalated, culminating in its peak in the seeds of apples between the 30th and 50th days of stratification, ultimately resulting in the elimination of dormancy.

In the peach, radicle elongation, and epicotyl elongation are the two distinct elements of germination that are influenced. No germination response happens in intact seeds throughout stage

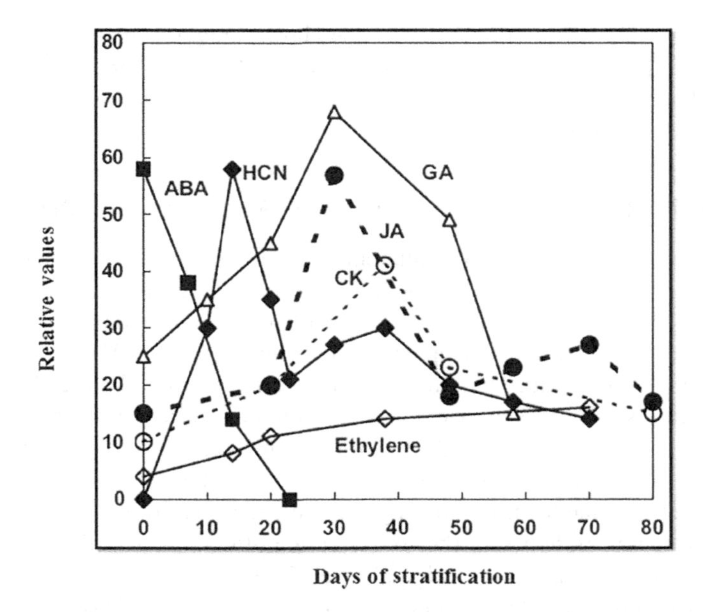

FIGURE 10.1 Schematic presentation of changes in relative levels of main growth regulators (ABA abscisic acid, ethylene rate of C_2H_4 emission, CK free cytokinins, GA free gibberellins A_4 and A_7, HCN free hydrogen cyanide, JA jasmonic acid) in apple seeds during the cold (5°C) stratification. (Modified from Lewak, 2011.)

I of stratification (0–30 days), whether they are chilled, transferred to 25°C at regular intervals, or presoaked in GA. In stage II (30–45 days), cold seeds do not sprout, but seeds that are moved to 25°C respond more favorably to germination and GA presoaking about doubles germination rates. Stage III (45–75 days) is when the seeds start to germinate quickly at cooling temperatures, and the seedlings start to grow normally. The responses reported are associated with hormone concentrations. All the coating of the seed and the cotyledons of newly picked peach seed, as well as other fruit crops including walnut, plums, apple, and hazelnut, contain a significant amount of ABA, as observed in studies by Martin et al. (1969) and Lin and Boe (1972). ABA, or abscisic acid, is the primary hormone responsible for seed dormancy in stone fruit seeds. Research conducted by Stein et al. (2021) has shown that as the stratification period increases, there is a notable reduction in ABA levels. ABA concentration decreases to almost nothing in stage I. Stage I intact seeds kept at chilling temperatures have low concentrations of gibberellin-like compounds, but stage II shows a sharp increase, suggesting that the ability to synthesize gibberellins is either present (Gianfagna and Rachmiel, 1986) or that there is a change from an inactive form to a free form. The introduction of paclobutrazol, an inhibitor of gibberellin synthesis, at the start of stage II significantly reduced the production of gibberellins but did not significantly affect the germination rate. Strong inhibition of the epicotyl and seedling elongation suggests that the radicle and epicotyl responses to chilling are distinct. These findings provide credence to the idea that cotyledons and testa of latent seeds both contain inhibitors (ABA). During the early phases of dormancy, they vanish (or are neutralized by cytokinins). Gibberellins are either created at the chilling temperature or changed into an accessible (or unbound) form, enabling radicle emergence (germination) to occur at warmer temperatures. Epicotyl elongation is a more regional phenomenon that either includes a different regulatory mechanism or has a greater threshold for gibberellins. Research on filbert (*Corylus avellana*) seeds has shown that the hormonal systems that influence germination are divided into those that promote and impede growth. The intact seed is dormant but the embryo is quiescent during the time of ripening. Both a large amount of abscisic acid and a noticeable amount of gibberellin have been found in the seed coat. The embryo is latent after the seed is dried after harvest, and the gibberellin levels drastically drop (Ross and Bradbeer, 1968). Germination requires several months of stratification.

During this chilling phase, gibberellin levels are low; however, when the seeds are placed in warm conditions, and germination occurs, gibberellin levels rise. The need for chilling treatment can be eliminated by applying gibberellic acid to the dormant seed. A combination of ABA and gibberellin prevents germination by offsetting the effects of GA.

10.2.2 Role of PGR's and Different Chemicals in Seed Dormancy Breaking

Adding specific chemicals and growth regulators to stratified seeds before sowing can enhance both the germination of seeds and their growth. Gibberellins, cytokinin (6-benzyl amino purine), jasmonic acid, and salicylic acid are plant growth regulators that have a role in breaking seed dormancy in temperate fruit, particularly in apples (Ranjan and Lewak, 1992).

Gibberellin: Kar (1979) obtained maximum germination in apples when seeds were treated with 50 ppm GA. Immature seeds, as well as mature dormant apple seeds, contain gibberellins A_4 and A_7. GA_4, GA_7, and GA_3 have demonstrated the ability to enhance the germination of apple embryos that have been isolated. The levels of Gibberellin A_7 (GA_7) in latent seeds were tenfold higher than the levels of GA_4 and remained constant throughout the entire period of cold stratification. During the first 4 weeks of treatment, however, the concentration of GA_4 increased considerably (by more than 1,000 times), then returned to its original level. The breaking of dormancy in apple seeds is also determined by GAs and their conjugates (glycosylated GAs releasing free hormones and GAs coupled with amino acids) (Lewak, 2011). Gibberellic acid has increased runner production inconsistently in strawberries. GA_3 foliar application at 50 mg/L in Gaviota strawberry plants, an increased runner which helps us to produce strawberry plants (Asadi et al., 2013).

According to Yildiz et al. (2007), jasmonic acid facilitates the germination of dormant apple seeds. The involvement of cytokinins in breaking embryonic dormancy in apple seeds is shown through alterations in endogenous hormone activity and their effect on the germination of isolated embryos (Zhang and Lespinasse, 1991). Salicylic acid (SA) has been associated with the regulation of seed dormancy and enhancement of seed germination through the reduction of oxidative damage (Chitnis et al., 2014). Applying GA_3, or a mixture of SA, GA_3, BAP, and JA during the process of seed stratification is a highly effective method for enhancing and accelerating seed germination and the growth of young seedlings. This results in a shorter apple breeding cycle, as demonstrated by Gornik et al. (2018).

Cyanide: Cyanide plays a role in regulating dormancy in apple seeds (Lewak, 2011). Hydrogen cyanide, in its unbound form, was detected in apple seeds that had undergone stratification, as well as in isolated dormant embryos that were being cultured. Moreover, HCN is unquestionably produced in apple seeds, just as in other plant materials, as a byproduct of ethylene production using amino cyclohexane carboxylic acid (ACC). When the substance HCN and its salts, like KCN, are administered to apple seeds or separated embryos, they exert a dual effect. Exposing apple embryos to gaseous HCN for a brief period promoted their germination and alleviated secondary dormancy signs, such as uneven development and cotyledon greening (Lewak, 2011).

Nitric oxide: Apple embryos that aren't in a state of dormancy and have been subjected to stratification or treated with NO or HCN exhibit rapid germination (Gniazdowska et al., 2007, 2010a, 2010b). Additionally, chloroplasts form in both cotyledons and are of equal size (Krasuska et al., 2015b). Apple seeds undergo cold stratification, which leads to an elevation in the production of hydrogen cyanide (HCN) (Lewak, 2011), reactive oxygen species (ROS), and reactive nitrogen species (RNS) (Dębska et al., 2013). On the contrary, the use of NO scavengers keeps embryos in a state of dormancy (Gniazdowska et al., 2007; Krasuska et al., 2016). The process of cold stratification seems to reduce seed dormancy by causing fluctuations in H_2O_2 levels and protein carbonylation. However, seed germination appears to depend on the presence of NO (Dębska et al., 2013). Reactive oxygen species (ROS) and reactive nitrogen species (RNS) are highly adaptable molecules that have significant functions in plant physiology, particularly in seed biology (Bailly et al., 2008; Šírová et al., 2011; Yu et al., 2014). The concept of a "nitrosative door" can be used to describe the dual function of RNS in seed germination. This concept is comparable to the "oxidative window" created by Bailly et al. (2008) for ROS. The "nitrosative door" opens when the intracellular level of NO reaches the proper concentration, allowing seed germination to occur. This idea was proposed by Krasuska et al. (2015a). The mechanism of action of RNS is centered around modifying amino acid residues in proteins after they have been translated. This modification can occur through processes that include nitration or S-nitrosation (Tichá et al., 2016; Yu et al., 2014). S-nitrosation of proteins or peptides occurs when a nitroso group is attached to a cysteine residue. Tichá et al. (2016) discovered that this modification occurring after protein synthesis has a notable influence on the structure and/or function of the protein.

Hydrogen peroxide (H_2O_2): Hydrogen peroxide (H_2O_2) has been observed to enhance the growth of seeds in numerous species (Diaz-Vivancos et al., 2013). The primary reasons for the stimulation of seed germination by H_2O_2 are as follows: the generation of O_2 to support the metabolism of mitochondria and respiration through the removal of H_2O_2 (Katzman et al., 2001), the promotion of seeds cracking, the oxidation of substances that inhibit germination (Ogawa et al., 2001), and the triggering of redox-sensitive proteins, leading to alterations at the proteome and transcriptome levels (Diaz-Vivancos et al., 2013). Regarding this matter, the activation of seed storage proteins and the decrease in ABA levels or its hindered movement from cotyledons to the embryo have been suggested as possible mechanisms that contribute to the improvement of seed germination by H_2O_2 (Barba-Espn et al., 2021). The primary hormone controlling seed dormancy in stone fruit seeds, ABA, has been found to significantly diminish with increasing stratification time (Sten et al., 2021; Leida et al., 2012). Applying exogenous H_2O_2 before stratification has been reported to enhance emergence in endocarp-less seeds of various Prunus species. H_2O_2 was found to greatly enhance both the percentage and speed of germination of seeds in the naturally occurring almond species *P. scoparia* and *P. communis*, as well as in sweet cherry (*P. avium*) (Imani et al., 2011; Sten et al., 2021). Key objectives of peach breeding projects include obtaining strong seedlings as well as quick and consistent seed germination. After 8 weeks of stratification, peach seeds without an endocarp were ingested, increasing germination rates and producing seedlings with healthy vegetative growth (Barba-Espin et al. 2022). The consumption of H_2O_2 in peach seedlings also affected ascorbate, glutathione, and phytohormones such as abscisic acid and jasmonic acid.

Other biochemical factors: Phloridzin (phloretin b-D-glucoside) is the most common monomeric phenol found in apple seeds (Lewak, 2011). Chlorogenic acid and hydroxylated cinnamic acids, together with phloridzin, were primarily found in the coat. The coat's high phenolic concentration has been proposed to contribute to the dormancy of the embryo. The existence of three primary polyamines (PAs), specifically putrescine, spermidine, and spermine, in the seeds of apples, was proven. During the cold-induced breaking of embryo dormancy, the levels of all these polyamines (PAs) decreased (stratification). When spermine, a polyamine that is most abundant in seeds, was given to a separate embryo, it significantly hindered germination. However, the other two polyamines had a positive effect by promoting germination. Spermine prevents ethylene synthesis, which helps to regulate dormancy maintenance, and the other PAs, which are not affected by ethylene, help to remove dormancy (Lewak, 2011). The amount of GSH (Glutathione) and S-nitroso glutathione increases in apple seeds after cold stratification, indicating that dormancy has been broken (Ciacka et al., 2019).

10.3 PROPAGATION THROUGH THE ASEXUAL (VEGETATIVE) METHOD

In particular for fruit crops, the vegetative propagation techniques include cutting (stem and leaf) and layering (stooling, trench, and air layering) (Davies et al., 2018). Successful vegetative propagation requires adventitious root and shoot development. All growth stimulants, such as auxins, cytokinins, gibberellins, ethylene, and abscisic acid, as well as other substances like growth inhibitors and retardants, polyamines, and phenolics, have a direct or indirect effect on the initiation of roots and shoots (Hartmann et al., 2015) (Table 10.2). It was found that IAA (Indole-3-acetic Acid) is a naturally occurring chemical that has a notable effect as an auxin.

TABLE 10.2
Effect of Plant Growth Regulators on Adventitious Root and Bud (and Shoot) Formation (Modified from Hartmann et al. 2015)

Plant Growth Regulators	Adventitious Root Formation	Adventitious Bud and Shoot Formation
Auxins	Ameliorate	Inhibit; low auxin high cytokinin ratio promote
Cytokinins	Inhibit; high auxin; low cytokinin ratio promote	Promote
Gibberellins	Inhibit	Inhibit; can enhance shoot elongation after organ formation
Ethylene	It can enhance the process of root formation in certain herbaceous plants when auxin is applied, but it typically does not have a direct effect on root development in woody plants	Not promotive
ABA	While inhibition might hinder rooting in certain species, it can enhance rooting when combined with auxin in other species	Although it was reported to enhance the adventitious bud production of a herbaceous species, it inhibits it
Ancillary compounds (Retardants/inhibitors, polyamines, jasmonate, brassinosteroids, phenolics)	When used together with auxin, it can enhance root development in some species	Non-stimulating; might inhibit shoot growth

Both IBA and NAA, which are synthetic compounds, are more efficient than naturally occurring or synthesized IAA in promoting root growth. IBA and NAA remain the two most commonly utilized auxins in tissue culture, layers, and stem-cutting propagation of plants. It has been consistently shown that auxin is essential for the initiation of adventitious roots on stems, and it has also been proven that the division of the initial cells of the first root relies on either external or internal auxin (Gasper et al., 1997). Although less common than IAA, indole-3-butyric acid also occurs naturally in plants (Ludwig-Muller and Epstein, 1994). IAA is capable of forming bonds of amide for conjugation, whereas IBA is capable of forming ester bonds. When IBA is administered to stem cuttings or micro cuttings of apples (Malus) to promote root growth, a portion of it is transformed into IAA, as observed in studies by Verma et al. (2017) and Patial et al. (2021). IBA can enhance tissue receptiveness to IAA and stimulate root growth by increasing the concentration of internal free IBA or by synergistically altering the effects of IAA or the natural synthesis of IAA within the plant. Cytokinin exerts the most significant influence on the start of buds and shoots from explants in tissue culture plants (Dobranszki and Teixeira da Silva 2010). Synthetic cytokinins, such as kinetin, thidiazuron (TDZ), and benzyl adenine (BA or BAP), are only used in fruit crops. Additionally, natural cytokinins like zeatin and zeatin riboside are also utilized in these crops (Kaushal et al., 2005; Mitic et al., 2012; Li et al., 2014). Typically, a greater proportion of auxin compared to cytokinin promotes the formation of adventitious roots, whereas a smaller proportion of auxin compared to cytokinin encourages the development of adventitious buds. An increased quantity of gibberellin consistently inhibited the formation of adventitious roots. This inhibition directly prevents the initial cell divisions necessary for the development of specialized stem tissues from reaching a meristematic state. Gibberellins control the synthesis of nucleic acids and proteins, and by disrupting these processes, especially transcription, they can inhibit root initiation. The inhibitory effect of gibberellins on root production is contingent upon the specific stage of root development. Ethylene can either promote, inhibit, or have no effect on the growth of adventitious roots. Elevated levels of auxin will also trigger the production of ethylene. The stimulation of root growth by ethylene is more common in undamaged plants than in cuttings (Hartmann et al., 2015). ABA plays a crucial role in promoting roots by counteracting the detrimental effects of gibberellins and cytokinins, which can hinder rooting. Additionally, ABA affects the ability of cuttings to tolerate water stress during propagation. Ancillary chemicals can alter the primary effects of hormones on roots, as well as the production of adventitious buds and shoots. The substances mentioned are growth retardants/inhibitors, polyamines, jasmonate, brassinosteroids, and phenolics. Growth retardants are commonly employed to decrease shoot growth. They have also been utilized to enhance rooting, usually in combination with exogenous auxin. This is because they impede the biosynthesis or activity of GA, which is typically inhibitory to rooting. Additionally, growth retardants reduce shoot growth, leading to reduced competition and increased availability of assimilates for rooting at cutting bases (Hartmann et al., 2015). Polyamines can function as secondary messengers in the process of rooting. The promotion of root growth in hazel and olive micro shoots is only observed when polyamines (putrescine, spermidine, and spermine) are present along with IBA and NAA, respectively (Rey et al. 1994).

Fruit plants are split into three categories in terms of how rooting is influenced by growth regulators. The first class is easy-to-root, the second class is moderately easy-to-root, and the third class is recalcitrant.

- Easy-to-root plants have auxin in addition to all the necessary endogenous compounds (root morphogens). Rapid root production happens when cuttings are created and placed in the right environmental conditions. Auxin is not typically necessary but may further improve rooting.
- Plants that are moderately easy to root have an abundance of naturally occurring substances that promote root development, but there is a limited supply of auxin. Auxin is necessary for enhancing root development.
- Recalcitrant plants exhibit a deficiency in either the presence of a rooting morphogen or the ability of cells to react to morphogens, regardless of whether natural auxin is present in large quantities or not. Externally applied auxin has negligible or no impact on roots.

10.3.1 Auxin Synergists – Rooting Co-factors

Hess utilized the mung bean, which is easily capable of producing roots, as a bioassay for assessing the biochemical impacts of rooting during the 1950s and 1960s. He referred to the non-auxin stimuli that promote root growth as rooting co-substances. This word was used to modify the rhizocaline theory, which suggests that biochemical factors other than auxin have a role in controlling root growth. The rooting co-factors were naturally occurring chemicals that seemed to enhance the rooting process in conjunction with indoleacetic acid through a synergistic effect. In 1986, Jarvis proposed a theoretical framework for combining the study of biochemistry and developmental anatomy in the process of adventitious root production. This was done by analyzing the four distinct stages of root growth, as shown in Figure 10.2. He hypothesized that the initial elevated concentration of auxin required during early rooting processes eventually hinders the formation of the primordium and its future growth. The regulation of endogenous auxin levels is crucial, with the IAA oxidase/peroxidase enzyme complex playing a major role. Specifically, IAA oxidase is responsible for metabolizing or breaking down auxin. He states that the activity of IAA oxidase is regulated by phenolics. Specifically, o-diphenols operate as inhibitors of IAA oxidase. However, when o-diphenols form complexes with borate, the activity of IAA oxidase increases. This leads to a decrease in the levels of IAA, which is beneficial for the latter stages of root development.

10.4 MICROPROPAGATION

The most crucial factor that controls the development and expansion of shoots in vitro is the selection of plant growth regulators that are utilized in the medium. The beginning and activity of axillary meristems, which are predominantly regulated by cytokinins, are necessary for shoot branching;

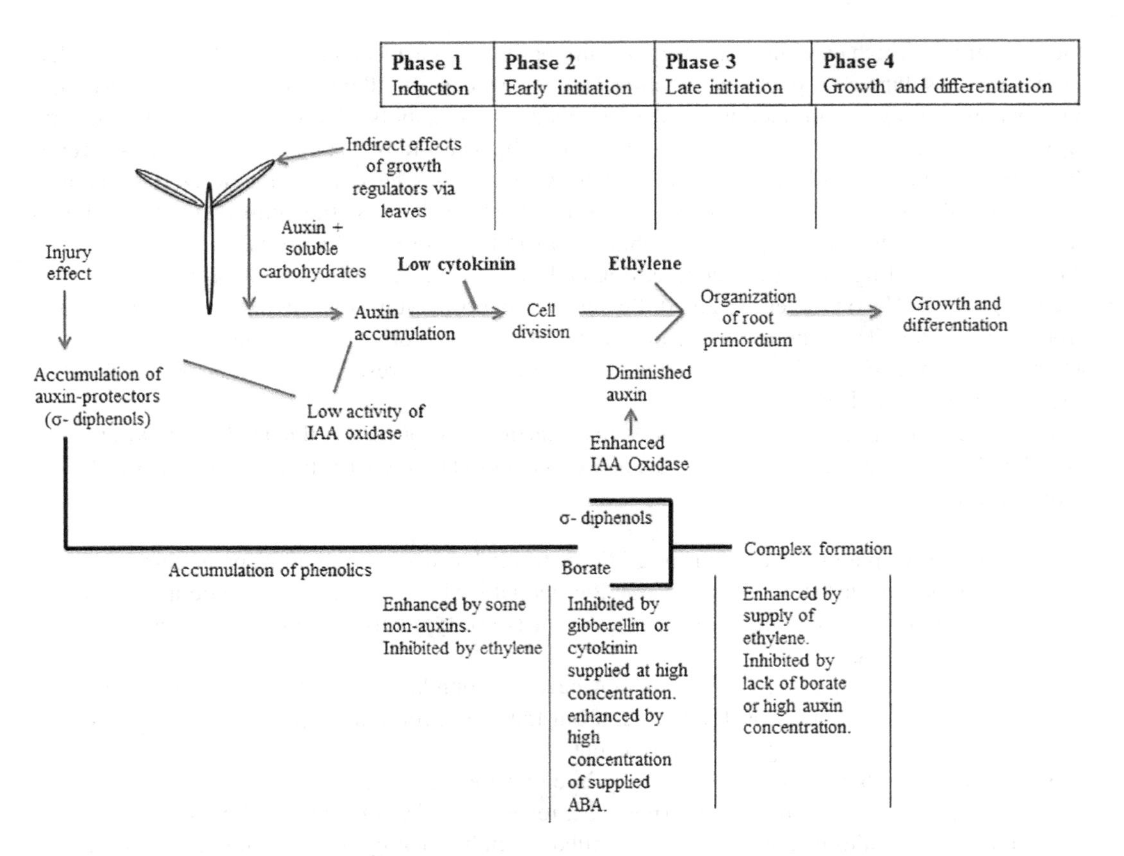

FIGURE 10.2 Hypothesized scheme showing the role of phenolics, IAA oxidase/peroxidase, borate, and phytohormones in the four development stages of adventitious root production. (Modified from Jarvis, 1986.)

yet, despite the indirect auxin action, they interact with auxins (Ward and Leyser, 2004). Auxins and cytokinins are the two most significant hormones utilized to regulate the development of organs and tissues. Usually stable, they are applied before the autoclave. Gibberellins have occasionally been utilized to encourage the elongation of shoots. Autoclaving gibberellins is not advised. The naturally occurring auxin, indole-3-acetic acid (IAA), is less stable compared to synthetic auxins like as NAA, IBA, and 2,4-D. The cytokinins utilized for the micropropagation of fruit crops encompass N6 – benzyl adenine (BA), kinetin, N6 – isopentyl-adenine (2iP), and Zeatin, with concentrations ranging from 0.1 to 10 mg/L (Sharma et al., 2000; Kaushal et al., 2005). BA and 2iP are the most often used substances due to their cost-effectiveness and effectiveness in promoting shoot growth. Thidiazuron and N-2-chloro-4-pyridyl N-phenyl urea (CPPU) exhibit cytokinin activity and are frequently employed in conjunction with conventional cytokinins (such as BA), but at a reduced dose of one-tenth. Thidiazuron (TDZ) is superior to N6-benzyl adenine (BA) in terms of shoot regeneration efficacy (Mitić et al., 2012; Li et al., 2014). The ideal concentration of TDZ varies depending on the genotypes of fruit crops (Magyar-Tábori et al., 2010; Mitić et al., 2012). Recent studies by Li et al. (2014) and Podwyszyńska et al. (2017) have developed procedures for inducing adventitious buds in various species and genotypes of Malus using TDZ. It is important to note that TDZ has been associated with abnormal shoot growth in various plant species (Dewir et al., 2018). The application of gibberellic acid (GA3) effectively promotes shoot elongation, although the specific dose needed may vary depending on the genotype (Dobranszki and Teixeira da Silva, 2010).

REFERENCES

Asadi, Z., Jafarpour, M., Golparvar, A.R., and Mohammadkhani, A. 2013. Effect of GA_3 application on fruit yield, flowering and vegetative characteristics on early yield of strawberry cv. Gaviota. *International Journal of Agriculture and Crop Sciences* 5(15):1716–1718.

Bailly, C., El-Maarouf-Bouteau, H., and Corbineau, F. 2008. From intracellular signaling networks to cell death: the dual role of reactive oxygen species in seed physiology. *Comptes Rendus Biologies* 331:806–814.

Barba-Espín, G., Hernández, J.A., Martínez-Andújar, C., and Díaz-Vivancos, P. 2022. Hydrogen peroxide imbibition following cold stratification promotes seed germination rate and uniformity in peach cv. GF305. *Seeds* 1:28–35.

Chitnis, V.R., Gao, F., Yao, Z., Jordan, M.C., Park, S., and Ayele, B.T. 2014. After-ripening induced transcriptional changes of hormonal genes in wheat seeds: the cases of brassinosteroids, ethylene, cytokinin, and salicylic acid. *Public Library of Science One* 9(1):e87543, 14p. https://doi.org/10.1371/journal.pone.0087543.

Ciacka, K., Krasuskaa, U., Otulak-Koziełb, K., and Gniazdowskaa, A. 2019. Dormancy removal by cold stratification increases glutathione and S-nitroso glutathione content in apple seeds. *Plant Physiology and Biochemistry* 138:112–120.

Davies, F.T., Geneve, R.L., and Wilson, S.B. 2018. *Hartmann & Kester's Plant Propagation Principle and Practices*, 9th edition, Pearson, New York.

Dębska, K., Krasuska, U., Budnicka, K., Bogatek, R., and Gniazdowska, A. 2013. Dormancy removal of apple seeds by cold stratification is associated with fluctuation in H_2O_2, NO production and protein carbonylation level. *Journal of Plant Physiology* 170:480–488.

Dewir, Y.H., Nurman, S., Naidoo, Y., and Teixeira da Silva, J.A. 2018. Thidiazuron-induced abnormalities in plant tissue cultures. *Plant Cell Reports* 37:1451–1470. https://doi.org/10.1007/s00299-018-2326-1.

Diaz-Vivancos, P., Barba-Espin, G., and Hernandez, J.A. 2013. Elucidating hormonal/ROS networks during seed germination: Insights and perspectives. *Plant Cell Reports* 32:1491–1502.

Dobranszki, J., and Teixeira da Silva, J.A. 2010. Micropropagation of apple a review. *Biotechnology Advances* 28:462–488.

El-Yazal, S.A.S., EI-Shew, A.A.El-M., and EI-Yazal, M.A.S. 2021. Impact of seed cold stratification on apricot germination and subsequent seedling growth as well as chemical constituents of seeds during stratification. *Horticulture International Journal* 5(4):151–157.

Gniazdowska, A., Dobrzyńska, U., Babańczyk, T., and Bogatek, R. 2007. Breaking the apple embryo dormancy by nitric oxide involves the stimulation of ethylene production. *Planta* 225:1051–1057.

Gniazdowska, A., Krasuska, U., and Bogatek, R. 2010a. Dormancy removal in apple embryos by nitric oxide or cyanide involves modifications in ethylene biosynthetic pathway. *Planta* 232:1397–1407.

Gniazdowska, A., Krasuska, U., Czajkowska, K., and Bogatek, R. 2010b. Nitric oxide, hydrogen cyanide and ethylene are required in the control of germination and undisturbed development of young apple seedlings. *Plant Growth Regulation* 61:75–84.

Gornik, K., Grzesik, M., Janas, R., Zurawich, E., Chojnowska, E., and Goralska, R. 2018. The effect of apple seed stratification with growth regulators on breaking the dormancy of seeds, the growth of seedlings and chlorophyll fluorescence. *Journal of Horticultural Research* 26(1):37–44.

Hartmann, H.T., Kester, D.E., Davies, F.T., and Geneve, R.L. 2015. *Hartmann and Kester's Plant propagation: Principles and Practices*, 8th edition. Pearson India Education Services Pvt. Ltd, New Delhi.

Imani, A., Rasouli, M., Tavakoli, R., Zarifi, E., Fatahi, R., Barba-Espin, G., and Martínez-Gómez, P. 2011. Optimization of seed germination in Prunus species combining hydrogen peroxide or gibberellic acid pre-treatment with stratification. *Seed Science and Technology* 39:204–207.

Jarvis, B.C. 1986. Endogenous control of adventitious rooting in nonwoody species. In: *New Root Formation in Pants and Cuttings*, Jackson, M.B. (eds.), Martinus Nijhoff Publishers, Dordrecht, pp. 191–222.

Kar, P.L. 1979. Studies on the physiology of dormancy and germination of a crab apple seed (*Malus baccata* Borkh.). Ph. D. Thesis, HPKV, College of Agriculture, Solan.

Katzman, L.S., Taylor, A.G., and Langhans, R.W. 2001. Seed enhancements to improve spinach germination. *HortScience* 36:501–523.

Kaushal, N., Modgil, M., Thakur, M., and Sharma, D.R. 2005. In vitro clonal multiplication of an apple rootstock by culture of shoot apices and axillary buds. *Indian Journal of Experimental Biology* 43:561–565.

Krasuska, U., Ciacka, K., and ryka, P., Bogatek, R., and Gniazdowska, A. 2015a. Nitrosative door in seed dormancy alleviation and germination. In: *Reactive Oxygen and Nitrogen Species Signaling and Communication in Plants*. Gupta, K.J., and Igamberdiev, A.U. (eds.), *Seria: Signaling and Communication in Plants 23*. Springer International Publishing, Switzerland, pp. 215–237.

Krasuska, U., Dębska, K., Otulak, K., Bogatek, R., and Gniazdowska, A. 2015b. Switch from heterotrophy to autotrophy of apple cotyledons depends on NO signal. *Planta* 242:1221–1236.

Krasuska, U., Ciacka, K., Orzechowski, S., Fettke, J., Bogatek, R., and Gniazdowska, A. 2016. Modification of the endogenous NO level influences apple embryos dormancy by alterations of nitrated and biotinylated protein patterns. *Planta* 244:877–891.

Leida, C., Conejero, A., Arbona, V., Gómez-Cadenas, A., Llácer, G., Badenes, M.L., and Ríos, G. 2012. Chilling-dependent release of seed and bud dormancy in peach associates to common changes in gene expression. *PLoS One* 7:e35777.

Lewak, St. 2011. Metabolic control of embryonic dormancy in apple seed: seven decades of research. *Acta Physiologiae Plantarum* 33:1–24.

Li, B.Q., Feng, C.H., Hu, L.Y., Wang, M.R., Chen, L., and Wang, Q.C. 2014. Shoot regeneration and cryopreservation of shoot tips of apple (*Malus*) by encapsulation–dehydration. *In Vitro Cellular and Developmental Biology Plant* 50:357–368.

Lin, C.F., and Boe, A.A. 1972. Effects of some endogenous and exogenous growth regulators on plum seed dormancy. *Journal of the American Society for Horticultural Sciences* 97:41–44.

Ludwig-Muller, J., and Epstein, E. 1994. Indole-3-butyric acid in *Arabidopsis thaliana* III. In vivo biosynthesis. *Journal of Plant Growth Regulation* 14:7–14.

Magyar-Tábori, K., Dobránszki, J., Bulley, S.M., Teixeira da Silva, J.A., and Hudák, I. 2010. The role of cytokinins in shoot organogenesis in apple. *Plant Cell, Tissue Organ Culture* 101:251–267. https://doi.org/10.1007/s11240-010-9696-6.

Martin, G.C., Forde, H., and Mason, M. 1969. Changes in endogenous growth substances in the embryo of *Juglans regia* during stratification. *Journal of the American Society for Horticultural Sciences* 94:13–17.

Mitic, N., Stanišić, M., Milojević, J., Tubić, L., Ćosić, T., Nikolić, R., Ninković, S., and Miletić, R. 2012. Optimization of in vitro regeneration from leaf explants of apple cultivars Golden Delicious and Melrose. *HortScience* 47:1117–1122.

Ogawa, K., and Iwabuchi, M.A. 2001. Mechanism for promoting the germination of Zinnia elegans seeds by hydrogen peroxide. *Plant Cell Physiology* 42:286–291.

Patial, S., Chandel, J.S., Sharma, N.C., and Verma, P. 2021. Influence of auxin on rooting in hardwood cuttings of apple (Malus×domestica borkh.) clonal rootstock 'M 116' under mist chamber conditions. *Indian Journal of Ecology* 48(2):429–433.

Podwyszyńska, M., Sowik, I., Machlańska, A., Kruczyńska, D., and Dyki, B. 2017. In vitro tetraploid induction of *Malus×domestica* Borkh. using leaf or shoot explants. *Scientia Horticulturae* 226:379–388.

Rey, M., Diaz-Sala, C., and Rodriguez, R. 1994. Exogenous polyamines improve the rooting of hazel micro shoots. *Plant Cell Tissue Organ Culture* 36:303–308.

Ross, J.D., and Bradbeer, J.W. 1968. Concentrations of gibberellin in chilled hazel seeds. *Nature* 220:85–86.

Sharma, M., Modgil, M., and Sharma, D.R. 2000. Successful propagation in vitro of apple rootstock MM106 and influence of phloroglucinol. *Indian Journal of Experimental Biology* 38:1236–1240.

Šírová, J., Sedlářová, M., Piterková, J., Luhová, L., and Petřivalský, M. 2011. The role of nitric oxide in the germination of plant seeds and pollen. *Plant Sciences* 181:560–572.

Stein, M., Serban, C., and McCord, P. 2021. Exogenous ethylene precursors and hydrogen peroxide aid in early seed dormancy release in sweet cherry. *Journal of the American Society for Horticultural Sciences* 146:50–55.

Tichá, T., Luhová, L., and Petřivalský, M. 2016. Functions and metabolism of S-nitrosothiols and S-nitrosylation of proteins in plants: the role of GSNOR. In: Lamattina, L., and García-Mata, C. (eds.), *Gasotransmitters in Plants*. Springer International Publishing, Switzerland, pp. 175–200.

Verma, P., Chauhan, P.S., and Chandel, J.S. 2017. Effect of plant growth promoting rhizobacteria (PGPR) and IBA treatments on rooting in cuttings of apple (*Malus×domestica* Borkh.) clonal rootstock Merton 793. *Journal of Applied and Natural Science* 9(2):1135–1138.

Ward, S.P., and Leyser, O. 2004. Shoot branching. *Current Opinion in Plant Biology* 7:73–78.

Yildiz, K., Yazici, C., and Muradoglu, F. 2007. Effect of jasmonic acid on germination dormant and non-dormant apple seeds. *Asian Journal of Chemistry* 19(2):1098–1102

Yu, M., Lamattina, L., Spoel, S.H., and Loake, G.J. 2014. Nitric oxide function in plant biology: a redox cue in deconvolution. *New Phytologist* 202:1142–1156.

Zhang, Y.X., and Lespinasse, Y. 1991. Removal of embryonic dormancy in apple (*Malus × domestic*a Borkh) by 6-benzylaminopurine. *Scientia Horticulturae* 46:215–223.

11 Embryogenesis

Divya Pandey, Sunny Sharma, and Umesh Sharma

11.1 INTRODUCTION

In many plant species, reproduction and a species' dispersal is accomplished by the use of seeds. In addition, seeds constitute a fundamental component of both human and animal nutrition. They possess an embryo inside of them that is protected by a seed coat, but this coat breaks apart when the seed germinates. Due to the close relationship between the seed and embryo, embryogenesis is frequently discussed within the context of germ growth. Embryos, though, are older than seeds in evolutionary terms. The embryonic origin demonstrates that the capacity of a plant to induce embryogenesis is not limited to the zygote, but is also present in other cells. Furthermore, it is crucial to broaden our perspective beyond zygotic development and explore alternative modes of embryo growth to fully understand the beginning of embryos and the molecular triggers involved. Extensive investigations in the field of biochemistry and genetics over the past few decades have provided valuable insights into the initial phases of embryo development. Subsequently, the regulatory framework for this crucial developmental event is examined, following an examination of the molecular data supporting the transcriptional control of embryo initiation. The life cycle of higher organisms is characterized by the presence of two distinct generations: the haploid gametophyte and the diploid sporophyte. The development of diploid sporophytes begins with fertilization and this process involves the formation of a zygote embryo and an endosperm nucleus (Yang and Zhang, 2011).

11.2 DIVERSE APPROACHES TO EMBRYOGENESIS

Embryos are believed to have constituted one of the earliest significant adaptations acquired by plants throughout the course of evolution. Their existence is indispensable for the reproductive processes of plants. The first plants capable of producing embryos, known as embryophytes, trace their origins back to a group of green algae. Over time, these embryophytes have diversified into numerous species, both currently existing and now extinct (Brooker, 2011; Radoeva and Weijers, 2014). Despite the current positioning of the embryo within the seed, paleobotanical evidence suggests that the latter is a relatively recent development (DiMichele et al., 1989; Radoeva and Weijers, 2014). Therefore, zygotic embryogenesis is not a characteristic that has been derived but rather a feature that has always been there. However, at present, seed plants predominate the plant world; this finding shows that seeds offer an advantage in terms of reproduction. Embryogenesis is the process through which a plant determines its overall body layout, regardless of where the embryo originated. There have been described several different methods of embryogenesis in addition to the embryogenesis that originates from the zygote. Somatic cells, microspores, cells involved in female reproduction, and extra-embryonic cells are all capable of inducing embryo development on their own (Radoeva and Weijers, 2014). When detailing the progression and morphogenetic events that occur throughout embryogenesis, the zygotic embryo is utilized as a model along with a few different types of alternative embryogenesis are briefly explored (Figure 11.1).

11.2.1 Zygotic Embryogenesis

The developmental progression of a plant from the stage of fertilization to the subsequent germination of its seeds is known as zygotic embryogenesis (ZE). Plant development involves several

DOI: 10.1201/9781003354055-11

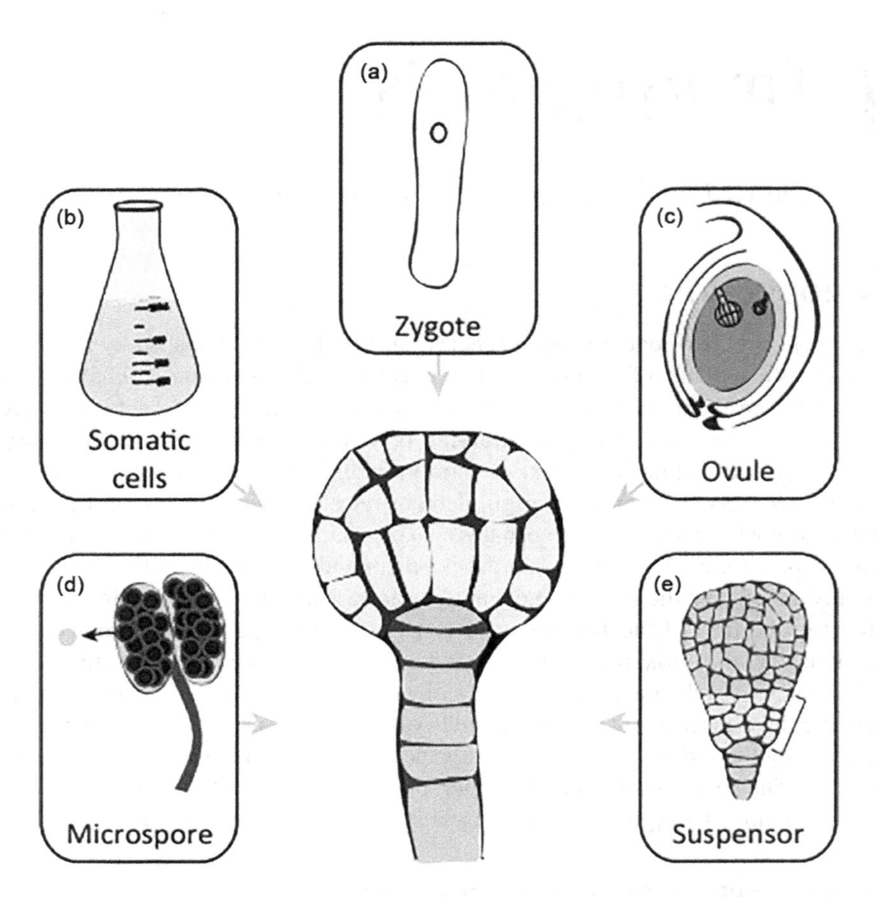

FIGURE 11.1 Different origins of plant embryos. (a) Zygotic embryogenesis, (b) Somatic embryogenesis, (c) Apomictic embryogenesis, (d) Microspore embryogenesis, and (e) Extra-embryonic (suspensor) embryogenesis.

processes, such as morphogenesis and maturation, which include the formation of the plant's fundamental structures, growth, nutrient buildup, and the shift to dormancy (Figure 11.2). Zygotic embryogenesis refers to the process in which an ovum is fertilized by a spermatozoon derived from pollen. In almost all higher plants, the initial stages of embryonic development exhibit a considerable degree of similarity (Radoeva and Weijers, 2014). After the process of fertilization, the zygote in numerous flowering plants, including *Arabidopsis thaliana*, undergoes elongation and asymmetric division. This division results in the formation of two distinct daughter cells: an apical cell, which is smaller in size, and a larger basal cell. These cells serve as markers for the apical-basal axis (De Smet et al., 2010). The phenomenon of polarity and asymmetry in zygote division does, however, differ from this general pattern in a few species (Radoeva and Weijers, 2014). In the *Arabidopsis* embryo, the establishment of the top and lower tiers of the proembryo occurs during the octant stage. This stage is characterized by three successive rounds of cell division of the apical cell, specifically two longitudinal divisions and one transverse division (Zhang and Laux, 2011). The proembryo then enters the dermatogen stage after all of its cells divide tangentially. The hypophysis, which eventually becomes a part of the proembryo, is recognized as the top suspensor cell in the following stage, known as the globular stage (Hamann et al., 1999). A single file with six to nine cells makes up the suspensor at this stage, which marks the end of its development. The characterization of shoot apical meristem (SAM) cells and cotyledons delineates the subsequent developmental phase known as the heart stage, wherein the proembryo undergoes a transition from its initial spherical morphology (Goldberg et al., 1994). The various tissue types that comprise the developing plant body originate from their respective precursors as embryogenesis advances (Mayer and Jurgens, 1998). The fully developed ovule, containing a mature embryo, transforms a seed. This seed remains dormant until favorable environmental conditions stimulate the process of germination, leading to the seedling emergence (Radoeva and Weijers, 2014).

11.2.2 Somatic Embryogenesis

Many somatic plant cells can start embryogenesis and thereafter regenerate a complete plant. The process described here is referred to as somatic embryogenesis. It was first observed and recorded in carrots and has since been observed in many other species over the past few decades (Smertenko and Bozhkov, 2014). This methodology has been used to study the basic regulatory processes involved in studies on plant zygotic embryogenesis. Additionally, it has been extensively utilized in the field of plant breeding and the in vitro cultivation of diverse fruit crops such as banana, papaya, grapes, date palm, and citrus, among others (Uma et al., 2021; Al-Shara et al., 2020; Jiao et al., 2018; Moradi et al., 2017; Cardoso et al., 2016). The induction of somatic embryos has been proposed as a complex process involving the acquisition of the ability to effectively react to inductive signals before induction. However, it is also feasible for somatic cells to be induced directly. In contrast to organogenesis in the in vitro propagation method, the bipolar nature of embryos generated through somatic embryogenesis enables the growth of plantlets without the need for a distinct stage of root development (Behera et al., 2022). Nonzygotic embryogenesis is a highly valuable model system for investigating the intricate processes involved in the physiological, morphological, biochemical, and molecular events that occur during embryonic development in higher plants. This phenomenon is observed in a diverse range of tissue types, making it an ideal subject for academic research. Numerous fundamental and practical elements of agriculture and plant sciences utilize in vitro monozygotic embryogenesis (Bhojwani and Dantu, 2013).

11.2.3 Apomictic Embryogenesis

Even the cells in the seed primordium are capable of starting embryogenesis when they are somatic cells. The process leads to embryogenesis, which is often referred to as asexual reproduction or apomixis, and subsequent seed development without fertilization. Apopmixis frequently skips meiosis in addition to avoiding fertilization, producing embryos that share the same genetic makeup as the parent plant. Apomixis is a highly captivating phenomenon that holds significant agronomic potential due to its ability to stabilize hybrid genotypes (Spillane et al., 2004). This particular form of asexual reproduction may be found in all plant species, and it can be divided into two primary categories: sporophytic (also known as adventitious embryony), and gametophytic (also known as diplospory and apospory) (Barcaccia and Albertini, 2013). These many kinds of apomixis are distinguished from one another by the kind of embryonic cells that are involved and the stage at which induction takes place. However, after induction, the course is quite similar to that of zygotic embryogenesis (Radoeva and Weijers, 2014).

11.2.4 Microspore Embryogenesis

Another reliable source for initiating the development of an embryo is microspore. Microspore embryogenesis, also known as androgenesis, refers to the phenomenon wherein microspores or pollen grains undergo development into haploid or doubly haploid embryos under controlled environmental conditions. Certain microspores deviate from the conventional pathway of gamete production and instead undergo embryogenesis due to an inductive stimulus, such as exposure to heat shock. Both morphological and metabolic changes occur in conjunction with this transition (Seguí-Simarro et al., 2008). The potential reproductive advantage associated with the ability to

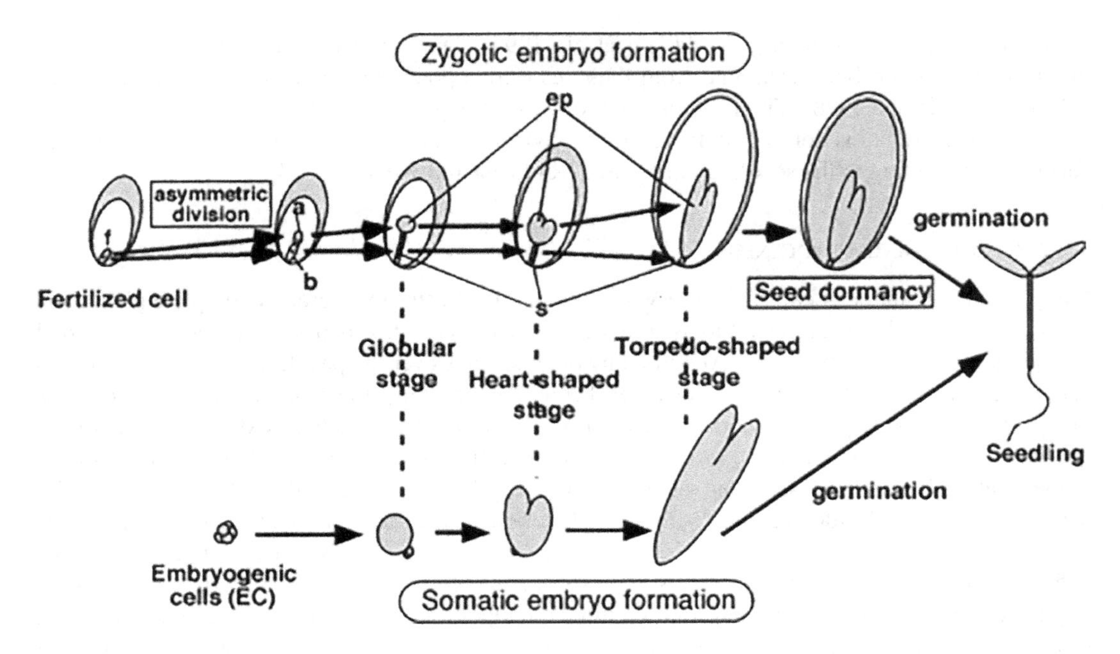

FIGURE 11.2 Different stages of zygotic embryogenesis and somatic embryogenesis.

undergo microspore embryogenesis, as well as the occurrence of analogous processes in natural settings, remains uncertain. Alternately, microspore embryogenesis might represent a holdover of a fundamental developmental potential given that embryogenesis evolved from evolutionary predecessors resembling spores (Taylor et al., 2009).

11.2.5 Extra-Embryonic (Suspensor) Embryogenesis

A different, less well-known embryogenesis method may be an old "reserve" system because it only occurs inside the boundaries of the fertilization product. In most plant species, an embryo is supported by a suspensor, which consists of a sequence of additional embryonic cells. Although the suspensor may be restricted to a few cells in certain species, it might consist of a considerably greater number of cells in others (Yeung and Meinke, 1993). The basal cell undergoes anticlinal division, leading to the formation of a suspensor consisting of six to nine cells that are extra-embryonic and eventually senesce, whereas the apical cell generates the embryo. Suspensor cells can divide periclinally under certain genetic conditions or after embryo ablation, resulting in the development of a second embryo with identical genetic composition. The relationship between suspensor-derived embryos and the ability of certain genes to induce embryogenesis from somatic cells is also unknown (Radoeva et al., 2020).

11.3 DIFFERENCES BETWEEN ZYGOTIC AND NONZYGOTIC EMBRYOGENESIS

Nonzygotic embryos or somatic embryos have secondary embryogenesis and asynchronous development (Bhojwani and Dantu, 2013). The significant difference in physical appearance between nonzygotic embryos grown in a laboratory setting and zygotic embryos found within seeds is the physical limitation placed on the latter (El-Esawi, 2016). The compressed morphology of zygotic embryos can be attributed to their consistently flattened form during the course of their developmental process. Nevertheless, somatic embryos enlarge and develop broader cotyledons and hypocotyls (Gray, 2011). Furthermore, compared to zygotic embryogenesis, somatic embryogenesis has a higher rate of developmental defects. Somatic embryos may exhibit structural abnormalities, which may include additional cotyledons and an underdeveloped apical meristem (Gray, 2011). Nonzygotic embryos, in contrast to zygotic embryos, do not possess a suspensor, which functions as the means for transporting essential nutrients. Additionally, monozygotic embryos frequently do not have a state of rest characterized by inactivity and minimal physiological processes (quiescent resting phase). In contrast, zygotic embryos of diverse crop species undergo a period of dormancy during the maturation of seeds. This resting phase serves as the principal mechanism that facilitates the preservation and utilization of seeds in agricultural practices. (Gray and Purohit, 1991). The developmental sequences of zygotic and nonzygotic embryos exhibit similarities. Monocots undergo globular, scutellar, and coleoptile phases, while dicots and conifers progress through globular, heart, torpedo, and cotyledonary stages. The embryo takes on a spherical shape during the early stages of development and is undifferentiated, yet it has a distinct epidermis. However, the enlargement of the coleoptilar region in monocots and the expansion of the cotyledon in dicots, mark the final stages of development (Gray, 2011).

The embryonic axis also develops more fully at the same time. The root apical meristem establishes itself in dicots. In monocots, the development of the embryo axis occurs in parallel alignment with the scutellum. The shoot apical meristem undergoes exogenous growth and is shielded by the coleoptile, whereas the meristematic apex of the root is deeply entrenched (Gray, 2011).

11.4 THE IMPORTANCE OF HORMONES IN THE PROCESS OF EMBRYOGENESIS

Significant focus has been given to hormones that have a central impact on the shaping of embryonic patterns, the process of maturation, the drying up of tissues, and the halting of development. This heightened interest stems from the substantial advancements made in comprehending their mechanisms of action within the framework of embryo development. This has resulted in a significant focus being directed towards the role that hormones play in embryonic development.

11.4.1 Function of Auxin

11.4.1.1 Embryonic Pattern Formation

Auxin, a plant hormone, is implicated in a diverse range of growth, developmental, and various physiological processes, exerting a substantial influence on each of these phenomena. A significant portion of these processes is reliant on the directed movement of auxin to the various tissues as well as organs within the plant body. The chemiosmotic hypothesis, also known as polar auxin transport, posits that the movement of auxin into and out of cells is facilitated by certain carriers. This transport is driven by the proton motive force, which operates across the plasma membrane. The hypothesis in this study is based on the idea that the proton motive force through the plasma membrane acts as the main driving factor for transport, as suggested by Fischer-Iglesias and Neuhaus in 2001. The hypothesis suggests that the transport's polarity is due to the uneven distribution of the efflux carrier on the basal side of the cells involved in the transport process. During embryonic development, it is important to notice that the shift from radial to bilateral symmetry is strongly associated with a redistribution of [3H], 5-N3IAA. When examining embryos with radial symmetry, we may witness the diffusion of auxin. Additionally, there is a possibility of active transport of auxin in various directions, such as from the inner lower part of the embryo towards the protoderm. In embryos with bilateral symmetry, there is a possibility of a change from a one-way active polarized transport to either a one-way or a two-way active polarized transport from the embryo's lower and inner proximal region toward the distal meristem (Bewley and Black, 1994).

11.4.2 Function of Cytokinin

11.4.2.1 Cell Division

The literature assessment conducted by Van Staden et al. (1982) reveals that a notable observation emerges regarding the relationship between rapid seed growth following fertilization and the occurrence of cell division and expansion. Specifically, there is a substantial elevation in cytokinin levels during this process. When food stores are built up, their levels fall to a lower level again. The augmentation of seed size is attributed to the heightened cell proliferation induced by cytokinin activity, leading to an expanded storage capacity (Fischer-Iglesias and Neuhaus, 2001). A strong association can be observed between endogenous concentrations of cytokinins and phases of cellular division and expansion during the maturation of seeds (Fischer-Iglesias and Neuhaus, 2001). The activity of cytokinin is responsible for increased cell number, which leads to increased seed size, which in turn results in increased storage capacity. It has been proposed that cytokinins operate as a trigger of mitosis since there is a link between the rate of cytokinin activity and the amount of tissue expansion. This suggests that the presence of cytokinin at a specific concentration is necessary for the beginning stages of cell division. During the developmental process of seeds, the endosperm undergoes various stages of maturation. One such stage is referred to as the milky or liquid stage at which the monocotyledons have a high level of cytokinin activity. In the case of dicotyledons, it appears that the situation is comparable. During the time that it is still there, The endosperm functions as a transient storage site for the nutrients that are being transported to the developing embryo (Aremu et al., 2020). This function may be connected to the high quantities of cytokinin that are seen in the tissue of the endosperm. Additionally, it has been postulated that cytokinins originating from the endosperm could indirectly influence the activity of the suspensor, hence The regulation of embryonic growth (Aremu et al., 2020). This is a proposed theory. Therefore, it may be deduced that the cytokinins produced by the endosperm or transferred to the embryo through the endosperm have a vital role in controlling the functioning of the suspensor cells, which are principally responsible for protein synthesis. CKs can be found in significant concentrations during the early stages of seed formation (Aremu et al., 2020). Prior research has demonstrated that there is a significant rise in their levels throughout the embryonic growth of seeds, namely during the enlargement of seed tissues. Afterward, these levels progressively decrease as the seeds mature.

11.4.3 Function of Gibberellic Acid (GA)

Gibberellic acid (GA) is a tetracyclic diterpenoid molecule that serves as an internal regulator in the growth and development of plants. Prior research has shown that the quantities of naturally occurring gibberellic acid (GA) in the suspensor are significantly greater than those seen in the embryo proper. These findings indicate that GA may play a role in the early stages of the embryo's growth (Fischer-Iglesias and Neuhaus, 2001). Gibberellic acid (GA) has a vital role in stimulating the expansion of both the cellular and the embryonic axis. The active gibberellins (GAs) are mostly present during the initial phases of seed development, while the inactive GAs are produced towards the end of seed maturation (Bewley and Black, 1994). The production of GA has been found to occur in embryonic seeds of different species.

11.4.4 Function of Abscisic Acid

The levels of the phytohormone abscisic acid (ABA) exhibit an increase in various monocot and dicot plant species during the intermediate and advanced phases of seed development, which aligns with the initiation of the process of maturation. These ABA concentrations reach their peak during this time (Rock and Quatrano, 1995). After that, the levels of ABA drop significantly, reaching extremely insufficient concentrations in desiccated seeds. (Rock and Quatrano, 1995). In most cases, the concentration of ABA reaches its peak at the same time as the growth of the seed, and then it rapidly decreases following the building of reserves and the start of the drying process. The accumulation of abscisic acid (ABA) during the middle and late stages has been observed to exhibit a dual peak in various species. The occurrence of this peak is observed within the intermediate period.

11.5 GENETIC REGULATION OF EMBRYOGENESIS

Numerous systems that regulate zygotic embryo development are also involved in the adult plant's subsequent tissue differentiation. To fully grasp the genetics of plant development, it is crucial to understand how this process is regulated.

According to Tvorogova and Lutova (2018), these principal steps are involved in the process of embryonic morphogenesis. After ovule fertilization, the embryo begins to develop. The procedure is typically broken down into many steps.

- During the two-cell embryo stage, the zygote undergoes elongation and subsequently splits asymmetrically. This division results in the formation of smaller apical daughter cells and larger basal daughter cells.
- The root apical meristem (RAM) consists of the apical cell, which forms the majority of the embryo, excluding its base. The suspensor, a filament of cells that links the embryo to the endosperm, is created by the horizontal division of the basal cell and its successive offspring. The root apical meristem originates from the suspensor cell, also known as the hypophysis, which is located adjacent to the embryonic tissue called the RAM.
- Octant stage: The apical cell divides in a sequence of well-timed steps. Four cells of the same size are initially produced after two rounds of longitudinal division. Transverse divisions are responsible for the creation of an eight-cell embryo with clearly defined upper and lower cell layers.
- Dermatogen, a stage of the 16-cell embryo: The apical and middle regions of the plant are already visible. They differ from each other in terms of their spatial arrangement, how specific genes are activated, and the resulting destiny of their cells. The four uppermost cells of the octant represent the apical area, from whence the shoot arises. The four cells positioned in the octant closest to the suspensor are suggestive of the core domain. The 16-cell embryo stage, also known as the dermatogen stage, is formed through the tangential division of octant cells, occurring parallel to the surface. This division results in the generation of eight cells on the exterior and eight cells on the interior. The term "protoderm" is used to describe the outermost layer of the embryo. The precursor to the epidermis is the protective tissue of the plant.
  - The proembryonic stage encompasses the period between fertilization and the emergence of the 16-cell embryo or until the transition to the establishment of the cotyledons, depending on the categorization system employed. After the 16-cell stage, the embryo progresses into a developmental phase referred to as the globular stage. In this context, the internal cells of the central domain undergo division along a plane that is aligned with the apical-basal axis of the embryo. This division process generates precursor cells for both the vascular tissues and the surrounding cortical cells.
  - The procambium is created later on during the ensuing development by vascular tissue precursor cells. During the postembryonic stage of plant development, a specific type of tissue undergoes differentiation to give rise to the phloem, which is responsible for transporting water and dissolved minerals from the shoot to the root. Additionally, this tissue also gives rise to the xylem, which transports water and various dissolved minerals in an upward direction. Furthermore, to a lesser extent, the cambium, which serves as the plant's primary lateral meristem, is also derived from this tissue.

- Furthermore, the precursor cells of vascular tissue have a secondary division along a plane that is parallel to the vertical axis of the embryo. The division leads to the creation of the pericycle, a layer of cells that surrounds the conducting structures of the root.
- Shoot apical meristem (SAM): It is imperative to acknowledge that solely the root and hypocotyl are responsible for the development of conducting tissues within the central region of the embryo. SAM serves as the origin of the shoot conducting system, which is mostly arranged during the postembryonic stage. As previously stated, the precursors of the root cortex differentiate and encompass the conducting tissues formed during the globular stage. During this developmental stage, the process of differentiation also takes place within the endoderm, the cortex's innermost layer. This layer, like the tissues that cover the roots, regulates how much water and minerals the roots take in. During this period, the development of the hypophysis, which is the suspensor cell closest to the embryo, takes place. Additionally, the SAM (shoot apical meristem) and RAM (root apical meristem), which are the precursor cells responsible for the formation of plant tissues, are distinguished.
- Later, when the cotyledons have developed, the plant moves on to the heart stage. The subsequent phase, known as "torpedo," is distinguished by the elongation of nascent cotyledons and the ongoing differentiation of the shoot apical meristem (SAM) and root apical meristem (RAM) (Tvorogova and Lutova, 2018).

11.6 FORMATION OF THE APICAL-BASAL POLARITY AXIS

The primary process that occurs during the early stages of embryonic development is the establishment of separate axes of polarity. The zygote experiences an initial division that is not symmetrical, resulting in the formation of smaller apical cells and larger basal cells. This, in turn, results in the initial emergence of the apical-basal polarity (Jeong et al., 2012). During subsequent developmental phases, the apical cell will generate a substantial majority of the plant's organs. The suspensor that connects the embryo to the endosperm is made up of the offspring of the basal cell. The embryo needs the suspensor for nutrition and hormones to develop in the correct orientation within the growing seed. During the process of zygotic embryogenesis, terminal differentiation occurs in most suspensor cells and ultimately dies as a result of autophagy (Smertenko et al., 2014). Subsequently, during the course of development, the hypophyseal cell will undergo differentiation and become an integral part of the apical meristem of the root.

11.7 CAUSE FOR ASYMMETRIC ZYGOTE DIVISION

In almost all flowering plants, the ovules undergo a process known as polarization before fertilization. This process positions the nucleus in the ovule's apical region, which is near the embryo sac's central cell, and places the vacuole in the ovule's basal region, which is more closely located to the micropyle. This polarity, on the other hand, is lost during the process through which the zygote is formed. At least in some species, the fertilized zygote continues to survive for some time as a symmetrical cell that retains the nucleus in the center. Later on, the zygote will undergo a process that will cause it to become polarized for the first time, and this will determine its first asymmetric division (Ueda et al., 2011). Multiple signaling mechanisms are responsible for conditioning the apical-basal polarity. MAPKK kinase YODA route is included in one of these pathways. By the YODA pathway (Figure 11.3), the receptor-like kinase SHORT SUSPENSOR stimulates YODA kinase, which, in turn, activates MPK3 and MPK6 kinases, which in turn triggers the work of TF GROUNDED through a process that is not yet fully understood. This transcription factor is essential for the elongation of the zygote, which is followed by the suspensor's development (Musielak and Bayer, 2014).

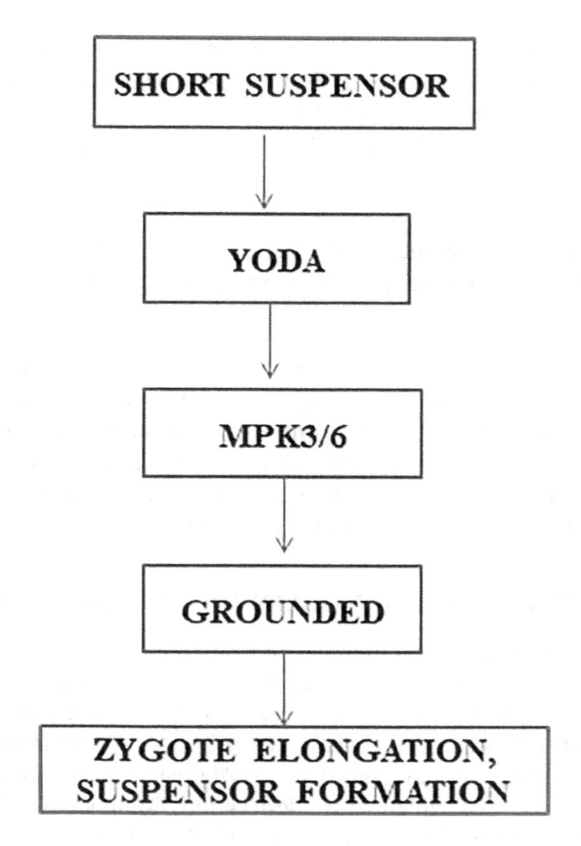

FIGURE 11.3 The YODA pathway.

11.8 RADIAL DIFFERENTIATION OF TISSUES

The division that occurs in the octant cells is periclinal, which implies that it happens parallel to the embryo's surface. As a result, eight outside and eight inner cells are created. According to De Smet et al. (2010), this stage is referred to as the dermatogen and has 16 cells. The outermost layer, known as the protoderm, undergoes subsequent development to form the epidermis in the fully developed plant. The division of all protoderm cells occurs in an anticlinal fashion, leading to an increase in the surface area of this particular layer while maintaining its structural integrity. This mechanism is influenced by the WOX2 gene, and it is conceivable that the WOX1 and WOX3 genes, which are homologs of WOX2, also contribute (Breuninger et al., 2008). This phase is accountable for the establishment of the reciprocal nature of expression between the markers of the outside and inside cells of the embryo.

11.9 SPECIFICATION OF ROOT AND SHOOT APICAL MERISTEMS

Transcription factors (TFs) from the HD-ZIP III and PLETHORA (PLT) families are essential for determining the unique characteristics of the root, which mostly arise from the central domain of the embryo. These TFs are also necessary for specifying the characteristics of the shoot, with the apical domain serving as the origin. During the globular stage, the genes PHAVOLUTA (PHV), PHABULOSA (PHB), REVOLUTA (REV), ARABIDOPSIS THALIANA HOMEOBOX 8 (ATHB8), and ATHB15, which belong to the HD-ZIP III set of genes, are active in the apical domain. The expression of these genes is restricted by the miR165/166 group of microRNA.

Conversely, genes from the PLT group (PLT1–4) are expressed in the central domain, specifically in the globule region closest to the suspensor. This region is located in a zone that is opposite to the zone where HD-ZIP III genes are expressed (Ten Hove et al., 2015). The TOPLESS gene, responsible for encoding the corepressor of the Aux/IAA BDL factor, suppresses the production of PLT genes.

11.10 DEVELOPMENT OF A BILATERAL AXIS OF SYMMETRY AND THE CREATION OF COTYLEDONS OCCUR

The transition of the embryo from the globular phase to the heart phase primarily involves the formation of a bilateral axis of symmetry and the development of the cotyledons. The PIN1 transporter predominantly generates auxin maximum, which is used to designate the locations where the cotyledon primordia are beginning to form. In this area, the auxin supply enters the plant via the protodermal cells and makes its way to the cotyledon-determining areas. The efflux of auxin occurs within the inner cellular layers of the primordium, which is where the zones that will be responsible for the creation of conducting tissues are determined (Benkova et al., 2003).

11.11 PASSAGE INTO THE GERMINATION AND DORMANCY PHASES

In addition to The entire process of its morphogenesis, the maturation of the embryo is a crucial stage during the process of embryogenesis. The process through which an embryo develops into an adult typically involves the embryo's growth, the collection of nutrients, and its transition into the dormant stage. Gibberellins, an antagonist of abscisic acid, are the primary hormones responsible for controlling seed maturity and germination. Abscisic acid works by inhibiting seed germination and increasing the seed's tolerance to desiccation. Compounds belonging to the LEAFY COTYLEDON group, including LEC1, LEC2, and FUS CA3 (FUS3), have received the most research attention. Amongst the transcription factors (TFs) various factors contribute to the maturing process. (Tvorogova and Lutova, 2018).

11.12 MAJOR TRANSCRIPTION GENES REGULATING PLANT EMBRYOGENESIS

The control of several phases of embryo development and maturation, including zygotic and somatic embryogenesis, is regulated by a group of 4 transcription factors (LEC1, LEC2, FUS3, and ABI3) that have been identified through the examination of Arabidopsis mutants with defective embryo formation. The factors encompass LEC1, LEC2, FUS3, and ABI3 (Baumbusch, 2006).

11.12.1 Leafy Cotyledon 1 (LEC1)

Lotan et al. discovered the gene LEC1 in 1998. This gene encodes the HAP3 component of the CCAAT-binding transcription factor. LEC1 expression is restricted to seeds. When LEC1 is produced in transgenic plants in a non-native environment, it triggers the development of structures that resemble somatic embryos. The expression of LEC1 is exclusive to seeds. The LEC1 mutation in plants leads to the formation of embryos with atypical characteristics, such as the presence of trichomes on the cotyledons. Additionally, this mutation causes a decrease in the ability of the embryos to resist drying out and a drop in the amount of storage proteins present (Vicient et al., 2000; Brocard-Gifford et al., 2003).

11.12.2 FUSCA 3 (FUS3)

The transcription of FUS3 genes, which encode a protein containing a B3 domain, is detectable in developing embryos starting at an early stage and continues until just before germination (Luerben et al., 1998). When comparing embryos of the wild type with fusca3 mutant (fus3) on the cotyledons, the embryos exhibit trichome formation, increased anthocyanin accumulation, and reduced seed storage protein accumulation. The results of Baumlein and Keith's (1994) research were published in the works of Kroj et al. (2003). Prior research has shown that the FUS3 protein has a direct affinity for binding to the RY motif, which is a regulatory element found in the promoter region of numerous genes specific to seeds. This relationship is essential in regulating the expression of these genes during embryogenesis. The work conducted by Gazzarrini et al. (2004) demonstrated that activating the FUS3 gene expression in the L1 region of the shoot apical meristem (SAM) using the AtML1 promoter resulted in the development of cotyledon-like structures in transgenic Arabidopsis shoot apical meristem.

11.12.3 Leafy Cotyledon 2 (LEC2)

The FUS3 protein and the B3-domain-containing protein that LEC2 encodes are very closely linked (Stone et al., 2001). The LEC2 cultivar's trichome-producing embryos result in an atypical suspensor shape of the plant's cotyledons. The expression of LEC2 is exclusive to the silique. When LEC genes are expressed in abnormal locations, it stimulates the growth of structures resembling somatic embryos and often gives seedlings traits similar to embryos.. LEC2 is a gene that was originally discovered in Arabidopsis. Upstream of target genes is where the LEC2 protein interacts with the RY motif (Braybrook et al., 2006). The LEC2 gene, which also controls the expression of the ABI3, FUS3, and LEC1 genes, is expressed in the leaves, which controls the accumulation of seed-specific lipids (Santos-Mendoza et al., 2005).

11.12.4 ABA-Insensitive 3 (ABI3)/Viviparous 1 (VP1)

The expression of ABI3/VP1 initiates early during zygotic embryogenesis and persists into the late stage of development. Furthermore, the involvement of LEC1, LEC2, FUS3, and ABI3 itself has been identified in the regulation of ABI3 expression (To et al., 2006).

11.13 CONCLUSION

Embryogenesis is a critical phase in the reproductive process of plants, commencing with the fusion of the zygote and concluding with the maturation of a fully formed embryo. This stage begins when the zygote is fertilized by an egg. The process is carried out according to predetermined patterns of cell division, which leads to the accurate formation of the body plan. An apical-basal axis, which connects the cotyledons to the shoot, hypocotyl, and root meristems, as well as a radial axis, which connects the epidermis, cortical/ground, and vascular tissues, define this layout. The formation of a functional shoot meristem is an important and well-described event that takes place during embryogenesis. During postembryonic growth, this meristem will be in charge of producing leaves, stems, and floral structures. It is a challenging procedure that a variety of factors regulate, such as proteins, transcription factors, and phytohormones. Gene expression is influenced by a variety of chemical signals, and the proper expression is necessary for the embryos to develop normally and quickly. Somatic and zygotic embryos of different plants have been used to investigate each factor impacting development. The expected outcome of the interaction between phytohormones and transcription factors is the manifestation of a specific stage of embryogenesis; novel factors regulating plant embryogenesis have recently been discovered. The mechanism of plant embryogenesis is still not completely understood, though. Recent results employing epigenetic techniques suggest that methylation of DNA and chromatin remodeling may possibly play a substantial role in plant embryogenesis. The relationships between these elements will be clarified and connected in further research, exposing the full regulatory mechanism for embryogenesis.

REFERENCES

Al-Shara, B., Taha, R.M., Mohamad, J., et al. 2020. Somatic embryogenesis and plantlet regeneration in the *Carica papaya* L. cv. Eksotika. *Plants* 9(3):360.

Aremu, A.O., Fawole, O.A., Makunga, N.P., et al.. 2020. Applications of cytokinins in horticultural fruit crops: Trends and future prospects. *Biomolecules* 10(9):1222.

Barcaccia, G. and Albertini, E. 2013. Apomixis in plant reproduction: A novel perspective on an old dilemma. *Plant Reproduction* 26:159–179.

Baumbusch, L.O. 2006 Genetic control of plant embryogenesis and embryo dormancy in Arabidopsis. In *Floriculture, Ornamental and Plant Biotechnology: Advances and Topical Issues,* eds. J. A. Teixeira da Silva, 417–428. London: Global Science Books.

Behera, P.P., Sivasankarreddy, K. and Prasanna, V.S.S.V. 2022. Somatic embryogenesis and plant regeneration in horticultural crops. In *Commercial Scale Tissue Culture for Horticulture and Plantation Crops*, eds. S. Gupta and P. Chaturvedi, 197–217. Singapore: Springer.

Benkova, E., Michniewicz, M., Sauer, M., et al. 2003. Efflux-dependent auxin gradients as a common module for plant organ formation. *Cell* 115:591–602.

Bewley, J.D. and Black, M.1994. Seed development and maturation-hormones in the developing seed. In *Seeds: Physiology of Development and Germination*, eds. J. D. Bewley and M. Black, 100–110. New York: Plenum Press.

Bhojwani, S.S. and Dantu, P.K. 2013. Somatic embryogenesis. In *Plant Tissue Culture: An Introductory Text*, eds. S.S. Bhojwani and P. K. Dantu, 75–92. New Delhi: Springer.

Braybrook, S.A., Stone, S.L., Park S., et al. 2006. Genes directly regulated by LEAFY COTYLEDON2 provide insight into the control of embryo maturation and somatic embryogenesis. *Proceedings of the National Academy of Sciences USA* 103:3468–3473.

Breuninger, H., Rikirsch, E., Hermann, M., et al. 2008. Differential expression of WOX genes mediates apical–basal axis formation in the Arabidopsis embryo. *Developmental Cell* 14:867–876.

Brocard-Gifford, I.M., Lynch, T.J. and Finkelstein, R.R. 2003. Regulatory networks in seeds integrating developmental, abscisic acid, sugar, and light signaling. *Plant Physiology* 131:78–92

Brooker, R.J. 2011. *Biology*. New York: McGraw-Hill.

Cardoso, J.C., Abdelgalel, A.M., Chiancone, B., et al. 2016. Gametic and somatic embryogenesis through in vitro anther culture of different Citrus genotypes. *Plant Biosystems-An International Journal Dealing with all Aspects of Plant Biology* 150(2):304–312.

De Smet, I. Steffen, L., Mayer, U., et al. 2010. Embryogenesis – The humble beginnings of plant life. *Plant Journal* 61:959–970.

DiMichele, W.A., Davis, J.I. and Olmstead, R.G. 1989. Origins of heterospory and the seed habit: The role of heterochrony. *Taxon* 38:1–11.

El-Esawi. 2016. Micropropagation technology and its applications for crop improvement, In *Plant Tissue Culture: Propagation, Conservation and Crop Improvement,* eds. M. Anis and N. Ahmad, 523–545. Singapore: Springer.

Fischer-Iglesias, C. and Neuhaus, G., 2001. Zygotic embryogenesis. In *Current Trends in the Embryology of Angiosperms*, eds. S. S. Bhojwani and W. Y. Soh, 223–247. Dordrecht: Springer.

Gazzarrini, S., Tsuchiya, Y., Lumba, S., et al. 2004. The transcription factor FUSCA3 controls developmental timing in Arabidopsis through the hormones gibberellin and abscisic acid. *Developmental Cell* 7:373–385.

Goldberg, R.B., Paiva, G.D. and Yadegari, R. 1994. Plant embryogenesis: Zygote to seed. *Science* 266:605–614.

Gray, D.J. 2011. Propagation from nonmeristematic tissues-nonzygotic embryogenesis. In *Plant Tissue Culture, Development and Biotechnology*, eds. R. N. Trigiano and D. J. Gray, 293–306. Boca Raton, FL: Taylor & Francis.

Gray, D.J. and Purohit, A. 1991. Somatic embryogenesis and the development of synthetic seed technology. *Critical Review in Plant Science* 10:33–61.

Hamann, T., Mayer, U. and Jurgens, G. 1999. The auxin-insensitive bodenlos mutation affects primary root formation and apical-basal patterning in the Arabidopsis embryo. *Development* 126:1387–1395.

Jeong, S., Volny, M. and Lukowitz, W. 2012. Axis formation in Arabidopsis—transcription factors tell their side of the story. *Current Opinion Plant Biology* 15:4–9.

Jiao, Y., Li, Z., Xu, K., et al. 2018. Study on improving plantlet development and embryo germination rates in in vitro embryo rescue of seedless grapevine. *New Zealand Journal of Crop and Horticultural Science* 46(1):39–53.

Kroj, T., Savino, G., Valon, C., et al. 2003. Regulation of storage protein gene expression in Arabidopsis. *Development* 130:6065–6073.

Lotan, T., Ohto, M., Yee, K.M., et al. 1998. Arabidopsis LEAFY COTYLEDON1 is sufficient to induce embryo development in vegetative cells. *Cell* 93:1195–1205.

Luerben H., Kirik, V., Herrmann, P., et al. 1998. FUSCA3 encodes a protein with a conserved VP1/AB13-like B3 domain which is of functional importance for the regulation of seed maturation in *Arabidopsis thaliana. The Plant Journal* 15:755–764.

Mayer, U. and Jurgens, G. 1998. Pattern formation in plant embryogenesis: A reassessment. *Seminars in Cell & Developmental Biology* 9(2):187–193.

Moradi, Z., Farahani, F., Sheidai, M., et al. 2017. Somaclonal variation in banana (*Musa acuminate* cv. Valery) regenerated plantlets from somatic embryogenesis: histological and cytogenetic approaches. *Caryologia* 70(1):1–6.

Musielak, T. and Bayer, M. 2014. YODA signalling in the early *Arabidopsis* embryo. *Biochemical Society Transactions* 42:408–412.

Radoeva, T., Albrecht, C., Piepers, M., et al. 2020. Suspensor-derived somatic embryogenesis in *Arabidopsis. Development* 147(13):dev188912. PMID 32554529.

Radoeva, T. and Weijers, D. 2014. A roadmap to embryo identity in plants. *Trends in Plant Science* 19(11):709.

Rock, C.D. and Quatrano, R.S. 1995. The role of hormones during seed development. In *Plant Hormones, Physiology, Biochemistry and Molecular Biology*, ed. P. J Davies, 671–698. Dordrecht: Kluwer Academic Publishers.

Santos-Mendoza, M., Dubreucq, B., Miquel, M., et al.. 2005. LEAFY COTYLEDON 2 activation is sufficient to trigger the accumulation of oil and seed specific mRNAs in Arabidopsis leaves. *FEBS Letters* 579:4666–4670.

Segui-Simarro, J.M. and Nuez, F. 2008. How microspores transform into haploid embryos: Changes associated with embryogenesis induction and microspore-derived embryogenesis. *Physiology Plant* 134:1–12.

Smertenko, A. and Bozhkov, P. 2014. Somatic embryogenesis: Life and death processes during apical–basal patterning. *Journal* of *Experimental Botany* 65:1343–1360.

Spillane, C., Curtis, M.D. and Grossniklaus, U. 2004. Apomixis technology development-virgin births in farmers' fields? *Nature Biotechnology* 22:687–691.

Stone, S.L., Kwong, L.W., Yee. K.M., et al. 2001. LEAFY COTYLEDON2 encodes a B3 domain transcription factor that induces embryo development. *Proceedings of the National Academy of Sciences* 87:11806–11811.

Taylor, T.N., Taylor, E.L. and Krings, M. 2009. *Paleobotany: The Biology and Evolution of Fossil Plants.* Amsterdam: Academic Press.

Ten Hove, C.A., Lu, K. and Weijers, D. 2015. Building a plant: Cell fate specification in the early Arabidopsis embryo. *Development* 142:420–430.

To, A., Valon, C., Savino, G., et al. 2006. A network of local and redundant gene regulation governs Arabidopsis seed maturation. *The Plant Cell* 18:1642–1651.

Tvorogova, V.E. and Lutova, L.A., 2018. Genetic regulation of zygotic embryogenesis in angiosperm plants. *Russian Journal of Plant Physiology,* 65(1):1–14.

Ueda, M., Zhang, Z., and Laux, T. 2011. Transcriptional activation of Arabidopsis axis patterning genes WOX8/9 links zygote polarity to embryo development. *Development* Cell 20:264–270.

Uma, S., Kumaravel, M., Backiyarani, S., Saraswathi, M.S., Durai, P. and Karthic, R., 2021. Somatic embryogenesis as a tool for reproduction of genetically stable plants in banana and confirmatory field trials. *Plant Cell, Tissue and Organ Culture* 147(1):181–188.

Van Staden, J., Davey, J.E. and Brown, N.A.C. 1982. Cytokinins in seed development and germination. In *The Physiology and Biochemistry of Seed Development,* Dormancy *and* Germination, eds. A. A. Khan, 137–156. Amsterdam: Elsevier Biomedical Press.

Vicient, C.M., Bies-Etheve, N. and Delseny, M. 2000 Changes in gene expression in the leafy cotyledon1 (lec1) and fusca3 (fus3) mutants of *Arabidopsis thaliana* L. *Journal of Experimental Botany* 51:995–1003.

Yang, X. and Zhang, X. 2011. Developmental and molecular aspects of nonzygotic (somatic) embryogenesis. In *Plant Tissue Culture, Development and Biotechnology*, eds. R. N. Trigiano and D. J. Gray, 307–325. Boca Raton, FL: Taylor and Francis.

Yeung, E.C. and Meinke, D.W.1993. Embryogenesis in angiosperms: Development of the suspensor. *Plant Cell* 5:1371–1381.

Zhang, Z. and Laux, T. 2011. The asymmetric division of the Arabidopsis zygote: from cell polarity to an embryo axis. *Sexual Plant Reproduction* 24:161–169.

12 Seed and Bud Dormancy

Shivali Sharma, Umesh Sharma, and Sunny Sharma

12.1 INTRODUCTION

A seed's dormancy is indicated by the existence of one or more barriers that prevent germination even in the presence of ideal humidity, temperature, and gas exchange conditions (Table 12.1). In the life cycle, dormancy is often considered an adaptation strategy that enables organisms to deal with unfavorable environmental circumstances that occur from time to time (Donohue et al., 2010). Dormancy is only one of several factors that might prevent a seed from germinating. Dormancy is the conviction that a little plant still has life, but that in order to start developing, it needs variables other than those that are external, such as water and temperature. But dormancy is more than simply a survival strategy. Woody plants in temperate regions have conditional responses to cooler regions (Cooke et al., 2012). Consequently, the accumulation of winter chill is necessary for optimal blooming since insufficient winter chill can lead to physiological issues such as delayed bud-burst and irregular flowering (Campoy et al., 2011). The seed stage is crucial for the long-term survival of higher plants as a species. The plant's dispersal mechanism can ensure its survival during the interval between seed maturation and the establishment of the next generation as a seedling after germination. The seed typically desiccated for its preservation, is equipped to endure prolonged durations in adverse environments. The seed remains dormant to enhance germination over a while.

In addition, dormancy inhibits germination before harvesting. Several research has been undertaken to comprehend the impact of different environmental conditions, such as the chemicals employed, on the process of germination. However, there is currently limited knowledge regarding the process by which the embryo emerges from the seed during germination, as well as how dormant seeds prevent the emergence of the embryo. An essential survival mechanism in plants is the capacity of seeds to defer germination until the appropriate timing and location. While the study of seed dormancy may provide challenges and intricacies, it is an essential aspect of plants' capacity to endure and adjust to their environment. Seed dormancy, as described by Upadhyaya et al. (1983), is a hereditary characteristic that can be influenced by environmental factors during the seed's growth. Plants that have undergone domestication for an extended period often exhibit reduced seed dormancy compared to wild or newly domesticated species. There is currently a significant focus on studying the types of plant species present in soil seed banks. For dicotyledonous plants, however, buds are the primary organs that create shoots, making them crucial for growth, reproduction, and the development of architectural structures. Depending on internal and external stimuli, buds can either commit to vegetative or floral growth at the moment of origination, or they can change later from vegetative to floral development. The majority of the mechanisms that regulate bud growth operate to halt growth by inducing and maintaining bud dormancy since growth is the default program for buds. For specific cultivars, determining the agroclimatic demands (AR; chilling and heat requirements) requires determining whether the buds or tree has completed the endo-dormancy phase and attained its CR. Various methodologies have been developed to quantify CR. To determine the date when endo-dormancy is broken, shoots from dormant trees were harvested during winter. These shoots were then exposed to controlled warm temperatures, and the chilling temperatures were measured until the time of shoot harvesting when it is believed that flower buds can start growing again. (Fadón et al., 2020). In order to assess cooling and heating accumulation, many models have been used in conjunction with hourly temperature data. According to (Campoy et al., 2019), the dynamic model is the one that quantifies chilling the most accurately. The Utah and

DOI: 10.1201/9781003354055-12

TABLE 12.1
Different Category of Dormancy

Category	Explanation
Quiescence:	When temperature, moisture, and photoperiods prevent buds from growing.
Correlative inhibition:	Bud growth is hindered by other plant parts, such as lateral bud dormancy, due to shoot terminal dominance.
Rest:	Dormant buds are due to inherent physiological blockages that impede growth even under optimum environmental conditions. Above-freezing chilling ends it.
Pre Dormancy	sometimes termed early rest. Various treatments can encourage organ growth in this early stage.
Mid dormancy	Full dormancy makes it hard to restart development using external factors.
Post dormancy	The stage of slow emergence from dormancy allows for an easy restart of growth.
Seed dormancy	A state where a seed fails to germinate despite favorable environmental conditions.
Bud dormancy	Perennial plants, such as trees and shrubs, undergo seasonal changes annually. Plant survival is challenging due to significant temperature changes, particularly in freezing circumstances. To withstand harsh conditions, growth zones, including apical buds, axillary buds, and underground structures like rhizomes and tubers experience "suspended growth" dormancy.
Primary dormancy	Internal factors might produce issues even when external circumstances are achieved. Hormone (ABA-GA imbalance), genetic factors
Secondary dormancy	They germinate quickly under favorable conditions but only under precise external conditions.
Exogenous dormancy	Conditions outside the embryo cause exogenous dormancy.
Physical dormancy	Impermeable to seed coverings. Low-moisture seeds. The embryo is usually quiescent. It is caused by the outer cell layer becoming water-impermeable. Physical dormancy occurs in legumes.
Mechanicaldormancy-	Seed coatings are too hard for embryo growth during germination. Causes include seed coat structure or fruit. Found in olive.
Chemical dormancy-	due to inhibitors in many fruits and seeds' outer coats. This occurs in fleshy fruits, hulls, and capsules of many dry fruits. Apples, citrus, grapes, desert vegetation. Endosperm and other tissues around the embryo may also contain it.
Endogenous dormancy	results from embryo conditions
Morphological dormancy-	The embryo is still developing at ripening. Need embryo growth following seed-plant separation.
Physiological dormancy	hinders embryo growth and seed germination until chemical changes occur. These compounds often hinder embryo growth, preventing it from breaking through the seed coat or other tissues. When gibberellic acid (GA3) or dry after-ripening or storage increases germination, physiological dormancy is indicated.
Combination (Intermediate) dormancy	Example Epicotyl dormancy For epicotyl, radicle, and hypocotyl, separate after ripening. Warm temperatures cause seed germination and root and hypocotyl development. Dormant epicotyl needs 1–3 months of cooling.
Double dormancy	Double dormancy is the term used to describe a combination of two or more forms of hibernation. It may be exo-endodormancy or morpho-physiological. require a cooling time for the embryo, a warming period for the root, and a cold period for the growth of the shoot.
Thermodormancy	High temperatures caused hibernation.
Photodormancy	extended exposure to too much light for seeds
Skotodormancy	needed light for germination if they are infused with darkness for a long time.

Weinberger models have also been used to compute CR in order to compare the outcomes to those of the earlier research.

There is a common misconception that seed dormancy solely refers to a state of rest when germination conditions are not optimal. Quiescence is a frequently used term to describe this state. Real dormancy, as described by Upadhyaya et al. (1983), refers to a condition where seeds are incapable of germinating, even in settings that are normally conducive to germination. In addition, seeds can enter a state of dormancy, when they do not sprout even when all necessary conditions are present, such as temperature, humidity, oxygen, and light. Seed dormancy can be defined as a condition in which a seed remains viable but requires specific conditions, such as optimal temperature and moisture levels, to be broken and allow germination to occur.

12.2 TYPES OF SEED DORMANCY

According to Hilhorst, it can be classified as:

Base of seed origin: Dormancy can be categorized into two primary classes based on the origin of the seed.

1. Primary dormancy
2. Secondary dormancy

Mode of seed structure: Seed dormancy can be categorized into two main groups based on the parts of the seed where the dormancy effects and control.

1. Embryo-induced-dormancy
2. Coat induced-dormancy

Mode of action: The classification system is founded on the dormancy effect, which has been proven to be dependent on the physioo-morphological traits of the seed.

12.2.1 Classification Based on Barrier Factors

One of the earliest classifications of dormancy is distinguished by three main types of dormancy based on barrier factors within the seed.

12.2.1.1 Exogenous Dormancy

Exogenous dormancy is frequently classified into three subgroups and is precipitated by external factors.

Physical dormancy: Physical dormancy is the consequence of the inability of water to pass through the multiple layers of mucilaginous outer cells. The endocarp of the seeds has hardened, thereby impeding the passage of water. This occurs when seeds are unable to exchange gases or liquids. Legume seeds are a prime illustration of physiologically inactive seeds, as they are unable to assimilate water due to their low moisture content and the seed coat. The seed coat can be cracked or chipped, allowing water to be absorbed. Impermeability is typically the result of an exterior cell layer composed of macrosclereid cells or a mucilaginous cell layer. The third cause of the seed coat's impermeability is a stiffened endocarp. During the final phases of seed development, seed covers that are impenetrable to gas and water are employed.

Mechanical dormancy: Mechanical dormancy occurs when the embryo is unable to develop during germination due to the rigidity of seed coats or other coverings. Previously, it was believed that a variety of animals employed this form of dormancy; however, it was ultimately determined that endogenous compounds were responsible for their hibernation phase. One of these endogenous realities is the physiological dormancy that results from insufficient embryonic development potential.

Chemical dormancy: The presence of growth regulators in the tissues surrounding the embryo is also referred to as chemical dormancy. There are numerous methods for removing them from the seed tissues, such as cleaning, soaking, and deactivating them. Another factor that inhibits seed germination can be eliminated by rainwater or snowmelt.

12.2.1.2 Endogenous Dormancy

Endogenous dormancy is categorized into three types: physiological dormancy, morphological dormancy, and combined (morpho-physiological) dormancy. This type of dormancy is usually initiated by forces originating from within the embryo itself.

Physiological dormancy: This sort of dormancy is the most prevalent in the realm of seed biology. By altering inhibitors, an embryo's growth is slowed, and seed germination is prevented. When a seed fails to satisfy certain physiological criteria for germination, it occurs. Physiological dormancy prevents the germination of seeds and the growth of embryos until particular chemical changes take place. Typically, inhibitors hinder the growth of embryos, preventing them from reaching a stage where they can escape their seed coat or various other organs. Germination rate is enhanced by the use of gibberellic acid (GA3), dry post-ripening, or dry storage, which are all indicative of physiological dormancy. Additionally, promoting germination can be advantageous by subjecting the seeds to either cold stratification or warm stratification for a period of up to 3 months. Alternatively, the time required for cold stratification can be reduced by allowing the seeds to undergo dry after-ripening. Following scarification, certain seeds exhibit enhanced germination, which is also indicative of physiological dormancy.

Morphological dormancy: Seeds exhibit the presence of separate cotyledons and hypocotyls, along with underdeveloped embryos. Embryos that are in morphological dormancy are often under optimal conditions and simply need time to develop and sprout, as they are not biologically inactive. Certain seeds possess fully formed embryos that require further expansion before germination, whilst others consist of embryos that have not yet undergone tissue differentiation before the fruit reaches maturity. Immature embryos: Certain plants can produce seeds even when the tissues of the embryos are not yet fully grown. The seeds may take a considerable amount of time, ranging from weeks to months, to undergo germination once they have reached maturity and absorbed water

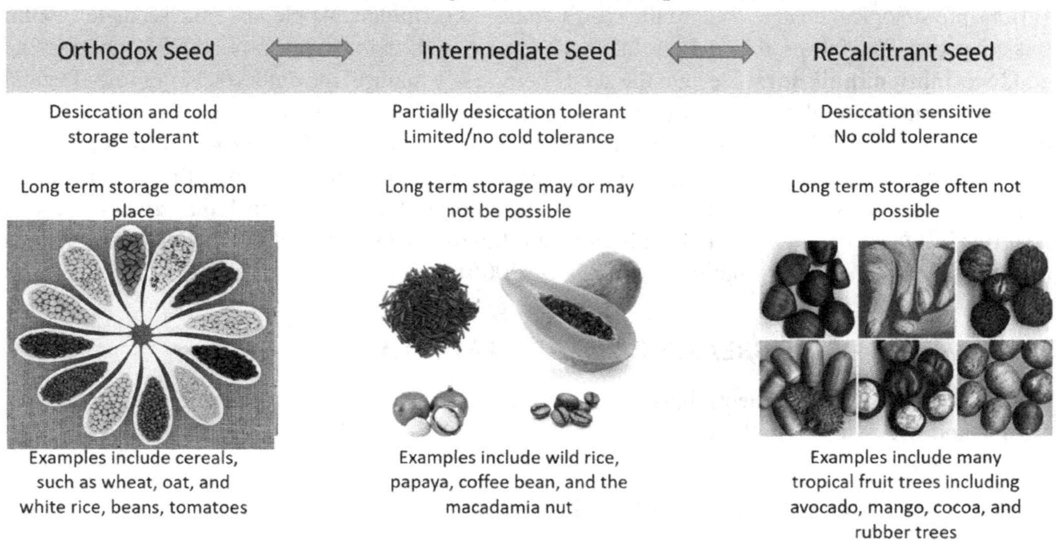

FIGURE 12.1 Different types of seeds.

while being on the ground. That is to say, the phase of development of the embryo is also its inactive phase.

Combined (Morpho-physiological) dormancy: Combination dormancy, which occurs in some seeds, is caused by both internal (physiological) and exogenous (physical) factors. Physiological components and differentiated embryos that are still growing in size are both included in this sort of dormancy. The dormancy of the seeds has to be broken, and the seeds also require time to grow and germinate. Seeds are dormant physiologically and morphologically. Morpho-physiological dormancy occurs when seeds with developing embryos show physiological hibernation. Therefore, dormancy-breaking procedures and a time to generate fully mature embryos are required for these seeds. Different types of seeds based on storage have been described in Figure 12.1

12.3 FACTORS RESPONSIBLE FOR SEED DORMANCY

Impermeability of seed coats to water: When mature, the seed coats of several species are completely water-impermeable. The seeds of several legumes, including the water lotus, morning glory, clovers, and alfalfa, have a high prevalence of this disease. Water must first get through the seed coverings before germination can take place.

Mechanically resistant seed coats: The seed coats of some seeds, such as those of mustard (*Brassica*), pigweed (*Amaranthus*), and shepherd's purse (*Capsella*), are so robust that they do not give in to the pressure of the enlarging embryo. These seeds' embryos don't become dormant and will develop right away if the seed covers are taken off.

Seed-coats impermeable to oxygen: Not equally dormant are the two cocklebur (*Xanthium*) fruit seeds. When a seed reaches maturity in a natural environment, it frequently emerges in the spring while the top seed does not until the following year. It has been discovered that the seed coverings' resistance to oxygen permeability is what makes these seeds dormant.

Rudimentary embryos: In certain plant species such as ginkgo, European ash, holly, and many orchids, the embryo becomes disorganized at seed expulsion and completes its development before germination (Cendán et al., 2013).

Dormant embryos: Despite favorable climatic conditions and fully formed embryos in ripe seeds, many species' seeds do not germinate. Dormancy in such seeds is caused by the physiological circumstances of the embryo. Even in the absence of seed coats, the embryos of these seeds will not undergo initial development. Throughout the latent phase, also known as after-ripening, the embryo experiences various physiological changes before the seed's ability to germinate. Apple and pine seeds fall within this collection. The fall-produced seeds in nature develop in the winter and sprout the following spring.

Germination inhibitors: Despite the seeds being fully mature, the embryos being fully formed, and the environmental conditions being ideal, the germination of seeds from various species is often unsuccessful. These seeds enter a state of dormancy due to the physiological state of the embryo. These seeds do not develop embryos once reaching maturity, even if the seed coats are removed. The endosperm (for instance, in *Iris*), the embryo (for instance, in *Xanthium*), or the seed coat (for instance, in *Cucurbita*) may all include inhibitors. Abscisic acid (ABA) is one of the most prevalent inhibitors of germination (Voisin et al., 2006).

12.4 METHODS OF BREAKING SEED DORMANCY

Seed researchers and engineers have employed diverse techniques to overcome seed dormancy. Simple and widely used methods are:

12.4.1 The Natural Breaking of Seed Dormancy

The state of dormancy terminates once the embryo is exposed to optimal conditions, including appropriate levels of humidity and temperature. Various factors, such as bacterial development, fluctuations in temperature, and the wear and tear from the digestive systems of birds and other seed-eating animals, might trigger the seed coat to break, making the seed permeable.

- The conclusion of the overripening phase is an additional method that is based on organic principles.
- The seed coat is releasing inhibitors.
- The deactivation of inhibitors through the application of thermal, cryogenic, and photonic energy.

The production of growth hormones can counteract the effects of inhibitors, as well as the leaching of excessive and extremely concentrated solutes from the seeds.

Scarification refers to any chemical or physical process that causes a weakening of the seed coat. The scarification approach is used when a hard-seen coat induces dormancy, as seen in legumes like Cajanus cajan (tur), gram, and other similar plants. This approach offers various methods to fracture the tough seed covering, such as:

The seeds of the plant are manually abraded using sandpaper. When manipulating seeds such as green gram, exercise caution to avoid damaging the seed's axis.

If the outer layer of the seed becomes excessively rigid, it is necessary to fully remove it, particularly in the case of woody seeds. Rubber, specifically from the Havea spp. the plant is utilized in the production of Indian teak wood.

Mechanical scarification: Scarification using mechanical methods involves the removal of the seed coat to break its dormant state. To safeguard the embryo and the subsequent seedling, it is necessary to intentionally break or scratch the seed coat at the appropriate location. The optimal location for mechanical scarification is the region of the seed coat directly above the tips of the cotyledons.

Acid scarification: Some spices and seeds can be effectively digested by immersing them in concentrated H_2SO_4 until the seed covering develops pits. It is advisable to examine the seeds at regular intervals, regardless of whether digestion occurs quickly or takes more than an hour. Seeds must undergo a process of thorough washing under running water after digestion in order to facilitate germination.

Treatment by soaking: Hard seed coats, such as cotton seeds and seeds from Indian temperature treatments: teak woods can have their seed coat impermeability removed by soaking them for 1–60 minutes in a strong or diluted H_2SO_4 solution.

Temperature treatments: Temperature treatments involve incubating the seed on the substrate at a low-temperature range of 0°C–5°C for 3–10 days. This process induces dormancy in the embryo factor, allowing it to achieve its ideal temperature. This element is essential for the process of germination to occur (For, e.g. the mustard plant). Before germination, specific seeds necessitate a brief period of incubation, lasting from a few hours to 1–5 days, at a temperature range from 40°C to 50°C. When employing this technique, make sure that the seed, such as paddy, has a moisture content that is below 15%. Employing hot water treatment as a means to alleviate the hardness of legume seeds is a viable and successful approach. The seeds are immersed in water at a temperature of 80°C for a duration of 1–5 minutes, depending on the seed type, before being placed for germination.

Light therapies: Certain seeds exhibit a lack of germination when deprived of light, hence requiring consistent or intermittent exposure to light for successful growth. Lettuce (Lactuca sativa) requires red light (660 nm) or white light for germination.

Growth regulator and other chemical treatments: Gibberellin (GA) and abscisic acid (ABA) both have significant effects on how dormancy is regulated, and the ratio of ABA to GA determines how dormancy is maintained. The manufacture of ABA is necessary to maintain dormancy because the synthesis of ABA at radicals prevents embryo development. An essential element that prevented the development of embryos was the presence of ABA in the endosperm of the Arabidopsis seed. Exogenous GA administration had an impact on the degree of dormancy in the seeds of *T. iliensis* and *T. tarbagataica*. GA is a positive regulator for seed germination. It is well known that cytokinins (KT) play a role in cell differentiation and are crucial for plant development. In particular, KT was able to end seed dormancy on her own. Lee also stated that KT enhanced the germination and dehiscence of ginseng seeds. The application of growth regulators and other chemical treatments can induce endogenous dormancy, as can the addition of germination inhibitors. Low concentrations of growth regulators such as gibberellins, cytokinins, and ethylene can be used to break seed dormancy. Gibberellins and cytokinins are the most commonly utilized growth regulators. To disrupt seed dormancy in sorghum seeds, presoaking seeds with GA3 at a dosage of 100ppm was used. To break the dormancy of the seeds in the plant, barley and tomato, KNO_3 (0.2%) and thiourea (0.5%–3%) are frequently employed.

There are several techniques for releasing seeds from their dormancy and reducing the amount of time they spend in it, allowing seeds to germinate more quickly. Scarification can break any dormancy that is brought on by one of the intrinsic problems with seed coverings. For example, compared to hand-harvested legume seeds, machine-thrashed seeds often exhibit a greater proportion of germination. Strong mineral acids have proven effective in breaking the dormancy of seeds brought on by resistant or impermeable seed coverings. To overcome dormancy, seeds can be immersed in certain substances, such as potassium nitrate, and thiourea, or particular plant hormones. Many seeds mature more quickly after being stored at low temperatures as opposed to higher temperatures. According to Finkelstein et al., 2002, the following are some general methods for releasing seeds from their dormancy:

Dry storage: For species whose dormancy is naturally brief, it is frequently sufficient to keep the sample for a brief period of time in a dry place.

Pre-chilling: Prior to reaching the appropriate temperature, the germination duplicates are placed in contact with a damp substrate and kept at a low temperature for a specific duration.

Pre-heating: The germination should undergo a heating process for a maximum of 7days at a temperature not exceeding 40°C, while ensuring there is adequate air circulation. Afterward, they should be transferred to the appropriate germination conditions. In certain situations, it may be necessary to extend the duration of the pre-heat phase.

Light: During the high-temperature phase and for at least 8hours every 24hours when seeds germinate at different temperatures, illumination should be given. Light output from a cool white bulb should range from 750 to 1,250 lux.

Potassium nitrate (KNO_3): To wet the germination substrate, mix 2g of KNO_3 with 1L of water to get a solution that is 0.2% KNO_3. At the start of the test, the substrate is saturated; however, after that, water is utilized to wet it.

Gibberellic acid (GA_3): A 500ppm solution of GA_3 can be used to wet the germination substrate. This solution is made by combining 500mg of GA_3 with 1L of water. In cases when dormancy is weak, 200ppm can be enough. If the issue is serious, you could use a 1,000ppm solution. A 0.01M buffer in distilled water can be used when the concentration is higher than 800ppm.

Pre-washing: Before performing the germination test, the seeds should be washed in running water at room temperature (25°C) to eliminate any naturally occurring chemicals in the seeds that serve as an inhibitor and affect germination. The seeds should be dried again at room temperature (25°C) after being washed.

Removal of the surrounding structures: Germination can be sped up by removing external elements like the lemma or the involucre of bristles.

Disinfection of the seed: If it is known that the seed has not previously had a fungicide treatment, it may be treated with one before being planted for germination.

Soaking: After soaking in water for 24–48hours, seeds with tough seed coats may germinate more quickly. After soaking, the seed should be sown for germination right away.

12.5 BUD DORMANCY

Bud dormancy refers to the temporary cessation of apparent growth in any plant structure that has a meristem.

12.5.1 Types of Dormancy

Based on seasonal/environmental, dormancy-imposing stimuli dormancy is classified into three types (Figure 12.2):

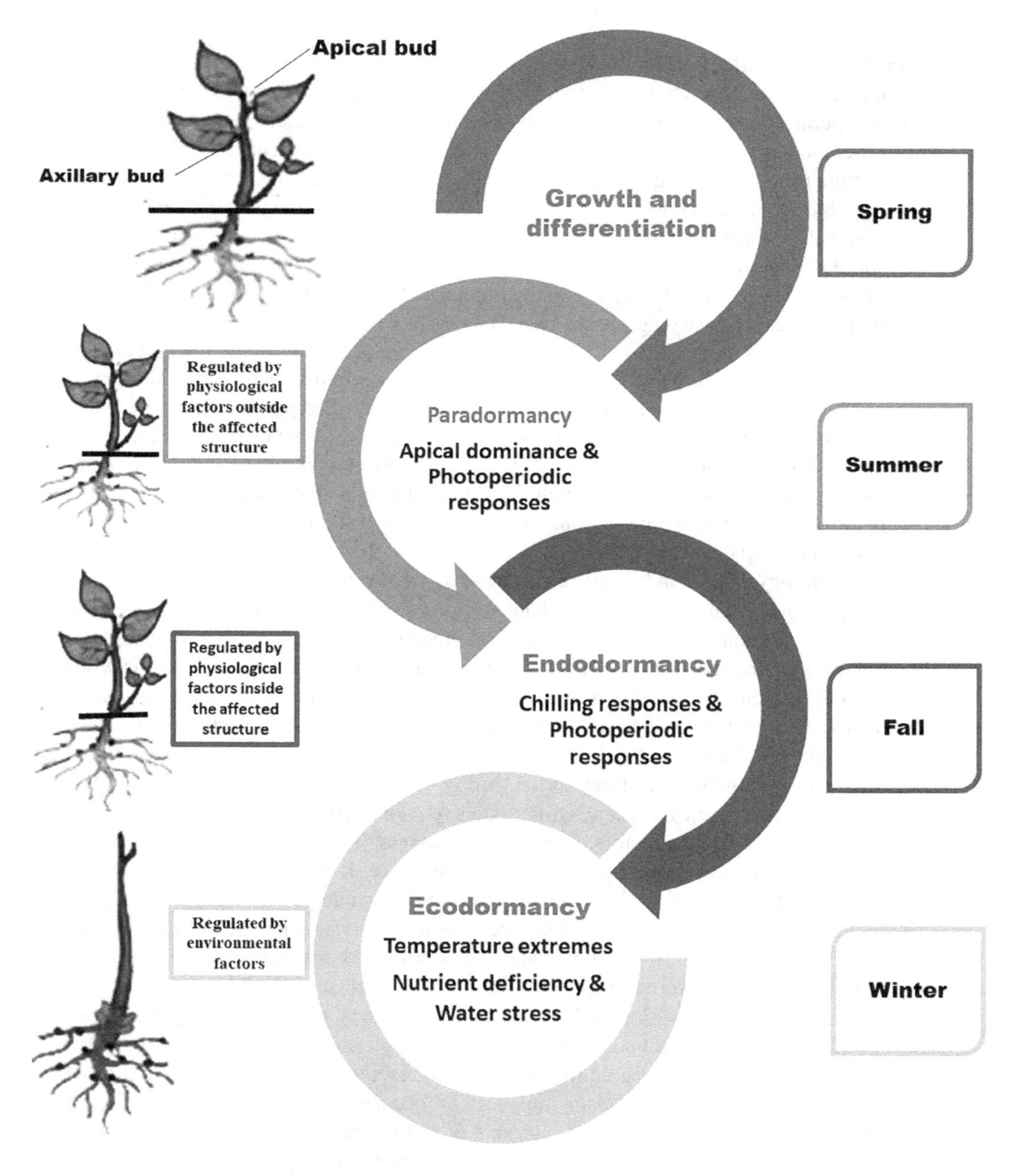

FIGURE 12.2 Signals and usual seasons associated with dormancy types in perennial plants.

i. **Para-dormancy** refers to the halting of growth in a plant, which is controlled by physiological forces within the plant but exterior to the structure that is impacted.
ii. **Endo-dormancy** refers to the stoppage of growth that is regulated by physiological processes within the affected structure.
iii. **Eco-dormancy** Endo-dormancy is the cessation of growth that is controlled by physiological mechanisms within the damaged structure.

12.5.2 Phases of Bud Dormancy

Bud dormancy can be classified into three phases:

i. **Bud dormancy induction**

During the transition from an active phase in the summer to a dormant condition in the fall, temperate trees undergo several changes. Bud set development ceases, leaflets lose their green color, and the tree enters a dormant phase. In places characterized by moderate temperatures and frequent cold autumns, the natural shedding of leaves can serve as a reliable indicator. However, in other regions, trees may continue to produce leaves even after their buds have already entered a state of dormancy. Bud emergence refers to the initial appearance of the scale leaves that will eventually encase the developing stalk. This process takes place during the late summer and early fall. Floral primordia of different species exhibit varying phases of development when they enter a state of hibernation (Diggle and Mulder, 2019). The stamens (consisting of filaments and the anthers with sporocytes in each locule) and the gynoecium (including stigma, style, ovary, and ovules) of Populus species are more advanced in their development compared to those of Prunus species. Species such as the apricot, the sour cherry, and the sweet cherry exhibit distinct verticils, but there is no observable evidence of ovule formation within the ovary (Fadón and Rodrigo, 2018). Flower development in Malus species is interrupted at an early stage during dormancy. During hibernation, the gynoecium and stamens in this group of plants become indistinguishable.

ii. **Endo-dormancy:** The condition is characterized by the inability to survive and thrive even in a favorable environment. The buds require a specific duration of chilling to exit their dormant state and restore their ability to respond to environmental cues that stimulate growth. This inhibitory mechanism hinders the process of bud burst throughout the winter when there are short periods of warm weather. Plant endogenous factors govern endo-dormancy. Temperature-based models are currently employed by scientists and farmers to determine the precise amount of cold exposure required to halt endo-dormancy. Moreover, these models are employed to forecast the timing of blooming for certain species of fruit trees. Despite their extensive usage, chill models remain rudimentary mathematical representations of the biological factors behind tree dormancy. The vegetative and reproductive meristems of the tree, which are particularly sensitive, are shielded by the buds. According to Fadón and Rodrigo (2018), flower primordia, which include petals, sepals, anthers, and pistils, can be easily distinguished from one another. Buds acquire the ability to endure extreme freezing conditions, which would cause significant harm or even death to both the bud and the tree if they were to happen during any other stage of growth.

iii. **Eco-dormancy:** The condition is defined by the incapacity to endure and prosper, even in a propitious setting. The buds necessitate a precise period of cooling to emerge from their inactive condition and regain their capacity to react to external stimuli that promote growth. This inhibitory mechanism impedes the occurrence of bud bursts all through wintertime whenever there are brief intervals of mild weather. Endo-dormancy is regulated by plant endogenous factors. Scientists and farmers already use temperature-based models to accurately predict the amount of cold exposure needed to stop endo-dormancy.

Furthermore, these models are utilized to predict the precise date of flowering for specific types of fruit trees. Although cold models are widely used, they still provide basic mathematical descriptions of the biological variables that cause tree dormancy. The buds protect the tree's vegetative and reproductive meristems, which are especially susceptible to damage. Fadón and Rodrigo (2018) state that floral primordia, which consists of petals, sepals, anthers, and pistils, can be readily differentiated from each other. Buds develop the capacity to withstand severe freezing temperatures, which would result in substantial damage or even mortality for both the bud and the tree if they were to occur at any other developmental phase.

The Growing Degree Hours model (GDH) measures growth-friendly temperatures by using parameters for basic, ideal, and critical temperature thresholds. Although during eco-dormancy trees don't show any overt evidence of activity, this model was later utilized in conjunction with cold models to predict tree bloom dates. To forecast the phenological stages of annual plants, this model was developed. Trees or buds gradually lose their ability to withstand cold before bursting into buds, making them more vulnerable to below-freezing temperatures. Internal changes take place during bud bursts before external ones (Fadón and Rodrigo, 2018; Fadón et al., 2015). Important sexual reproduction processes, including pistil growth, pollen production, and anther meiosis are happening simultaneously inside the buds at this time. Deciduous trees, as opposed to evergreen species, rely on reserves accumulated from the previous growth season for all of these activities. Storage also becomes essential for a good start to the season in histerant species, where the reproductive buds (i.e., flowers) restart growth before the vegetative buds (i.e., leaves). Warm temperatures during this moment, like they did at a previous stage, promote growth. Heat models like increasing degree days (GDD) or rising degree hours (GDH) can thus be used.

12.5.3 Techniques for Overcoming Bud Dormancy

For the timing of blooming and fruiting, and consequently for the survival of a plant species in its geographic area, the biological process of breaking bud dormancy is essential (Ionescu et al., 2017). Bud dormancy, a crucial plant stage in deciduous fruit trees, is largely impacted by the winter's cold temperatures and the accumulation of specific cold hours, which ultimately lead to bud growth as a result of the spring's mild temperatures (Zhu et al., 2021).

i. In order for seeds and buds to emerge from dormancy, phytohormones like IBA, IAA, GAs, and ABA are required.
ii. The balance of hormones and ROS controls bud endo-dormancy; if sufficient cooling requirements are met, buds resume growth (Zhu et al., 2021).

Dormancy breaking requires a multitude of oxidative and reductive processes (Considine and Foyer, 2014).

i. **Chemical substances:** Dormant buds need varying amounts of freezing to emerge from the endo-dormant condition (Endo-Dormancy), depending on the species and cultivar. Uneven bud break, insufficient vegetative growth, incomplete anthesis, and poor flower development are just a few of the physiological issues that inadequate chilling hours can lead to (Khalil-Ur-Rehman, 2019).

   As a result, a fruit tree lacking enough chilling hours in a region with mild winters may benefit from the exogenous administration of particular chemical substances between September and January, such as urea (10%), zinc sulfate (5%), and copper sulfate (10%) (Khalil-Ur-Rehman, 2019). According to (Ionescu et al., 2017), these substances are usually categorized as S- and N-based chemical components S- and N-based substances (such as potassium nitrate, sodium azide, and urea).

ii. **Organic compounds:** Organic chemicals such as mineral oils can be employed to stimulate trees to exit their dormant state (Table 12.2). Organic chemicals have been employed to interrupt bud dormancy due to environmental concerns and the toxicity of certain chemical substances. These chemicals include extracts of cinnamon, coffee, ginger, clove, colocynth, olive, garlic, red chilies, nigella, and turmeric (Ahmed et al., 2014). These organic components contain abundant volatile chemicals, tannins, phenols, antioxidants, vitamins, and minerals. The recent expansion of organic farming has led to a higher need for organic molecules that become active after a period of inactivity, serving as harmless alternatives to chemicals. According to Dos et al. (2020), Garlic extract (*Allium sativum* L.) is an exceptional natural ingredient for reversing dormancy. Using garlic extract alone at a concentration of 5%–10% or in combination with mineral oil at a concentration of 2%–4% was found to be highly efficient in inducing the end of dormancy in grape, apple, and kiwifruit. The study conducted by Kotb et al. in 2019 revealed that the application of garlic extract (at a concentration of 1%) on the leaves of peach trees resulted in the initiation of bud break 5 days earlier than usual. The stimulating characteristics of garlic extract are attributed to its primary sulfur components, namely di-allyl-sulfides (mono-, di-, and tri-) and dimethyl disulfide. The study conducted by Orrantia-Araujo et al. (2019) showed that grapevine dormancy can be effectively broken using fresh garlic extract, garlic oil, or di-allyl-sulfide without causing any harm to the seedlings. Garlic extract contains a significant amount of sulfur molecules, which leads to a reduction in sulfur production and the subsequent creation of cysteine. The process of cysteine metabolism leads to the synthesis of glutathione, which serves to neutralize reactive oxygen species (ROS) and free radicals. Reduced glutathione can enhance the process of increasing the transcription of 1,3-b-d-glucanase, an enzyme believed to play a crucial role in breaking dormancy. Administering a lower amount of exogenous glutathione increased grapevine bud break. According to observations made by Tohbe et al. in 1998, S-methyl cysteine sulfoxide caused 100% of the buds from several table grape varieties to end their dormant state. Upon application of onion extract to apple leaves, there was an increase in the levels of free amino acids, hydrogen peroxide, proline, auxins, and anthocyanin. However, the activity of the enzyme catalase and the total amount of free phenol decreased. Bluprins, a product developed by Biolchim Spa in Bologna, Italy, has the potential to stimulate bud break and promote consistent flowering in grapes, cherries, and kiwifruit. This is achieved by the use of a combination of amino acids, polysaccharides, nitrogen, and calcium. In a comparative study conducted over 4 years, Fenili et al. (2018) found that all treatments exhibited significantly better results than the control plots. However, the combination solution of mineral oil (3.5%) + HC (0.35%) was more effective than Bluprins in inducing blossoming in the apple.

iii. **Phytohormone:** It has been shown that plant hormones play a part in the intricate orchestration of bud dormancy and that these hormones are interdependent. Chemicals produced by plants called phytohormones control most, if not all, of the developmental aspects of plant life. However, a number of advancements in the mechanistic aspects of hormone signaling at the cellular and molecular levels have made hormone sensing, signal transduction, and signal interaction in numerous significant hormones clearer.

iv. **Growth regulators are used to disrupt bud dormancy:**

Recent findings indicate that the levels of naturally occurring ABA rise during the initiation of dormancy and decline at the termination of dormancy (the transition from endodormancy to ecodormancy) (Figure 12.3). According to Zheng et al. (2015), the amount of ABA in grapevine buds increases by up to three times during the early stages of dormancy and then gradually drops until dormancy is released. This was shown by measuring the rate of bud break from node cuttings under controlled settings. Chmielewski

TABLE 12.2
Overcoming of Dormancy in Fruit Crops

Fruit Crop	Dose Concentration	Time of Application	Remarks
Apple cv. Starking Delicious	1.0% Dormex + 2.0% mineral oil	Before completion of chilling requirement (1208 CU)	• Induced early and uniform flowering
Cold-stored strawberry runner	0.1%(HCN) + 3% (KNO_3)		• Breaking the dormancy
Pear cv. Buggugosha	2%–4% Dormex	40–50 days before the expected bud break	• Enhanced bud burst by 2 weeks • Enhanced flowering by 10 days
Kiwifruit cv. Hayward	4% Dormex + 2% mineral oil	45 days before natural bud breaks immediately after pruning	• Advancement of bud break by 9–10 days • Flower bud emergence by 9 days • Full bloom by 9–10 days • Fruit set was advanced by 11–12 days
Pistachio nut	4% Dormex		• Advanced the normal bud break by 15 days • Flowering by 11 days • Improved natural pollination by synchronization of male and female flowers

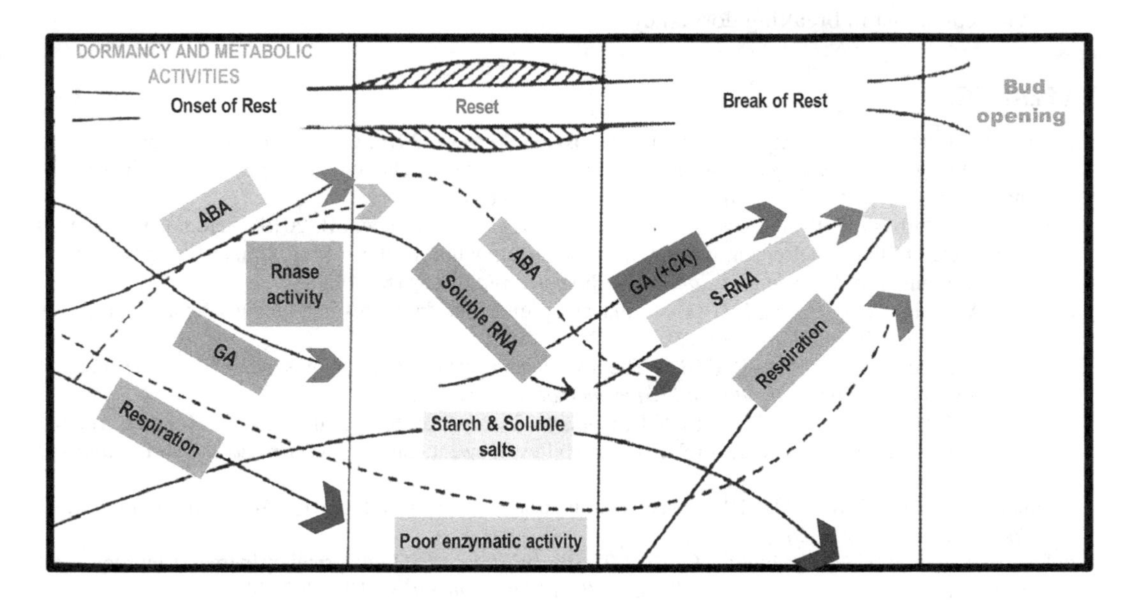

FIGURE 12.3 Metabolic activities during dormancy.

et al. (2018) observed comparable outcomes in other arboreal species, such as the peach, pear, and sweet cherry. Before dormancy is established, the SD photoperiod causes ABA content to increase, and that chilling buildup leads it to decrease. The potential that ABA plays a role in modifying environmental signals as well as starting and progressing the dormancy cycle is raised by the simultaneous connection between ABA levels and dormancy depth. There is now more knowledge available on the effects of exogenous ABA therapy on bud dormancy. In some woody species, exogenous ABA administration was shown to promote the onset of dormancy and delay bud break.

Gibberellins (GAs) play a vital role as a plant hormone in the regulation of bud dormancy. Previous studies have consistently shown large fluctuations in the levels of bioactive GAs both before and during dormancy (Cooke et al., 2012). In general, GA levels decrease during the process of inducing dormancy and increase following the release of dormancy or bud burst. Several arboreal species, such as Japanese apricots, grapevine, and sweet cherry, have demonstrated comparable responses to GA concentrations. During the initial stages of dormancy, a decrease in GA levels leads to bud set and growth halt. However, the process of growth cessation can be reversed by introducing exogenous GA, even when subjected to SD treatment (Molmann et al., 2005). The results indicate that inhibiting the synthesis of GA through drugs may occur after the processes for dormancy onset mediated by SD photoperiodic perception.

IAA, a potent stimulator of growth, has been linked to the onset of dormancy in several species. Exogenous auxin has been demonstrated to facilitate the degradation of dormancy callose and promote the reestablishment of symplastic pathways in the phloem of the magnolia, a crucial process for bud break. According to Nagar and Sood's (2006) study, the levels of IAA in the tea gradually rise from the release of dormancy until the buds start to open in the spring. IAA levels exhibit a sustained decrease over the whole inactive period. According to Zhang et al. (2018), the Chinese fir and Chinese plum have exhibited elevated amounts of IAA upon the termination of dormancy. According to Wolbang and Ross (2001), the transportation of IAA in a polar manner promotes the synthesis of GA, which is crucial for development. Therefore, it is possible that IAA, when combined with GA, can assist in breaking dormancy.

REFERENCES

Ahmed, F.F., Ibrahim, H.I.M., Abada, M.A.M., and Osman, M.M.M. 2014. Using plant extracts and chemical rest breakages for breaking bud dormancy and improving the productivity of superior grapevines growing under hot climates. *World Rural Observations*, *6*(3):8–18.

Campoy, J. A., Darbyshire, R., Dirlewanger, E., Quero-García, J., & Wenden, B. 2019. Yield potential definition of the chilling requirement reveals likely underestimation of the risk of climate change on winter chill accumulation. *International Journal of Biometeorology*, *63*, 183–192.

Campoy, J.A., Ruiz, D., and Egea, J. 2011. Dormancy in temperate fruit trees in a global warming context: A review. *Scientia Horticulturae*, *130*(2):357–72.

Cendán, C., Sampedro, L., and Zas, R. (2013). The maternal environment determines the timing of germination in *Pinus pinaster*. *Environmental and Experimental Botany*, *94*:66–72.

Chmielewski, F.M., Baldermann, S., Götz, K.P., Homann, T., Gödeke, K., Schumacher, F., Huschek, G., and Rawel, H.M. 2018. Abscisic acid-related metabolites in sweet cherry buds (*Prunus avium* L.). *Journal of Horticulture*, *5*, 237.

Considine, M.J., and Foyer, C.H. 2014. Redox regulation of plant development. *Antioxidants and Redox Signaling*, *21*(9):1305–26.

Cooke, J.E., Eriksson, M.E., and Junttila, O. 2012. The dynamic nature of bud dormancy in trees: Environmental control and molecular mechanisms. *Plant, Cell and Environment*, *35*(10):1707–28.

Diggle, P.K., and Mulder, C.P. 2019. Diverse developmental responses to warming temperatures underlie changes in flowering phenologies. *Integrative and Comparative Biology, 59*(3):559–70.

Donohue, K., Rubio de Casas, R., Burghardt, L., Kovach, K., and Willis, C.G. 2010. Germination, postgermination adaptation, and species ecological ranges. *Annual Review of Ecology, Evolution, and Systematics*, *41*:293–19.

Fadón, E., and Rodrigo, J. (2018). Unveiling winter dormancy through empirical experiments. *Environmental and Experimental Botany*, *152*:28–36.

Fadón, E., Herrero, M., and Rodrigo, J. (2015). Flower development in sweet cherry framed in the BBCH scale. *Scientia Horticulturae*, *192*:141–147.

Fadón, E., Herrera, S., Guerrero, B. I., Guerra, M. E., and Rodrigo, J. 2020. Chilling and heat requirements of temperate stone fruit trees (*Prunus sp.*). *Agronomy*, *10*(3):409.

Fenili, C. L., Petri, J. L., Sezerino, A. A., Martin, M. S. D., Gabardo, G. C., & Daniel, E. D. S. 2018. Bluprins® as alternative bud break promoter for 'Maxi Gala' and 'Fuji Suprema' apple trees. *Journal of Experimental Agriculture International*, *26*(2), 1–13.

Finkelstein, R. R., Gampala, S. S., & Rock, C. D. 2002. Abscisic acid signaling in seeds and seedlings. *The Plant Cell*, *14*(suppl_1), S15–S45.

Ionescu, I.A., Møller, B.L., and Sánchez-Pérez, R. 2017. Chemical control of flowering time. *Journal of Experimental Botany*, *68*(3):369–382.

Khalil-Ur-Rehman, M., Wang, W., Dong, Y., Faheem, M., Xu, Y., Gao, Z., and Tao, J. 2019. Comparative transcriptomic and proteomic analysis to deeply investigate the role of hydrogen cyanamide in grape bud dormancy. *International Journal of Molecular Sciences*, *20*(14):3528.

Kotb, H.R.M., El-Abd, M.A.M., and Salama, A. 2019. Response of "White Robin" peach trees cv. (*Prunus persica* L.) to cultivation under plastic covering conditions and foliar application by hydrogen cyanamide and garlic extract. *Journal of Plant Production*, *10*:1187–94.

Molmann, J. A., Asante, D. K., Jensen, J. B., Krane, M. N., Ernstsen, A., Junttila, O., & Olsen, J. E. 2005. Low night temperature and inhibition of gibberellin biosynthesis override phytochrome action and induce bud set and cold acclimation, but not dormancy in PHYA overexpressors and wild-type of hybrid aspen. *Plant, Cell & Environment*, *28*(12):1579–1588.

Nagar, P. K., & Sood, S. 2005. Changes in endogenous auxins during winter dormancy in tea (Camellia sinensis L.) O. Kuntze. *Acta Physiologiae Plantarum*, *28*, 165–169.

Orrantia-Araujo, M.A., Martínez-Téllez, M.A., Corrales-Maldonado, C., Rivera-Domínguez, M., and Vargas-Arispuro, I. 2019. Changes in glutathione and glutathione disulfide content in dormant grapevine buds treated with garlic compound mix to break dormancy. *Scientia Horticulturae*, *246*:407–410.

Tohbe, M., Mochioka, R., Horiuchi, S., Ogata, T., Shiozaki, S., and Kurooka, H. 1998. Roles of ACC and glutathione during breaking of dormancy in grapevine buds by high temperature treatment. *Journal of the Japanese Society for Horticultural Science*, *67*:897–901.

Upadhyaya, M.K., Naylor, J.M., and Simpson, G.M. 1983. The physiological basis of seed dormancy in Avena fatua. II. On the involvement of alternative respiration in the stimulation of germination by sodium azide. *Physiologia Plantarum*, *58*(1): 119–123.

Voisin, A. S., Reidy, B., Parent, B., Rolland, G., Redondo, E., Gerentes, D., ... & Muller, B. 2006. Are ABA, ethylene or their interaction involved in the response of leaf growth to soil water deficit? An analysis using naturally occurring variation or genetic transformation of ABA production in maize. *Plant, Cell & Environment*, *29*(9), 1829–1840.

Wolbang, C. M., & Ross, J. J. 2001. Auxin promotes gibberellin biosynthesis in decapitated tobacco plants. *Planta*, *214*, 153–157.

Zhang, Z., Zhuo, X., Zhao, K., Zheng, T., Han, Y., Yuan, C., and Zhang, Q. 2018. Transcriptome profiles reveal the crucial roles of hormone and sugar in the bud dormancy of *Prunus mume*. *Scientific Reports*, *8*(1): 1–15.

Zheng, C., Halaly, T., Acheampong, A. K., Takebayashi, Y., Jikumaru, Y., Kamiya, Y., and Or, E. 2015. Abscisic acid (ABA) regulates grape bud dormancy, and dormancy release stimuli may act through modification of ABA metabolism. *Journal of Experimental Botany*, *66*(5): 1527–1542.

Zhu, Y., Liu, X., Gao, Y., Li, K., and Guo, W. 2021. Transcriptome-based identification of AP2/ERF family genes and their cold-regulated expression during the dormancy phase transition of Chinese cherry flower buds. *Scientia Horticulturae*, *275*:109666.

13 Physiology of Flowering

Trina Adhikary, Aeshna Sinha, and Pankaj Das

13.1 INTRODUCTION

The phase of flowering in any plant is the most mysterious and complex. It is a significant physiological occurrence that establishes the base for fruit production. A thorough understanding of the flowering phenomena in perennial fruit crops not only aids in developing management strategies for flowering-related issues but also raises the productivity and profitability of commercial fruit cultures. The efficient use of various production factors, which is primarily dominated by solar use efficiency, is necessary to induce flowering (Ravishankar et al., 2014). This method incorporates the process of extracting carbon dioxide and converting it into photosynthate. It encompasses many sink-source connections, including carbon gradients within the plant system that are ultimately established in the chain. It monitors a broad spectrum of metabolic processes controlled by genes that are up- or down-regulated or passively support flowering. The various processes associated with flowering phenology appear complex and frequently defy an effective approach to managing the flowering and fruit set events are used to improve the productivity of outcomes. Long-term studies have focused on the physiological processes influenced by environmental cues that enhance the yield of perennial fruit crops. Studies have shown that tree training, pruning, genetics, rootstocks, translocation of carbon, transformation of that carbon, growth regulators, and other orchard management techniques can all have a substantial impact on tree yields (Sinha, 2004; Dhillon and Bhat, 2012).

For perennial fruit crops to reap the rewards from the potential advantages mentioned earlier, it is essential to comprehend the "whole plant physiology". Although many temperate perennial fruit crops can be used as a preliminary step for problem-solving, tropical, subtropical, cold arid, and desert fruits have yet to be fully explored for management techniques. Focusing too much attention on shoot morphogenesis concerns the detriment of rhizosphere activities, particularly the root signals, which appear to be the main cause of this. They are important in regulating shoot behavior and constructing intelligent shoots-roots communication that eventually leads to reproductive phenophases. The event of flowering followed several steps viz., (i) Induction of flowering (Differentiation of floral primordia), (ii) Organization of floral parts (dedifferentiation to develop individual flower parts) (iii) Maturation of floral parts (growth & development of floral parts including male and female gametes) (iv) Anthesis (Opening of the flower) (Figure 13.1). Flowers are specialized structures made up of reproductive plant elements that operate according to distinct functional principles.

Fruits provide numerous prospects for lucrative fruit growth in the sub-continent in addition to being sources of nutritional and nutraceutical elements and occupying a major position in the family food basket from a health standpoint. Fruit cultivation diversity produces a range of yields. Variations in flowering phenology are frequently blamed for low production in different crops. There are still some knowledge gaps surrounding the phenology of flowering, which must be filled.

13.2 CHANGES OCCURRING IN TARGET TISSUES

13.2.1 Biochemical Changes

(i) Advancement of compounds that impede chloroplast function. (ii) Transitioning away from the floral stimulus originating from leaves. (iii) Augmentation of overall RNA synthesis and RNA synthesis

 DOI: 10.1201/9781003354055-13

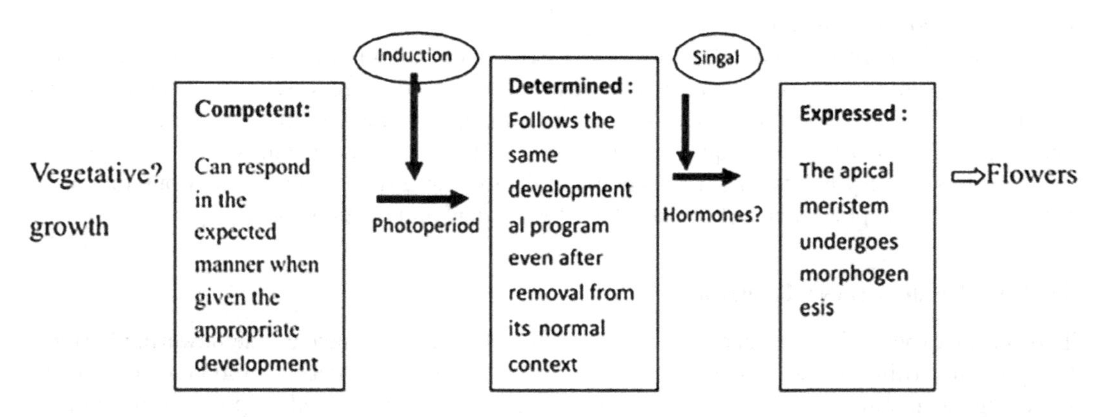

FIGURE 13.1 The event of flowering (Source: Metzger, 1987).

dependent on chromatins. (iv) The process of DNA synthesis is enhanced. There is an increase in mitotic activity. The quantity of mitochondria experiences a rise. (vii) The level of starch falls.

13.2.2 Metabolic Changes

(i) The height of the meristem increases. (ii) The size of the cells increases. (iii) The duration of the mitotic phase (second mitotic peak) becomes longer. (iv) Cells transition from the photosynthetic phase to the mitotic phase. (v) Initiation of flower buds

The initiation of flower bud primordia formation is stimulated by the synthesis of new DNA and an elevation in mitotic activity, leading to the production of cells required for development. Morphogenesis refers to the process in which the flower bud primordia, also known as the morphogenetic phase, develops (Dhillon and Bhat, 2012).

13.3 MECHANISMS OF FLOWERING

The enigma of flowering has been explained by a number of theories, including:

13.3.1 Phytochrome Theory

Phytochrome is a photo-receptor pigment that is non-photosynthetic and exists in two forms, P_r form, and P_{fr} form. P_{660} light is absorbed by P_r, whereas P_{730} light is absorbed by the P_{fr}. Both forms can be converted into the other. According to theory, phytochrome regulates plant flowering. P_{730} encourages flowering in plants with long days while suppressing it in plants with short days. On the other hand, P_{660} inhibits flowering in plants with long days and encourages it in those with short days. Sunlight serves as red light. P_{fr} variant is dominant as the light period draws to a close or in the evening after sundown, whereas P_r variant predominates close to dawn or near the conclusion of the dark period. P_{fr} form changes to P_r form throughout a lengthy night. Short-day plants' ability to blossom is hampered if the lengthy period of darkness is broken by red light because not enough P_r form is produced. The P_r/P_{fr} ratio affects flowering in short-day plants.

Short-day plant flowering is inhibited by a low P_r/P_{fr} ratio, whereas long-day plant flowering is encouraged. A low P_r/P_{fr} ratio inhibits flower production in long-day type plants while a high ratio between P_r and P_{fr} favors it in short-day type plants.

13.3.2 Bunning's Hypothesis

Bunning (1958) mentioned about two crucial phases, i.e., the photophilous phase and the skotophilous phase, which also regulate the flowering. They are the two endogenous rhythms (oscillators) that exist in plants. The skotophilous phase occurs when catabolic processes like the breakdown of starch take place, and the photophilous phase, when an anabolic activity like flowering takes place. Unlike long-day plants, which require a minimum of 15 hours of daylight and some light during the dark phase, short-day plants only need 9 hours of daylight during the light phase. Long-day plants blossom due to the preparation of the oscillator for passage into the photophilous phase.

13.3.3 Chailakhyan's Hypothesis

This theory explains how the chemicals gibberellin and anthesins make up the flowering hormone florigen. According to the theory, two phases are required for flowering in annual seed plants. First, higher amounts of GA in leaves were associated with increased respiration and glucose metabolism. During the second phase, there is a need forrobust nitrogen metabolism, characterized by higher concentrations of anthesins in the leaves and nucleic acid metabolites in the stem buds. Extended periods of daylight promote the initial stage, but shorter periods of daylight are more advantageous for the subsequent stage, which depends on the as-yet-unidentified compound anthesin. According to Chailakhyan (1937), Vernalin is thought to be formed in plants at cold temperatures. It becomes gibberellins in conditions of prolonged daylight. Plants that have lengthy days contain anthesin. Long-day plants begin to flower when anthesin and vernalin are present. However, the vernalin is transformed to gibberellin during short-day conditions, preventing flowering. Gibberellin can induce flower production in long-day, unvernalized plants because these plants already have anthesin. Short-day plants lack anthesin, hence gibberellin is useless at causing flowers to grow (Figure 13.2).

Purvis (1961) postulated that substance A is converted to an unstable B after chilling, which can then be converted to a stable compound D (Vernalin) at suitable temperatures. When exposed to high temperatures, B is converted to C (devernalisation). When exposed to the appropriate amount of light, Vernalin undergoes a conversion process and becomes F (florigen), which is a hormone responsible for triggering the development of flowers.

13.4 PHOTOPERIODISM

The physiological reaction of plants to light duration is known as photoperiodism. It is the most crucial element in the beginning of flowering. Once the biological clock reaches a specific duration, the leaf begins producing a stimulus that triggers the growth of a floral bud in the shoot apex, causing the vegetative shoot meristem to change into a flower bud (Dhillon and Bhat, 2012). Plants can

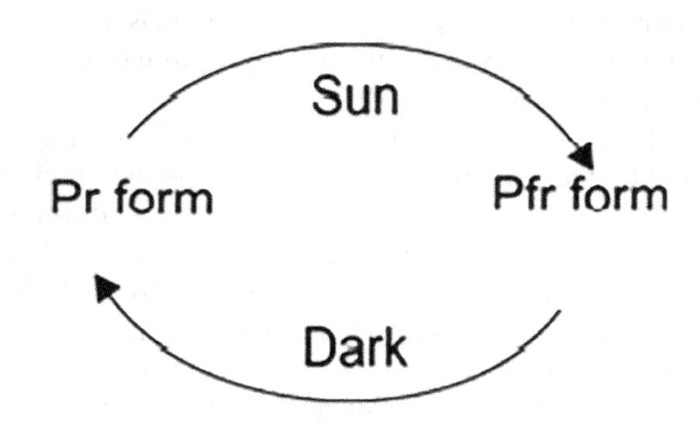

FIGURE 13.2 Schematic diagram of Chailakhyan's hypothesis.

TABLE 13.1
Response of SDP and LDP to Photoperiods

Length of Day/Night	Duration (hours) Day	Night	SDP	LDP
Long day Short night	16	8	No flowering	Flowering
Long day Long night	16	16	Flowering	No flowering
Short day Short night	8	8	No flowering	Flowering
Short day Long night	8	16	Flowering	No flowering
Continuous day	24	-	No flowering	Flowering

TABLE 13.2
Difference between SDPs and LDPs

Short-Day Plants (SDPs)	Long-Day Plants (LDPs)
Able to flourish in environments of short days	Flower under the long fay condition
Flowering is prevented by a flash of light that ends a lengthy period of darkness.	No disruption of the long dark period by a burst of light prevents blossoming.
During the light period, interruptions do not prevent flowering.	A break during the light time prevents flowering
Under continuous darkness, flowering can occur if food is supplied	Such conditions cannot induce flowering

be categorized into three groups based on their response to the duration of daylight, known as the photoperiod.

i. **Short-day plants (SDP):** These kinds of plants only blossom when exposed to day lengths below a particular critical maximum. An essential day lasts 11–15 hours. Tropical plants, such as pineapple, strawberries, and others, tend to have short days (Tables 13.1 and 13.2).
ii. **Day-neutral plants (DNP):** Whatever the photoperiod, these plants undergo a vegetative phase of development before blooming. The duration of the day has no bearing on flowering, which can happen at any time of the year, as in the case of citrus and guava.
iii. **Long-day plants (LDP):** When these plants are exposed to day durations that are longer than a crucial minimum, they start to bloom. These plants continue their vegetative growth after the critical period. Long-day plants produce the majority of fruits in the temperate zone. The ideal day is between 12 and 14 hours long. When the lengthy nighttime phase in plants with short days is broken by brief exposure to light, the plants do not blossom. Similar to how short nights affect short-day plants, short nights affect long-day plants. When the lengthy nighttime phase is broken by a brief period of exposure to light, the short-day plants do not blossom. Similar to how short nights affect short days, long-day plants respond to long nights (Tables 13.1 and 13.2).

13.5 PHOTOPERIODIC INDUCTION

It is defined as the flowering initiation in a plant following a required amount of time spent in the presence of light. The photoinductive cycle is a word used to describe a certain period of exposure to light during a day. It lasts for 24 hours. These cycles, which collectively are known as photoinduction, can be one or more than one depending on the specific species of plants.

13.5.1 Site of Photoinduction Perception

Leaves are the sites of photoinduction; for example, defoliated plants inhibit flowering even after receiving the right amount of sunlight (Chailakhyan, 1937).

13.5.2 Transmission of Stimulus

The stimulus is sent from the leaves to the desired place (in the axils, at the branch apices, or in flowering regions) via the phloem at a velocity of 10–24 mm/hour. However, photosynthates move at a speed of 1,000 mm/hour. Transmission rates can vary from plant to plant, but for LDP, they are 30 cm/hour, whereas 2.4 mm/hour for SDP.

13.6 NATURE OF STIMULUS

Based on the following facts, Chailakhyan (1937) hypothesized that the stimulus is a hormone and termed it florigen:

- Transmission through the phloem.
- Transmission of stimulus from a photoinduced branch to a non-photoinduced branch through grafting.
- Initiation of flowering in a non-photoinduced plant by the leaf extracts of a photoinduced plant, e.g., xanthium. However, some scientists like Sachs and Hackett (1983), disagreed with the florigen theory and thought that photoinduction causes nutrients to be diverted from meristems to floral initiation (Figure 13.3).

13.7 PHYTOCHROME AND FLOWERING

i. **Phytochrome:**
   - Phytochrome, also known as rhodopsin, is a homodimer consisting of two identical protein molecules, each of which is connected to a light-absorbing molecule.

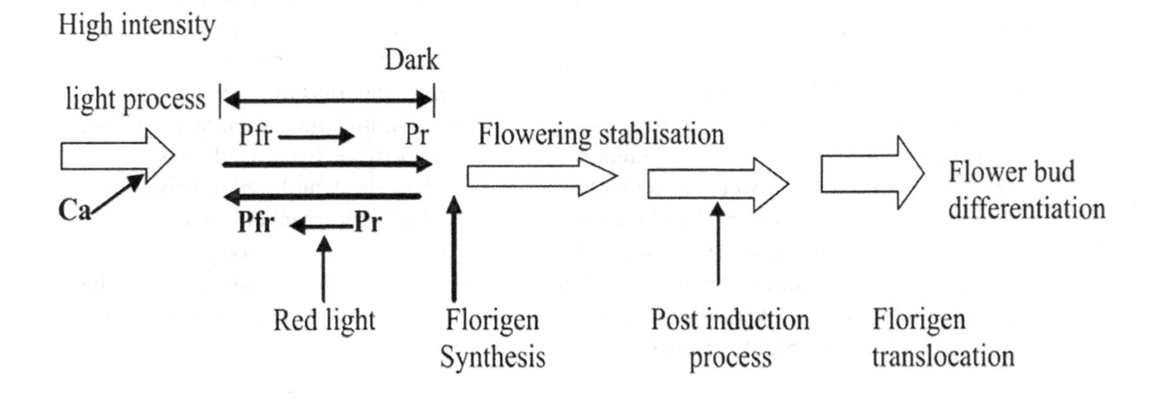

FIGURE 13.3 Physiology of flowering in Xanthium.

- Plants make 5 phytochromes: PhyA, Phy B, C, D, and E.
- Although some of the functions performed by the many phytochromes are redundant, several of them also appear to have special properties.
- Two interconvertible types of phytochromes are present. PR since it absorbs red light (660 nm) Far Red (730 nm) light is absorbed by PFR.
- The connections are as follows: Red light is transformed into PFR when it is absorbed by PR. Far-red light that is absorbed by PFR is transformed into PR. PFR reverts to PR on its own in the dark.

The cocklebur plant experiment by Bonner and Hamner (1938) explained the behavior of phytochrome:

- At dusk, all of the phytochromes are PFR because sunlight has a higher portion of red (660 nm) content than far red (730 nm) light.
- The PFR switches back to PR at night.
- The flowering signal cannot be released without the PR form.
- As a result, the cocklebur requires 8.5 hours of darkness to convert all the PFR existing at dusk into PR and complete the further reactions that result in the release of the blooming signal (florigen).
- A flash of 660 nm light can stop this process in its tracks, reverting the PR to PFR and undoing the previous night's effort (C).
- After exposure to far-red light, the pigment is again (730 nm), and the process of releasing florigen can be finished (D).
- By preventing the spontaneous conversion of PFR to PR, exposure to intense far-red light early in the evening allows for an approximately 2-hour time change (E).

ii. **Working of phytochrome:**

- When PR is changed into PFR by sunlight (660 nm), the PFR transfers from the cytoplasm to the nucleus.
- Then, it forms a bond with the PIF3 protein (phytochrome-interacting factor 3).
- Many transcription factors, including PIF3, are helix-loop-helix proteins.
- The two-part complex binds to and activates promoters with the sequence CACGTG GTGCAC.
- Genes that themselves encode different transcription factors contain these promoters.
- When the plant is exposed to light, several genes that are expressed begin to be transcribed as a result of these other transcription factors.
- Far-red light exposure changes the PFR back to PR, which separates from PIF3 and returns to the cytoplasm. According to research on phytochrome's function in etiolation, The active form is PFR, while the inactive form is PR. Indeed, PR is necessary for long-night/short-day plants like the cocklebur to blossom.

The diagram (Figure 13.4) demonstrates how the final exposure can either start or stop flowering. This suggests that a pigment system functions as a photoreceptor in plants. Additionally, the fact that photoinduction and deduction occur at different wavelengths suggests the involvement of two

660nm 730nm 660 nm 730 nm

SDP → No flowering → Initiation of flowering → No flowering → Initiation of flowering

Long night Long night Long night Long night

FIGURE 13.4 Pigment system functions as photoreceptors on the plants (Source: Bonner and Hamner, 1930).

distinct pigments, one for induction and one for deinduction, or two variants of the same pigment that change conformation in response to different light qualities and affect induction or deinduction in the appropriate ways. Phytochrome was discovered in 1959 by Butler et al. and is located in plant cell plasma membranes. It is a ubiquitous component of green plants and has a variety of biological roles.

iii. **Light and phytochrome**

Plant membranes are likely where phytochrome, a biliprotein pigment, is found. In response to the light it receives, it alters in form and absorption. A form of phytochrome called P_r (660 nm) absorbs primarily red light before changing into P_{fr} (660 nm), which absorbs primarily in the far-red range (730 nm). It's unclear exactly what part phytochrome plays in the development of flowers. It should be noted that while this pigment is not the stimulus for flowering, it may help to stimulate the stimulus for flowering. P_{fr} is highly concentrated after the light phase in SDP, and the ratio of P_{fr} to P_r is such that it hinders the development of a flowering stimulus. If the dark period lasts for a long time, P_{fr} either dies or transforms back into P_r. This stage's P_{fr} to P_r ratio causes the production of a flowering stimulus by starting the necessary activities. When red light momentarily interrupts the dark period, P_{fr} and P_r are formed, changing their relative proportions and delaying the emergence of flowering stimuli. In contrast, a high P_{fr} to P_r ratio is required by LDP to stimulate flowering. After a hard day, we reach this ratio. After a lengthy night, P_{fr} is converted back to P_r, preventing the creation of a blooming stimulus. In situations where red light interrupts the night, P_r is converted to P_{fr}. Here, a blooming stimulus is produced due to the high ratio of the two pigments. Phytochrome was identified by Hendricks and Borthwick (1955) as the photoreceptive pigment that triggers flowering.

The flowering response process consists of four steps: (i) Leaf phytochromes' perception of the stimuli. (ii) The production of the florigen, a hormone that causes flowers to bloom, is changed to a new pattern of metabolism in the leaves. (iii) The transfer of florigen to the primordial bud. (iv) The emergence of floral buds from vegetative buds.

13.8 FLOWERING AND PHOTOPERIODISM IN FRUIT CROPS

In the mango, cool temperatures instead of short photoperiods cause bloom induction, whereas warm temperatures instead of long photoperiods cause flowering inhibition. The usual observation is that the mango tree's eastern side, which gets more sunshine hours and flowers a few days earlier than other sides, illustrates how light affects the mango's reproductive process. In Kanyakumari, a variety like Neelum flowers twice, while in north Indian circumstances, it only does so once. The growth of strawberry blooms is influenced by the length of the daily light cycle. Shorter light hours tend to favor flower development over runner formation, while longer days tend to favor runner formation over flower production. Although short days encourage flower production independent of temperature, day duration is somewhat influenced by temperature. Generally speaking, any drop in temperature shortens the day, allowing flowers to form. In the Korona strawberry, 12 and 13.5 hours photoperiods were similarly effective in producing yield, but a 13.5 hours photoperiod maintained a more vigorous vegetative development during treatment. In their 2005 study of eight strawberry cultivars, Serce and Hancock (2005) found that Frederick 9 and CFRA 368 produced the same amount of blooms both during short and long days. The genotypes of day-neutral, such as RH 30 and LHSO 4, generated runners under short-day conditions. Contrarily, the strawberry cultivar "Fort Laramie" is heat-tolerant in flowers.

In Coorg Honey Dew observed that lengthy days and high temperatures encouraged the formation of female flowers. Similar findings were made by Storey (1986), who discovered that the long day treatment—a photoperiod cycle of sixteen hours of sunlight and eight hours of darkness—favored femaleness in Co. 1 papaya. Cavichioli et al. (2006) investigated four production methods, including artificial light, irrigation, and shade, as well as artificial light and a system that relies on natural lighting (2006). The findings demonstrated that irrigation and artificial light both boosted the yellow passion fruit's blossom and fruit production as well as overall yield. The irrigation decreased the number of flowers in the shaded treatment but did not affect the artificial light treatment's fructification, flowering, or yield. The number of blooms was decreased by shade, both with and without irrigation. The proportion of fructification increased after the artificial light treatment. Plants exposed to light had much earlier flowering. Similar to this, light therapy enhanced the number of blooms and fruit sets. No diurnal temperature difference or short days are necessary for a pineapple to naturally flower. Instead, reaching a certain minimum size and a cool nighttime temperature controls when spontaneous blossoming will take place. After reaching a certain vegetable growth stage 11–12 months after planting and the development of at least 40 leaves, pineapple plants often blossom.,

A pineapple only ever produces one fruit throughout its lifetime. Citrus flowering is not quantitatively affected by photoperiod (Lenz, 1964). Guava is a day-neutral plant, too. Although photoperiod appears to have no effect on floral initiation in fruit crops, including apples, cherries, peaches, and plums, it nevertheless affects the vegetative growth of these species. In a controlled environment using artificial light, According to Tromp (1984), reducing the amount of light for the initial 7 weeks of flowering growth diminished flowering by 50% the next year, but cutting back from weeks seven to sixteen had a much smaller and non-significant impact.

13.9 FLORIGEN CONCEPT

The location of photoperiodic perception is the leaf, while the location of floral bud production is the shoot apex. This indicates that a signal known as a blooming stimulus must travel from the leaf through the shoot apex. In most cases, the flow of molecules that resemble plant hormones is what allows information to be sent to plants. The florigen (flowering hormone) theory was proposed by Chailakhyan in 1937. It states that a hormone created in leaves exposed to the appropriate photoperiodic cycle and transported through phloem activates the flowering genes at the shoot apex. It is yet uncertain what florigen is biochemically. Only grafting experiments may yet be used to infer the existence of florigen. The identification and characterization of florigen have been one of the most crucial objectives in the study of flowering.

Florigen may not be universal because confusion has resulted from the various ways in which the florigen notion is presently interpreted. The idea that florigen is a single element found frequently in plants has been misconstrued. The results of the grafting trials suggested that florigen was shared by many species since they demonstrated that the flowering stimulus could be passed between different plant species. Gibberellins are not regarded as florigens since they do not trigger flowering in short-day plants, although they can activate flowering in some long-day plants under noninductive conditions. However, there is no justification for assuming that florigen must be widely used. Gibberellins may be the florigen of long-day plants, while chemicals other than gibberellins may be the florigen of short-day plants. Because grafting is only possible amongst taxonomically near relatives, the results of the grafting studies indicated above should not be generalized. Brain (1959) proposed that GA-like hormones may have a function in the induction of flowering. He claims that under the influence of red or infrared light, CO_2 is transformed into an inactive or less active processor, producing a GA-like hormone that may create florigen. When exposed to far-red light, the hormone that resembles GA can be transformed into its precursor. Dark treatment can do the same thing, but it takes a very long time. This theory by Brain is based on the interaction between long-day plants treated with GA and short-day circumstances, including the effects of red and far-red light or even total darkness. Numerous fruits include growth-regulating compounds that have caused flowering. According to widespread consensus, high levels of auxins or auxin-like compounds may stimulate flowering by either diminishing the efficiency of gibberellin (GA) or by reducing the permeability of cell membranes, notably the plasmalemma (Baxter, 1970).

Although it has been hypothesized that florigen has existed for about 70 years, it has not yet been isolated. According to some plant biologists, florigen has not been isolated because it does not exist, and they do not rely on its existence to explain the controlling mechanism of the flowering process. Alternative hypotheses include the presence of several components, a flowering inhibitor, and electrical impulses.

13.9.1 Multiple Factors

According to the florigen theory, a single hormonal agent acts only to promote flowering. However, several substances that aren't just for controlling blooming can control hormones. Endogenous concentrations of cytokinins, putrescine, sucrose, and calcium ions rise in some plant species and travel to the shoot apex when conditions are favorable for flowering. These data imply that flowering takes place in the shoot apex when these chemicals interact. Therefore, each of these substances may be essential for the induction of flowering. The multifactorial idea is what is meant by this. The florigen idea is not predicated on a single universal component. It need not presume any particular chemicals. But it might be a combination of two or more substances, perhaps even well-known substances with various uses, as typical plant hormones. In this regard, the multifactorial notion is comparable to the florigen concept.

13.10 VERNALIZATION AND FLOWERING

Other environmental factors than the length of the night affect flowering. Another crucial factor is temperature. Some plants need to flower in a low-temperature environment (between 0°C and 10°C) for a few days to a few weeks. The term vernalization describes the induction of flowering in plants at a low temperature. Although it is seen in many long-day plants, photoperiodism is not related to it. The most efficient temperature is found to be 4°C, and the time of chilling might vary from species to species, ranging from four days to a few months.

Young leaves and a vigorous apical meristem serve as the vernalization site. While stimulus reception occurs in the apical meristem of the shoot in biennials and perennials, it occurs in the embryo axis of annuals. Biennial plants grow vegetatively during the first year before flowering the following summer after vernalization over the winter. But perennial plants require a vernalization cycle every winter. Meristems serve as both stimulation and transmission sites. Vernalin is thought to be transferred via grafting from vernalized plants to non-vernalized plants.

13.10.1 Mechanism of Vernalization

Two hypotheses have been put out to explain how vernalization works.

i. **Phasic development hypothesis**

   Annual seed plants grow via several physiological phases that happen in a predetermined order so that one phase ends and the next starts. These have been determined to be the key phases:

   a. **Thermostage:** The length of the vegetative phase, which varies on the plant and its surroundings, is determined by low temperature (0.14°C), moisture, and aeration.
   b. **Photostage:** It is also known as photoperiodism and demands a high temperature. Vernalin aids in the production of florigen in this instance.
   c. **Third Stage:** The development of gametes and sex organs depends on it.

ii. **Hormonal Hypothesis:** It shares Chalikhyan's theory.

   Significance of vernalization:

   a. Shortening the vegetative phase to produce many crops in a single year.

b. Protection from freezing injury by sowing winter crops in spring.
c. Tropical plants can be grown in temperate climates even if they need long summers (short summers).

13.11 TRANSLOCATION OF FLOWERING STIMULUS

The flowering stimulus travels from the leaf to the apical apices via phloem cells. The results of causing photoperiodic stimuli might be localized or widespread depending on the category of plant. Transmission of stimuli from one plant to another also happens, for example, through grafting.

13.12 FACTORS FOR INDUCTION OF FLOWERING IN FRUIT CROPS

The primary factors influencing the induction of flowering can be broadly divided into five groups:

i. **Environmental factors**
   Photoperiod, temperature, water relations, location, etc.
ii. **Internal factors**
   Nutritional elements, carbohydrate status, phytohormones, and nitrogen metabolism. Environmental cues are incorporated by plants for both development and, more specifically, flowering. The floral transition is the most significant phenomenon in the flowering process that is controlled by external conditions.
iii. **Soil factors:** Nutrient status, soil structure, texture, moisture content, etc.
iv. **Horticultural traits:** Age, crop load, rootstocks, type of shoots, etc.
v. **Cultural practices and use of chemicals:** Smudging, girdling, root, and shoot pruning, use of TIBA, SADH, PBZ, CCC, etc.

13.13 CONCLUSION

- The vegetative apex is converted into a reproductive structure as the flower bud develops.
- Induction, evocation, and initiation are the three processes that take place when buds begin to flower.
- The flat apical meristem first shows signs of differentiation by domeding, followed by partitioning of the core meristem and the emergence of the pith meristem.
- In fruit crops, photoperiodism and hormonal balance are more important determinants for flowering than the C:N ratio. Long-day plants' flowering is regulated by low temperatures, while the flowering of mango and citrus is enhanced by water stress.
- Many researchers have proposed several theories to explain the mechanism of flowering, however, none of them are still tenable.
- Numerous cultural activities, such as branch bending, defoliation, chemical treatments, nutrition control, girdling, and pruning, can also affect the flowering of fruit crops. The impact of growth regulators on flowering is dramatic. GA inhibits it, whereas auxins encourage it. NAA is used to stimulate flowering in mango and pineapple.
- Ethephon plays a significant role in controlling pineapple blossom. Different growth inhibitors, such as Alar, Cycocel, TIBA (antiauxin), SADH, and PBZ, promote flowering by preventing the synthesis of GA. To encourage flowering in mango trees during the "off" year, paclobutrazol has been frequently employed.
- Although numerous flowering-related genes have been identified, the underlying molecular pathways remain poorly understood. Thus, blossoming is a complicated phenomenon that is controlled by a variety of circumstances.

REFERENCES

Baxter, P. 1970. The flowering process – a new theory. In *Plant Growth Substances*. Ed., DJ Carr. CRC Press, Boca Raton, FL, pp. pp 775–779.

Bonner, J., and Hamner, K. 1938. Photoperiodism in relation to hormones as factor in floral initiation and development. *Bot Gaz* 100:338–431. https://bcs.whfreeman.com/WebPub/Biology/hillis1e/Animated%20 Tutorials/at2702/pol_2702_scr.html

Brian, P. W. (1959). Effects of gibberellins on plant growth and development. *Biological Reviews*, 34(1), 37–77.

Bunning, E. 1958. *Die physiologische Uhr I Aufl.* Springer, Berlin.

Butler, W.L., Norris, K.H., Siegelman, H.W., and Hendricks, S.B. 1959. Detection, assay and preliminary purification of the pigment controlling photoresponsive development of plants. *Proceedings of the National Academy of Sciences of the United States of America* 45:1703–1708.

Cavichioli, J.C., Ruggiero, C., Volpe, C.A., Paulo, E.M., Fagundes, J.L., and Kasai, F.S. 2006. Flowering and fructification of yellow passion fruit submitted to artificial light, irrigation and shade. *Revista-Brasileira-de-Fruticultura* 28(1): 92–96.

Chailakhyan, M.C. 1937. Concerning the hormonal nature of plant development processes. *Comptes Rendus de l'Academie des Science de* l' *USSR* 16: 227–230.

Dhillon, W.S., and Bhat, Z.A. 2012. *Fruit Tree Physiology*. Narendra Publishing House, New Delhi.

Hendricks, S.B., and Borthwick, H.A. 1955. A reversible photoreaction controlling seed germination. *Proceedings of the National Academy of Sciences of the United States of America* 38(8): 662–666.

Lenz, F. 1964. Day length and temperature effect on citrus cuttings. *Australian Horticulture Research Newsletter* 11: 27–28.

Metzger, J.D. 1987. Hormone and reproductive development. In *Plant Hormone and their Role in Plant Growth and Development*, Ed., PJ Davies. Springer, The Hague, pp. 432–441.

Purvis, O.N. 1961. *The Physiological Analysis of Vernalisation*. Springer Verlag, Berlin, pp. 76–122.

Ravishankar, H., Singh V.K., Misra A.K., and Mishra, M. 2014. *Souvenir National Seminar-cum-Workshop on Physiology of Flowering in Perennial Fruit Crops*. Central Institute for Subtropical Horticulture (ICAR) Rehmankhera, Lucknow.

Sachs, R.M., and Hackett, W.P. 1983. Source sink relationships and flowering. In Beltsville Symposia in Agricultural Research. Vol. 6. *Strategies of Plant Reproduction*. Ed., WJ Meudt. Allanheld, Osmun, pp. 263–272.

Serce, S. and Hancock, J.F. 2005. The temperature and photoperiod regulation of flowering and runnering in the strawberries, Fragaria chiloensis, F. virginiana, *and* F. *x ananassa*. *Scientia Horticultarae* 103: 167–177.

Sinha, R.K. 2004. Physiology of flowering. In *Modern Plant Physiology*, Ed., Sinha R. K.. Narosa Publishing House, New Delhi, pp. 513–521.

Storey, W.B. 1986. *Carica papaya*. In *Handbook of Flowering*, Vol 2, Ed., Halevy, A. H. CRC Press Inc., Boca Raton, FL, pp. 1–5.

Tromp, J. 1984. Flower-bud formation in apple as affected by air and root temperature, air humidity, light intensity and day length. *Acta Horticulturae* 149: 39–47.

14 Regulation of Flowering and Off-Season Production of Horticultural Crops

Stuti Pathak, Susmita Das, and Sunny Sharma

14.1 INTRODUCTION

In India, horticulture has evolved into an essential part of agriculture, providing farmers with various crop diversification options. Protected farming of crops with significant value has grown into the most important approach to ensuring greater yield, better quality, financial benefits, and off-season production in the face of competitive economic integration, scarce land, and changing climates (Kaushal and Singh, 2019). Flowering is a critical step in fructification, as the absence of flowers results in the absence of fruit. Apart from their significance in determining crop yield, certain events occurring during flower formation and fruit set have an effect on fruitlet development, final fruit size, and quality, and thus on returns (Singh et al., 2019). Thus, it is critical to understand how crops flower and fruit. Since both moisture and temperature stress are essential for blossoming, unrestricted development leads to greater foliage development than reproductive growth (Cassin et al., 1969). Regulated crops are required to prevent market gluts, guarantee a steady supply of crops, and ensure off-season availability. Extended periods of stress—which can be brought on by variables like temperature and humidity extremes—delay flowering. According to Singh and Chadha (1988), applying stress led to uniform flushing, and the degree of flushing was correlated with the stress level as determined by the plant's proportion of water before relief. Commercial vegetable crops like tomatoes, capsicums, and cucumbers can have their flowering times appropriately controlled, either early or by using a variety of blossoming and fruit regulation tools and techniques, including mechanical, ecological, and chemical-based methods. To improve crop production in

TABLE 14.1
Effect of Heavy Fruit Set and Low Fruit Set in Fruit Crops and their Management

	Heavy Fruit Set	Low Fruit Set
Effects	Smaller sized fruits Limb breakage Decreased return bloom Alternate bearing	Low yield
Reasons	Ideal conditions at bloom Light or no pruning	Frost injury Low temperatures or rains Lack of pollination
Strategies	Crop thinning Pruning	Bloom delay (chemicals or evaporative cooling) Improving fruit set (chemicals) Pollinizers/pollinator

DOI: 10.1201/9781003354055-14

TABLE 14.2
Bahar Regulation in Guava

Bahar	Time of flowering	Time of Harvesting
Ambe bahar	February to March	Rainy season- July to September.
Mrig bahar	June to July	Winter season- Nov to Jan
Hast bahar	October to November	Summer season- Feb to April

TABLE 14.3
Effect of Different PGR's on the Modulation of Fruit Crops

Fruit Crop	Dose Concentration	Time of Application	Remarks
Le Conte pear	Sucrose @ 10%	At full growth stage	• Enhanced fruit set by 6.11% • Number of mature fruits per meter shoot length by 2.89 • Increased yield by 16.8%
	GA3 application @ 5 ppm and 1,000 ppm chlormequat		• Improved fruit set and better-quality fruits
Plum	GA at 50 and	10th day after full bloom	• Increase fruit set from 3.1% to 17.7% and
	1,000 ppm of 2,4,5 – T (fenoprop)		• 23.0 fruits per 100 flowers
Strawberry	NAA (50 ppm)	At flower initiation stage.	• Improve fruit set

tropical and subtropical fruit crops, a variety of techniques are currently used, including selective and non-selective pruning, crop removal, drought stress (verdelli method), branch girdling, and plant growth regulators (PGRs). Tropical fruit production utilizes a number of methods, including selective pruning, mechanical pruning, crop sacrifice, drought stress, and plant growth regulators to maximize production Tables 14.1–14.3.

14.2 PRINCIPLE AND OBJECTIVES OF FLOWER REGULATION

The primary concept behind flower regulation is to alter the crop's normal blossoming and bearing cycles to increase fruit yield, quality, and revenue (Boora et al., 2016). The concept relies on the observation that most crop flowers are only formed on immature, succulent vegetative growths that are rapidly merging. Flower regulation's primary purpose is to compel the plant to rest and produce an abundance of blossoms and fruits within a specific season (Bhakti, 2016). Reducing the cost of cultivation is important because continuous blossoming would produce light crops all year round, which would require expensive monitoring and marketing (Lal et al., 2017). It also ensures a consistent, high-quality crop and maximizes earnings for the grower.

14.3 IMPORTANCE OF PROTECTED CULTIVATION

Protected horticulture, as the most efficient method of overcoming climatic variation, has the potential to meet the needs of small growers by multiplying yields while also greatly improving the quality of the food to meet market demand in the offseason (Kumar et al., 2014). It is an agro-technology that entails covering the crop in order to regulate macro- and microenvironments, hence promoting optimal plant growth and development, extending the length of growth, inducing earliness, and improving yield and quality (Gruda and Tanny, 2015). Protected cultivation of fruits is intended to provide early harvesting and protect them from late frost, late summer rains, disease, and pests.

During winters: The main systems used for greenhouse climate control are heating, ventilation, and CO_2 enrichment.

During summers: The main systems used are ventilation (natural or forced), shading (by screens or whitening), CO_2 enrichment, and cooling (by fan and pad systems).

14.4 FACTORS AFFECTING FLOWERING UNDER PROTECTED CONDITIONS

14.4.1 Temperature

Because the steel structure of the poly home is wrapped with polythene, the interior temperature can reach 40°C. The ventilation system, as well as cooling pads and fans, are utilized to regulate the temperature within the poly home (Pickens and Sibley, 2019). However, the fruit set is determined by the 24-hour mean temperature and the temperature variation between day and night.

14.4.2 Relative Humidity

The relative humidity (RH) value indicates the amount of water vapor in the air. The capacity of air to contain water vapor is temperature-dependent (Kong and Singh, 2011). We should maintain an appropriate humidity level for vegetables in order to ensure their health and optimal growth. Vegetables require a humidity level of 60%–65%. Transpiration is a critical plant function that assists the plant in cooling itself, obtaining nutrients from the root system, and allocating resources within the plant. Transpiration determines the greatest efficiency of photosynthesis, the efficiency with which nutrients are transported into the plants, and the efficiency with which these nutrients are dispersed for plant growth (Perez, 2019). Relative humidity minimizes evaporation loss from plants, resulting in optimal nutrient usage. Additionally, it preserves the turgidity of cells, which is beneficial for enzymatic activity, resulting in a better yield.

14.4.3 Light

Photosynthesis occurs exclusively in visible light. Light is managed with poly-house technology in such a way that plants receive the greatest amount of visible light and the remainder is reflected back. By and large, the more intense the light, the greater the rate of photosynthesis and transpiration (increased humidity) in the greenhouse, as well as the solar heat gain. It has been demonstrated that decreasing light levels from 10,000 to 25,000 lux delays floral start. The greenhouse absorbs a percentage of the light that falls on it, allowing up to 80% of the light to reach the crop around midday and an average of 68% throughout the day (Brown, 2006).

14.4.4 Carbon Dioxide

CO_2 content in our surrounding environment is 0.03%, or 300 parts per million (Lindsey, 2020). Photosynthesis is how plants utilize this CO_2. At night in a poly home, photosynthesis ceases, but CO_2 is released through respiration. Because this CO_2 remains concentrated around plants at night, in comparison to the outdoors, polyhouses always have a higher CO_2 concentration. This CO_2 is then utilized for quick photosynthesis by plants growing in polyhouses. It has been demonstrated that when a poly house contains 1,000 ppm of CO_2, vegetable yield increases four to fivefold over normal circumstances. The optimal CO_2 concentration should be between 350 and 1,000 ppm (Bao et al., 2018).

14.4.5 Wind Movement

If a polyhouse has a higher humidity level, the risk of illness and pests increases. Under these conditions, the poly-side house's vents are opened to improve airflow within the structure. As a result of the increased air movement, the humidity level lowers and the risk of disease also decreases.

14.4.6 Nutrients Required by Plants

Plant chemical analysis indicates the existence of more than 90 elements, but only 17 have been identified as necessary for the effective proliferation and development of plants based on the essentiality criterion. Recent findings have revealed that the growth of some plants necessitates the presence of several elements such as vanadium, silicon, sodium, and nickel. The carbon to nitrogen ratio, or C/N ratio, of soils is the ratio of organic carbon to nitrogen. The C/N ratio of residue has an effect on the pace of organic matter breakdown. The organisms that break down residues require nitrogen (and other important components) in addition to carbon; if the residue has insufficient nitrogen, decomposition is slowed. Additionally, if the residue contains little nitrogen, microbes will absorb inorganic nitrogen in the soil to meet their nitrogen requirements, competing with plants for nitrogen and lowering the quantity of soil nitrogen available for plant growth (Aczel, 2019). The optimal supply of nutrients aided early flowering in cucumber due to vigorous vine growth and fruit maturity is largely dependent on fertilizer application, as previously demonstrated that 100% fertigation can induce early harvest in polyhouse conditions (Janapriya et al., 2010).

14.5 REGULATION OF FLOWERING AND FRUITING IN COMMERCIAL CROPS

14.5.1 Flowering

Typically, tomato blossom clusters are not pruned until they have three to four fully developed fruits. However, anomalous flowers (such as the huge captivated flower) must be destroyed immediately upon detection. This flower will bear a fruit with a cat's face (Bhakti, 2016). When a young capsicum plant has 7–13 leaves, it begins flowering approximately 2–6 weeks after planting. The ideal temperature range for flowering should be between 20°C and 21°C on average, both day and night (Paradiso and Pascale, 2014). Decreased nocturnal temperatures diminish the ability of pollen in capsicum flowers to remain viable, modify the structure of the flowers, and decrease the efficiency of self-pollination. Flowers that undergo development under a night temperature below 18°C tend to yield fruits with an elongated, pointed blossom-end, resembling a 'tail'. Lowering the temperature during blooming leads to a decrease in the number of fruits with four compartments, or even fruits with only two compartments, which is considered undesirable. A flower cultivated under frigid conditions (below 10°C at night) yields a small, compressed fruit. Blossom end rot is promoted by temperatures over 28°C (Elitzur et al., 2009).

14.5.2 Pollination

The tomato flower's female organs are encased within the male organs (five anthers form a cone around the female organ). Upon reaching maturity, anthers undergo dehiscence, allowing the release of pollen into the surrounding environment (Melissa, 2021). Once fully developed, the anthers will have a vibrant yellow hue, and the flower will maintain its ability to be pollinated within 48 hours. The tomato fruit's size and weight are positively connected with the amount of pollen transported to the female portion of the flower. Within the tomato, utilize pollinators such as bumble bees or honey bees inside the greenhouse to ensure a set of high-quality crops. Recent research demonstrates that stingless bees are also an effective alternative to honeybees for pollinating a variety of greenhouse crops of considerable economic and social importance, including strawberries, tomatoes, and sweet peppers.

Cucumber pollen grains are big and sticky, they require an external substance to facilitate pollen transmission between flowers. Cucumbers are primarily pollinated by insects, primarily honeybees. In greenhouse circumstances, pollination is not required for growing gynoecious kinds of slicing cucumber. However, monoecious species rely on pollination within protected structures, which is primarily performed by honeybees. Colonies are adapted to their environment, and bees forage for food and other products up to a few kilometers away from the hive. Nonetheless, greenhouse crops are pollinated by native and Africanized bees.

14.5.3 Parthenocarpy

Auxin caused the development of seedless fruits in cucumbers and watermelons. PCPA at concentrations of 50–100 ppm induced parthenocarpy in tomato and brinjal. Additionally, the application of 2,4-D at a concentration of 0.25% in lanolin pastes to the cut end of styles or as foliar sprays to newly opened flower clusters, has been found to promote parthenocarpy, as reported by Prajapati et al. in 2015. Plant growth regulators facilitate fruit production by parthenocarpy, which occurs in the absence of fertilization. Staminate flowers were induced in a parthenocarpic cucumber line by applying the plant growth regulator GA3 at a concentration of 1,500 ppm and silver nitrate at a concentration of 200–300 ppm. This was done by four sprays at a 4-day interval, as described by Singh and Ram in 2004.

14.5.4 Training and Pruning

The training approach places a premium on the plant's capacity to obtain sufficient sunlight for growth. Additionally, it is critical to maintain adequate air circulation around the plant to minimize the danger of pest infestation. The interaction between the source and sink influences the growth habit, fruit-bearing pattern, and seed yield of tomatoes. Tomatoes have determinate, semi-determinate, or determinate growth behavior (Grant, 2021). Indeterminate varieties/hybrids are preferred for greenhouse hybrid seed production (Pavani et al., 2020). These plants can be grown for an extended period of time and produce numerous fruit trusses. Tomatoes grown in greenhouses are clipped to a single stem. Plastic string is used to support the plants. With a short, non-slip loop, secure one end of the twine loosely to the bottom of the plants. The other end is connected to an overhead supporting wire located between 1.8 and 2.5 m above the plant row. Twist the string around the plant as it grows in one or two gentle rotations for each fruit cluster. When plants develop larger and heavier, "twister" or plastic snap-on clips may be used to secure the plant to the string. The string is untied and lowered frequently when the plant reaches the wire, enabling the lower section of the plant to lie on the ground. Approximately 6–7 weeks before harvesting is to cease, the plant's growth tip is typically clipped off or topped. Numerous training methods are used, like vertical, arch, V-shape, S-hook, and lateral.

De-shooting: By pinching off any side shoots, prune tomato plants to a single stem. Avoid using a knife. This should be done at least once a week; side shoots should be removed while they are very little. Remove no side shoots above the most recent blossom cluster (Novak and Maskova, 1979).

De-leafing: Remove the two to three leaves beneath ripe fruit clusters that begin to yellow and wilt. Before removing leaves, ensure that the fruit has reached the mature green stage (Garderner, 2015).

Thinning of tomato flowers and fruit is optional. Prune flower clusters and fruits to promote tomato growth. Prune the remaining blooms in clusters once three to four well-formed fruits have developed on each cluster. Eliminate any aberrant flowers or misshapen fruits to create less competition for producing fruits. Tipping (removal of the apical bud) of tomato plants resulted in a 6-day delay in maturity, while leaf pruning resulted in a 6-day delay in maturity. Tanaka and Fujita (1974) showed that photosynthates flow from sources surpass sink demands and that partial leaf removal compensates for the decreased net assimilation rate of remaining leaves; consequently, fruit growth is unaffected. Pruning impacts carbon partitioning and consequently the ratio, and so the fruitfulness, of tomato.

In pepper, the flower at the base of the first branch is a crown bud. The terminal blossom is not allowed to mature and is plucked immediately upon presentation. Each plant should have two main stems following the cutting or pinching of the remaining branches, leaving two leaves and one blossom at each internode. These two stems are trained on strings to the main wire that runs the length of the row at an 8–9-foot height. The stems are trellised either loosely or tightly around the strings. String is used to secure the stems using rings or plastic clips. During the era of rapid growth, training, and pruning should be performed every 3 or even 2 weeks. According to Resh (1996), trimming peppers grown in greenhouses enhances light interception, fruit set, and fruit quality due to the lower branch count. In general, no or minimal pruning leads to excessive vegetative growth on tiny fruiting plants. The removal of fruit from bell pepper plants' initial flowering node 10 days after the fruit set had no effect on the partitioning of dry mass to fruit on the plant's upper nodes. The first flowering node fruit operates as a significant sink for photosynthates (10.2%) for the first 20 days after flowering and then becomes a weaker sink.

Several training strategies for greenhouse cucumber exist. The fundamental objective guiding the development of a training system is to enhance leaf interception of sunlight uniformly throughout the home. Pruning and training of the canopy, as well as proper spatial arrangements, have been highlighted as critical management methods for maximizing marketable yields from greenhouse crops. The system that is chosen will be determined mostly by the greenhouse facility, the production method, and the grower's desire. The umbrella system is the most often used pruning method for vertical cordon or V-cordon-trained plants.

The Umbrella System is straightforward, labor-intensive, and easily understood. Secure the cucumber plant to a 7-foot-tall vertical wire. Pinch out the top growth point. Provide support for every fruit that develops on the main stem's lower portion. Eliminate any side branches from the primary stem's leaf axis. Train the uppermost two lateral branches to droop down on opposite sides of the main trunk. Permit these to reach a height that is two-thirds of the total length of the primary stem. Once the crops on the first laterals have been harvested, it is necessary to cut them back to a strong shoot to allow the subsequent laterals to assume control. Continuously do the process of renewal for the side branches, as this method will guarantee productivity. Before day 10, after the blossom, the fruit has a slight taper at the stem end. But by day 14, the fruit becomes uniformly cylindrical. Typically, commercially adequate fruit size is achieved around the 11th day after the flowering period opens in the spring season.

14.6 SPECIAL PRACTICES

14.6.1 Tomato

To limit further growth, prune the terminal development node above the uppermost flower cluster (the final cluster to be pollinated) on tomato plants approximately 45 days before the intended crop cessation date (Guan, 2018).

Vibrators powered by batteries: Vibrators are little devices used in tomatoes that are powered by a weak electrical current from a battery. For a few seconds, vibrate the flowers by stroking the stem of the flower cluster. This tool's powerful vibration will disperse more than enough pollen to fertilize the majority of the ovules in the ovary (Marin, 2019). Pollinate the flowers every other day on bright days when the greenhouse's humidity level is between 60% and 80%. Avoid touching the cluster stem and avoiding the flower itself since this will result in a hole being made in the developing fruit. Pollinating 700 plants in a 30 96-foot greenhouse takes around 30 minutes three times a week (Rutledge, 2015).

In tomato, greenhouse tomatoes can be pollinated with a normal-speed household leaf blower with the airflow directed at the flower clusters. Three times every week, utilize this instrument. When compared to the electric vibrator, it takes half the time to pollinate the same number of plants.

However, the quantity of seeds per fruit was less and the fruit size and weight were lesser than when the vibrator was used to pollinate the fruit. In general, expect a 5% drop in yield while using this device (Delaplane, 2013).

Bumblebees: By pollinating one or two greenhouses with bees, we can save time. Bumblebees are excellent greenhouse tomato pollinators. Each bee will visit the blossom and vibrate it briefly to collect pollen for feeding. As a result of this procedure, the stigma of the flower is showered with a significant amount of pollen, resulting in effective pollination and fertilization of nearly all the ovules in the ovary (Potts et al., 2014). Bee pollination is supposed to result in a larger fruit with a heavier weight (Junqueira and Augusto, 2016). Bees work from sunrise to dusk; they do not take extended vacations or days off (Lallensack, 2018). Each bee is thought to be capable of pollinating up to 350 flowers (Stein et al., 2017). Pollination by bumblebee produced the most fruit of any approach.

14.6.2 Mangoes

Mango flower induction takes place throughout the dry season. Unlike other varieties of fruit trees, the mango tree requires 2–3 months without rain to produce flowers. Additionally, the mango tree is capable of flowering even before the start of the rainy season. It is relatively simple to induce off-season flowering of mango. Using paclobutrazol followed by a foliar application of KNO_3 is an efficient method for inducing flowering in the mango variety 'Gadung 21'. Applying paclobutrazol to the soil is also an efficient method for inducing mango flowers. Nevertheless, the prolonged residual impact of paclobutrazol after being applied to the soil (for up to 3 years) leads to a new issue, namely, the aberrant growth of the tree. Directly applying paclobutrazol through a foliar application on the tree canopy or shoots/leaves can effectively address the issue of residual effects. Applying paclobutrazol through soil drenching, foliar spray, or bud spray, then applying a foliar spray of 40 g/L KNO3 1 month later, did not result in a decrease in the percentage of bud break or total buds. Applying paclobutrazol through a soil soak and foliar spray, followed by a foliar spray of 40 g/L KNO3 1 month later, increased the overall bud count. Nevertheless, the utilization of paclobutrazol resulted in a decrease in the number of vegetative buds while simultaneously promoting the growth of generative buds or flowers. The control tree did not develop any flowers, with only 1.3 flowers per tree or 3% of all the shoots. Trees subjected to paclobutrazol treatment exhibited blossom production on 16%–56% of the branches. The procedure greatly augmented the quantity of inflorescences. Soil drenching treatment was the most efficient strategy, although the foliar spray method was also effective. The first flower appeared after the use of paclobutrazol at a rate of 2.0 g a.i./tree through bud spray (directly spraying paclobutrazol at each terminal bud), and at a rate of 1.0 g a.i./tree through foliar spray. The latest flower appeared after the application of paclobutrazol at rates of 1.0 and 4.0 g a.i./tree through bud spray. Applying paclobutrazol at a rate of 1.0 g a.i./tree through soil soaking or foliar spray resulted in normal inflorescences, similar to the control group. However, when applied at rates of 2.0 g and 4.0 g a.i./tree, the length of the inflorescence somewhat decreased. The application of paclobutrazol through bud spray resulted in growth suppression, which began at a dosage of 1.0 g a.i./tree and reached its peak level at a dosage of 4.0 g a.i./tree.

14.6.3 Mangosteen

Mangosteen flowering is triggered by drought stress during the dry season, and the blooms appear shortly after the rainy season. By applying a soil drench of 2 g of paclobutrazol, dissolved in 1 L of water, around the trunk, the trees were induced to flower 46 days following the treatment. The act of wrapping a wire (with a diameter of 3 cm) around the trunk of the mangosteen tree at heights of 30 and 40 cm above the earth resulted in the tree being strangled. Additionally, this action caused the tree to start flowering 46 days sooner than usual. These two treatments resulted in a threefold increase in flower and fruit yield compared to the control.

14.6.4 Rambutan

Usage of paclobutrazol did not enhance blooming in rambutan when compared to mango, citrus, mangosteen, and durian. To resolve the issue, an alternative approach (known as the ringing technique) was required to stimulate off-season rambutan flowering. During the ringing procedure, a 2-cm strip of bark was cut from around the trunk. Shortly after completion, the injured trunk was wrapped with black plastic tape. After approximately 1 month after ringing, a callus bridge develops between the lower and higher sections of the ringed bark. Ringing rambutan trees can induce blooming and fruit production during non-seasonal periods. Applying KNO3 1 month after the ringing process resulted in a 10–20 day increase in flowering time compared to untreated plants, as observed by Poerwanto and Kubota in 2003.

14.6.5 Kiwifruit

14.6.5.1 Fruit Quality

- Thus, heavy crop creates a severe competition between the fruits for water, nutrients, and photosynthates, which leads to the production of small-sized fruits.
- Therefore, to harvest quality crops of good size, hand thinning is essential, as chemical thinning is ineffective.
- Flowers of fruit thinning (20%) to the extent of retaining 5–6 fruits/flowering shoot produces more fruits of A grade without any adverse effect on total yield. In hand thinning only laterals flowers or fruits are removed.
- To enhance fruit growth and size is CPPU (named Sitofex or Caplit, SKW, Trostbeg, Germany and KYOWA Hakko, Japan).
- This is an N-(2-chloro-4-pyridyl)-N-phenyl urea derivative characterized by a strong cytokinin-like activity capable of increasing fruit size and yield and enhancing harvest maturity.
- The standard concentration used ranges between 5 and 20 ppm and the optimum application time is 2–3 weeks after full bloom. It has proven a useful tool to improve fruiting performance.

14.6.5.2 Bud Break

- Inadequate chilling can be countered by using hydrogen cyanamide (Dormex or HI-Cane, SKWT Trostberg, Germany) and fatty acid polymer, Armobreak (AKZO- Nobel, the Netherland).
- The main effects induced by Dormex are an increase in bud break and bud fertility, a synchronization of flowering of male and female vines, and a reduction in the duration of flowering.
- Dormex is usually applied at a concentration between 2% and 5%, 40–50 days before the expected bud break.

14.6.6 Strawberry

- The plant bio-regulators are used commercially to modify the physiological processes of strawberries.
- GA3 is a potential bio-regulator for improving the vegetative growth and runner production of strawberries.
- Likewise, soil application of PP333 in strawberries is also beneficial for increased runner production.
- The application of triacontanol also plays an important role in maximizing the leaf area, leaf number, and runner production.

- For breaking the dormancy of cold stored runners, these should be treated with 0.1% hydrogen cynaide (HCN) and 3% potassium nitrate.
- Flower thinning is an important operation in strawberries for enhancing the yield.
- The foliar application of ethephon (500 ppm) at the flower initiation stage is effective for the thinning of flowers.
- The application of GA3 (50 ppm) before initiation of flowering was equally effective to suppress flowering.
- The fruit set in strawberry can be improved with foliar application of NAA (50 ppm) at the flower initiation stage.
- The application of NAA (0.05–0.10 mg/L) to emasculated flowers resulted in parthenocarpic development of fruit.

REFERENCES

Aczel MR (2019) What is the nitrogen cycle and why is it key to life? *Frontiers for Young Minds* 7: 41.

Bao J, Lu WH, Zhao J, and Bi XT (2018) Greenhouses for CO_2 sequestration from atmosphere. *Carbon Resource Conservation* 1: 183–190.

Bhakti P (2016) *Growth Regulation Practices in Important Fruits Crops*, Navsari Agriculture University, Navasri, Gujarat. Accessed on September 8, 2021.

Boora RS, Dhaliwal HS, and Arora NK (2016) Crop regulation in guava-A review. *Agricultural Reviews* 37: 1–9.

Brown JW (2006) *Light in the Greenhouse: How Much Is Enough?* Cropking Incorporated. Accessed on September 9, 2021.

Cassin J, Bourdeaut J, Fourgerl A, Furon V, Gillard JP, Montagut G, and Morevilli C (1969) The influence of climate upon the blooming of citrus in tropical area. Proceedings of the First International Citrus Symposium, University of California Riverside, California, U.S.A., 315–23.

Delaplane KS, Dag A, Danka RG, Freitas BM, Garibaldi LA, Mark Goodwin R, and Hormaza JI (2013) Standard methods for pollination research with Apismellifera. *Journal of Apicultural Research* 52(4): 1–28.

Elitzur T, Nahum H, Borovsky Y, Pekker I, Eshed Y, and Paran I (2009) Co-ordinated regulation of flowering time, plant architecture and growth by *Fasciculate*: The pepper orthologue of *self pruning*, *Journal of Experimental Botany* 60(3): 869–880.

Garderner L (2015) Gardening Myth: De-leafing Tomato Plants. *Laidback garderner*. Accessed on September 5, 2021.

Grant BL (2021) *Determinate vs Indeterminate Tomatoes: How to Distinguish a Determinate from An Indeterminate Tomato*. Gardening Know How. Accessed on September 5, 2021.

Gruda N and Tanny J (2015) Protected crops e recent advances, innovative technologies, and future challenges. *Acta Horticulturae* 1107: 271–278.

Guan W (2018) *Prune Determinate Tomatoes*. Purdue University. Accessed on September 12, 2021.

Janapriya S, Palanisamy D, and Ramaswamy MV (2010) Soilless media and fertigation for naturally ventilated polyhouse production of cucumber (*Cucumis sativus* L.) Cv. Green Long. *International Journal of Agriculture Environment and Biotechnology* 3(2): 199–205.

Junqueira CN and Augusto SC (2016) Bigger and sweeter passion fruits: Effect of pollinator enhancement on fruit production and quality. *Apidologi* 48:131–140.

Kaushal S and Singh V (2019) Potentials and prospects of protected cultivation under Hilly Conditions. *Journal of Pharmacognosy and Phytochemistry* 8(1): 1433–1438.

Kong F and Singh RP (2011) *Chemical Deterioration and Physical Instability of Food and Beverages*. Woodhead Publishing Series in Food Science, Technology and Nutrition, Woodhead Publishing: New Delhi. pp. 29–62.

Kumar S, Patel NB, and Saravaiya SN (2014) Response of parthenocarpic cucumber to fertilizers and training systems under NVPH. *International Journal of Current Research* 6(8): 8051–8057.

Lal N, Sahu N, Marboh ES, Gupta AK, and Patel RK (2017) A review on crop regulation in fruit crops. *International Journal of Current Microbiology and Applied Sciences* 6(7): 4032–4043

Lallensack R (2018) Busy bees take a break during total solar eclipses. *Smithsonian Magazine*. Accessed on September 7, 2021.

Lindsey R (2020) *Climate Change: Atmospheric Carbon Dioxide*. Climate.gov science and Information for a climate-smart nation.

Marin MV (2019) Buzz pollination: Studying bee vibrations on flowers. *National Library of Medicine* 224(3): 1068–1074.

Melissa P (2021) Stamen. *Encyclopedia Britannica*. Accessed on September 10, 2021.

Novak FJ and Maskova I (1979) Apical shoot tip culture of tomato. *Scientia Horticulturae* 4: 337–344.

Paradiso R and Pascale SD (2014) Effects of plant size, temperature, and light intensity on flowering of phalaenopsis hybrids in mediterranean greenhouses. *The Scientific World Journal* 420807ID. http://dx.doi.org/10.1155/2014/420807

Pavani K, Jena C, Divya VV, and Mallikarjunarao K (2020) Cultivation Technology of Tomato in Greenhouse. In *Protected* Cultivation *and Smart Agriculture*, NEW DELHI PUBLISHERS: New Delhi. pp. 121–129.

Perez JCD (2019) *Postharvest Physiology and Biochemistry of Fruits and Vegetables*. Woodhead Publishing: New Delhi, 157–173 pp.

Pickens J and Sibley J (2019) *Greenhouse Cooling: An Overview of Fan and Pad Systems*. Crop Production Alabama A and M and Auburn Universities.

Poerwanto R and Kubota N (2003). Effects of ringing and application of dormancy breaking substance on off-season production of rambutan. *Proceedings of 2nd Seminar toward Harmonization between Development and Environmental Conservation in Biological Production*, pp. 187–191. Graduate School of Agricultural and Life Sciences, The University of Tokyo

Potts SG, Breeze T, and Herren BG (2014) Crop pollination. In *Encyclopedia of Agriculture and Food Systems*, pp. 408–418. https://doi.org/10.1016/b978-0-444-52512-3.00020-6

Prajapati S, Jamkar T, Singh OP, Raypuriya N, Mandloi R, and Jain PK (2015) Plant growth regulators in vegetable production: An overview. *Plant Archives* 15(2): 619–626.

Resh HM (1996) *Hydroponic Food Production*, 5th edn, Woodridge Press Publ. Co., Santa Barbara, CA.

Rutledge AV (2015) *Commercial Greenhouse Tomato Production*. Agricultural Extension Service the University of Tennessee, United States.

Singh HP and Chadha KL (1988) Regulation of flushing and flowering in acid lime (*Citrus aurantifolia* swingle) through stress management. *Progressive Horticulture* 20: 1–6.

Singh J, Vishwakarma G, Singh RK, and Pandey K (2019) Crop Regulation in Fruit Crops. In *Hi-Tech Horticulture Improved Production Techniques*. New India Publishing Agency (NIPA), New Delhi.

Stein K, Coulibaly D, and Stenchly K (2017) Bee pollination increases yield quantity and quality of cash crops in Burkina Faso, West Africa. *Scientific Reports* 7: 17691.

Tanaka A and Fujita K (1974) Nutriophysiological studies on the tomato plant. IV. Source-sink relationship and structure of the source-sink unit. *Soil Science and Plant Nutrition* 20: 305–315.

15 Flower Drop and Thinning

Akriti Chauhan and Dinesh S. Thakur

15.1 INTRODUCTION

Plant components physically separate from the main body of the plant through a process known as flower drop or abscission due to the dissolution of the cell wall or growth of subtending parts accompanied by the loss of floral and fruiting parts (Nautiyal et al., 2022). The process of flower drop/abscission is highly regulated and occurs in reaction to the environmental factors the plant was exposed to. For instance, in response to environmental cues that signal the impending winter, the leaves of perennial trees drop during the autumn. (Thomas et al., 2003). Whereas, in some plant species, abnormalities in flower structure cause the flowers to drop, e.g., all flowers containing aborted pistils shed in peach and 83% of abscised blooms in *Citrus* species shed because of apparent defects (Kaska, 1989). In reaction to injury or disease invasion, floral abscission also occurs in some perennial crop species. In other cases, the floral components are eliminated by the plant after pollination, having served their purpose of luring pollinators, as in the case of walnut, where the male flowers produced in inflorescence (catkins) fall after pollen spilling (Von Mohl, 1860). Various aspects, including the climatic factors (temperature, wind, rainfall, etc.), high abscisic acid (ABA) content, low photosynthate, ethylene production in a specific amount, inadequate fertilization, abscission-related signal, low Indole-3-acetic acid (IAA), auxin deficiency and heavy crop load were indicated as main causes of flower drop in fruit plant species (Table 15.1).

TABLE 15.1
Different Fruit Crops with Various Cause of Fruit Drop

Crop	Causes
Malus domestica	Inadequate pollination
	Fungal diseases
	Frost damage
Musa Paradisica	Water stress
	Nutrient deficiencies
	Pests and diseases
Vitis vinfera	Poor weather at flowering
Citrus sinesis	Insufficient water
	Excessive heat
	Pests and diseases
Mangifera india	Lack of pollination
	Drought stress
	Fungal diseases
Fragaria ananassa	Poor pollination
	Water stress
	Nutrient imbalances
Ananas comosus	Water stress
	Inadequate pollination
	Nutrient deficiencies

(Continued)

DOI: 10.1201/9781003354055-15

TABLE 15.1 *(Continued)*
Different Fruit Crops with Various Cause of Fruit Drop

Crop	Causes
Carica papaya	Inadequate pollination
	Water stress
	Nutrient deficiencies
Prunus avium	Frost damage
	Inadequate pollination
	Fungal diseases
Pyrus communis	Poor pollination
	Water stress
	Nutrient deficiencies
Prunus domestica	Inadequate pollination
	Frost damage
	Pests and diseases
Persea americana	Water stress
	Wind damage
	Fungal infections
Actinida deliciosa	Insufficient pollination
	Frost damage
	Nutrient imbalances
Citrus limon	Water stress
	Nutrient deficiencies
	Pests and diseases
Punica granatum	Poor pollination
	Water stress
	Fungal infections
Rubus idaeus	Inadequate pollination
	Water stress
	Pests and diseases
Vaccinium spp.	Frost damage
	Poor soil conditions
	Fungal infections
Rubus spp.	Water stress
	Insufficient pollination
	Pests and diseases
Psidium guajava	Lack of pollination
	Drought stress
	Fungal diseases

15.2 PHYSIOLOGY BEHIND FLOWER ABSCISSION IN FRUIT CROP SPECIES

In some fruit crop species, flower abscission may be slightly influenced by the competition for nutrients between the blooms and the nearby young leaves. It has been proposed that it corresponds to the competition for carbohydrates in walnuts. (Deng et al., 1991). Abscission usually requires de novo protein synthesis, however, the presence of actinomycin D, an inhibitor of mRNA synthesis, reduced petal abscission in certain crop species (Van Doom and Stead, 1997). A positive correlation was found between abscission zone weakening and an increase in peroxidase activity, which was mainly due to de novo protein synthesis (Henry et al., 1974). Insufficient transport or absorption by abscission zone cells may be the cause of the actinomycin D effect's absence in some systems or the mRNA may have been produced much earlier. In general, insoluble wall pectin levels drop, and soluble pectic acid levels rise before abscission. This causes hydrolytic enzymes to break down pectin and its associations with other molecules. This in turn leads to cell wall degradation and hence abscission. An increase in cellulase activity has also been reported to be associated with leaf and flower abscission. The role of free calcium and changes of acidity in the cell walls is yet unclear but several reports suggested that calcium ions were lost from the walls of the abscission zone cell (Osborne, 1989). The middle lamellae and primary cellulose walls dissolve, causing the abscission layer's cells to split from one another under the impact of the enzymes pectinase and cellulase activity. An abscission zone develops as a result of a series of morphological and physiological changes that result in cytolysis as the weakened cells of the zone are unable to maintain the extending plant, and it abscises under its weight or as a result of an external stimulus (Esau, 1976).

15.3 ROLE OF EXTERNAL (ENVIRONMENTAL) FACTORS IN FLOWER ABSCISSION

15.3.1 Temperature

During the early plant growth stages, flower abscission can be initiated in response to cultural and environmental factors or events, such as temperature, rainfall, wind, disease, and pathogens, that affect pollination and fertilization and lead to the deliberate shedding of organs that are no longer needed for the plant to operate, as evidenced by the bloom after assistance in pollination (Khandaker et al., 2016). Periodically unfavorable circumstances change the physiological status, which prompts a response that stimulates enzymes that previously existed and changes gene expression (Figure 15.1). It is now clear that, in addition to nutritional status and moisture stress, other environmental cues also have an impact on the floral initiation phenomena. Several workers have linked temperature, light, and carbohydrates to the abscission process (Byers et al., 1990).The role of temperature, although it does not have any direct effect, has been reported to be contradictory on flower initiation, where warm temperatures advance the process, however, cool temperatures retards it (Abbott, 1984). However, some reports suggested that an increase in temperature results in an accelerated metabolism that subsequently increases the rate of energy reserves utilization by the tissue and hence progresses rapidly towards abscission (Guinn, 1974; Borochov et al., 1985). The pace of pollen tube growth and the process of flower initiation are greatly influenced by temperature and pollen supply. The amount of the stigmatic surface that germinated pollen covers on apples and pears depends on the temperature of the surrounding area at the time of pollination (Jackson, 2003). In the "Spartan" cultivar of apple, pollen had a higher germination percentage at 8°C to 100°C than that of "Cox's Orange Pippin". In addition, the rate of pollen tube growth is positively correlated with rising temperature and negatively correlated with ovule longevity. Pollen grains germinate freely at levels above 10°C but are inhibited from doing so at temperatures below 4.4°C, which cuts down the time necessary for pollination and causes the flower to wilt. (Goldway et al., 2012). Elevated temperatures are known to stimulate flower abscission in several crop species (Cochran, 1936; Monterroso and Wien, 1990; Konsens et al., 1991). In mango, environmental factors influence the flowering process where the very high and very low temperature during flowering poses a risk to pollen and the tree fails to flower or the flower formed will drop later on (Dag et al., 2000). In a study to determine the effect of temperature in the abscission of plum flowers, it was found that abscission occurs as the consequence of temperature deviations from the egg cells' optimal range, which was shown to be highest at 20°C and less evident at lower temperatures (5°C, 10°C, and 15°C). Similar has been found in apples and pears (Racsko et al., 2007). A significant rate of floral bud abscission was seen in numerous crops when they were exposed to high temperatures (Hendriks, 1990) (Table 15.2).

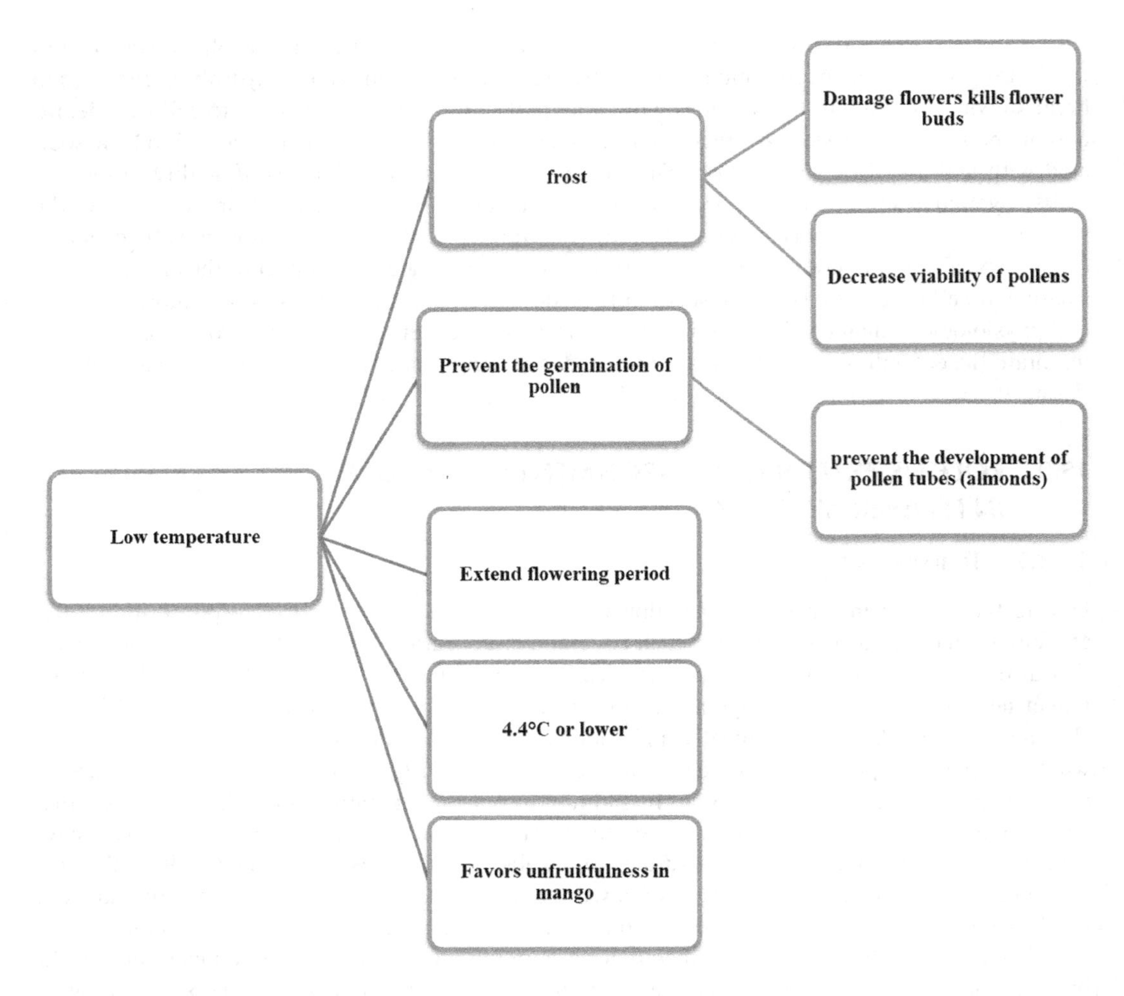

FIGURE 15.1 Effect of Low temperature on fruit crops.

TABLE 15.2
Differet Causes of Fruit Abortion in Different Crops

Common Name	Reason of Abortion
Apple	Defective embryo, defective ovules
Almond	Defective embryo sac, gynoecium abnormality
Grapes	Degeneration of nucleus
Kiwifruit	Pollen degeneration
Strawberry	Lower bud abortion, defective pistil
Sour cherry	Defective embryo
Mandarin	Abnormal pistil
Pecan nut	Defective pistil
Plum	Degeneration of pistil
Peach	Degeneration of nucleus, embryo abortion
Olive	Pistil abortion
Litchi	Embryo abortion

15.3.2 Pollination

The two most significant events that occur after a fruit crop species is established are flowering and fruit set (Khandaker et al., 2016). As fertilization is a prerequisite for preventing flower drop in several perennial crop species (Becquerel, 1907), flower shedding, in general, happens due to the lack of pollination and fertilization; however, the presence of a single embryo in some of the crop species prevented flower fall (Gartner, 1844). Clifford and Sedgley (1994) reported that in certain species, pollination alone prevented flower to fall in the absence of fertilization. However, this does not necessarily imply that only unpollinated flowers exists. It has been suggested that in a given crop species, the fruits from the base flowers have a geographical and temporal advantage and are more likely to mature than those from later-pollinated flowers (Bushnell, 1920; Herbert, 1979). In order to reach the younger fruits, flowers, and buds further along the inflorescence, leaf and root assimilates must pass the lower fruits. When these assimilates/resources are scarce, the reproductive structures that are most distant from the source are shed first (Wyatt, 1980). Three flowers are pollinated over 5 days on each inflorescence in a controlled series of trials with *Catalpa speciosa* blooms, and the results revealed that the last flowers fertilized have a much lower probability of setting (Stephenson, 1979).

15.3.3 Wind

In several anemophilous crop species, flower abscission is exacerbated by wind and weather conditions (Nagy and Kovacs, 2004). Wind-pollinated species also tend to abscise pollinated flowers. According to a 2-year study on Quercus alba, between 16% and 57% of the fallen flowers had pollen on them (Williamson, 1966). The pear cultivar 'Hardy' is especially susceptible to the wind. Where the local winds are strong, it has been advised that windbreaks or hedges be used to safeguard the plantations of pomaceous fruit species (Racsko et al., 2007). The effects of wind have been summarized in Figure 15.2.

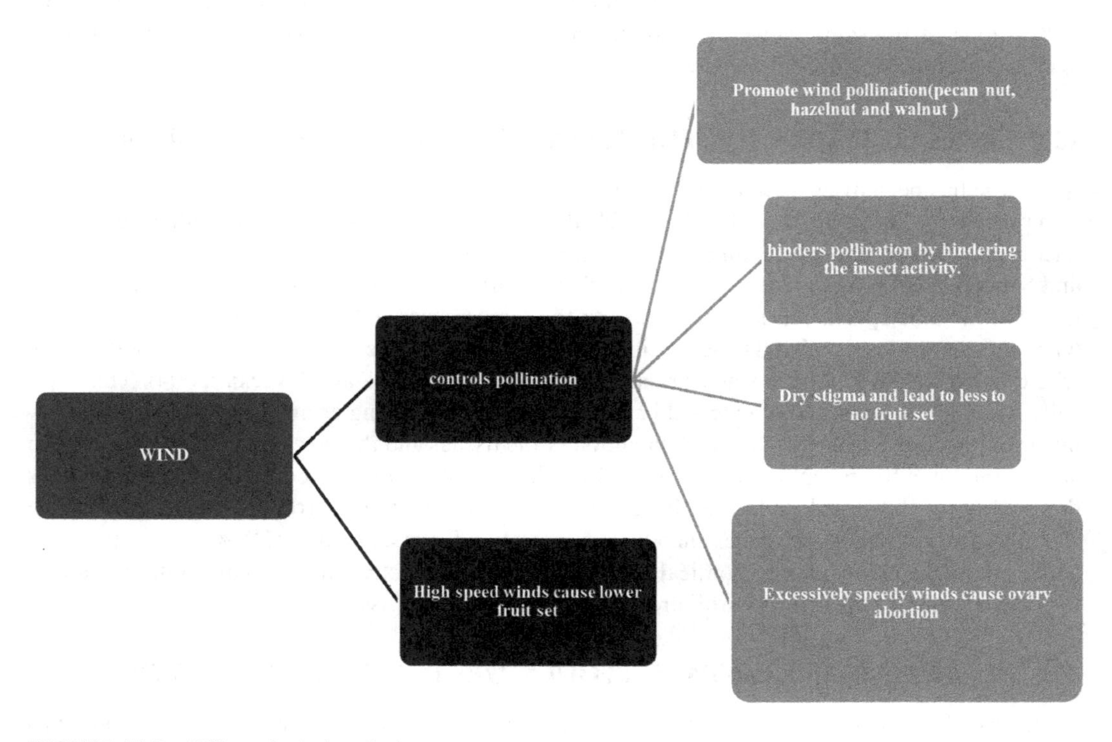

FIGURE 15.2 Effect of wind on fruit crops.

15.3.4 Diseases or Pathogens

Pathogens alter the metabolism of plants in a characteristic way that leads to flower abscission (Racsko et al., 2007). It has been reported that the majority of diseases that damage flowers encourage and hasten the abscission process. During a pathogen attack, the plant's defense is unsuccessful, and to stop the spread of ailments, it gains from shedding the organ. Changes in the scared tissues' enzymatic patterns can be seen, which are comparable to senescence's signs and often lead to flower/fruit drops. *Monilia* infections frequently result in the abscission of vegetative portions in perennial fruit crops. In hazel, as a result of monilia-infection fruits get shrivelled and mummified within the hard pericarp during the dry weather of July. In peaches, *Taphrina* is registered as the most important pathogen responsible for early flower drop (Singh et al., 2005). In walnuts, *Xanthomonas* leads to the flower drop, where the male inflorescences (catkins) become brown or black, followed by the deformed flower parts which abscise prematurely (Maria et al., 1997). In the Citrus *species*, *Colletotrichum acutatum* shows its symptoms on the petals, followed by turning the flowers brown (Timmer and Brown, 2000), which sheds later on (Li et al., 2003).

15.3.5 Other Factors

Certain other factors can contribute to the abscission process like relative humidity, light, carbon dioxide concentration, etc. Humidity prevents nutrients like calcium and magnesium from being absorbed, which increases the likelihood of flowers falling off prematurely, especially in high-humidity circumstances where it may reduce transpiration (Hendriks, 2001). No incidence of flower abscission was observed at lower humidity conditions (Zieslin and Gottesman, 1983). The role of carbon dioxide is still unknown because it accelerates photosynthesis and increases reserves of food, which could occasionally defer abscission but is not always the case (Stiling et al., 2002). In water-stress conditions, an increase in the production of ethylene and abscisic acid (ABA) because of water conduction within the desiccated xylem can speed up senescence (Reid, 1991). Additionally, it might cause stomatal closure, which would trigger the synthesis of ABA and hasten abscission. Rehydration of the same, however, leads to an increase in ethylene production which ultimately leads to abscission (Abeles et al., 1992).

15.4 ROLE OF INTERNAL (GENETIC) FACTORS IN FLOWER ABSCISSION

Very little has been discovered about the function of genetic (internal) factors in the flower abscission process, however, it is said that flowers shed when it is more energetically advantageous to produce a new flower than to sustain an old one as part of their resource allocation strategy (Ashman and Schoen, 1994). Abscission of flower and flower parts usually happens due to cell wall dissolution, but in some species, it does occur due to the forces generated by the growing fruit. Internal (Genetic) factors/forces related to abscission include unsuccessful fertilization or lack of pollination, where flowers shed to reduce water or carbohydrate loss (Erdelska and Ovecka, 2004). This typically occurs in the absence of external or inhibiting stimuli, serving as an internal control mechanism. Depending on the physiological condition of the tissues and the type of cells that make up the abscission zone, the receptors in the internal mechanism should be in a configuration that allows the binding of the signal molecule. To recognize stimuli and react correctly, they engage in the signal transduction pathway, genes encoding transcription factors, and other DNA-binding proteins. (Ascough et al., 2005). The anatomical variables suggest that active middle lamella dissolution or the degeneration of the primary wall are the causes of floral abscission.

15.5 ROLE OF PLANT GROWTH REGULATORS IN FLOWER ABSCISSION

Plant growth regulators serve as messengers, required at low concentrations in small amounts to regulate the flowering and other developmental processes (Khandaker et al., 2013). Hormonal relations zand physiology of flower abscission have been studied only in a few species which indicates a role of both auxins and ethylene. Ethylene advances the process of flower abscission. However, when the flower remains unfertilized, it leads to a decrease in the level of endogenous auxins due to an increase in ethylene sensitivity, which may act as a trigger of abscission. Other than these two, it is yet unclear how other growth regulators affect floral abscission.

15.5.1 Auxin

According to reports, abscission tends to be triggered by an effect that is lowered by applying auxin at the cut end by eliminating the distal leaf part, and continuous production of auxin by the distal vegetative parts of plants, such as blooms, flowers, and flower parts, assists in avoiding abscission (Osborne, 1989). Flower removal resulted in rapid pedicel abscission in several crop species; however, applying NAA on the cut surface lessened this impact (Roberts et al., 1984; Wien and Zhang, 1991). Auxins, depending on their concentration, can either reduce or promote the flower abscission process (Abeles et al., 1992). The reduction and reversal of abscission in rose pedicels upon auxin application demonstrates that auxin can lessen the reactivity of abscission zone cells to ethylene (Goszczynska and Zieslin, 1993).

The delay of apple abscission is one of the earliest responses physiologically linked to auxins. Exogenously applied auxins were reported to prevent or delay flower abscission in apple (Van Overbeek, 1952) and cherry plants (Addicott and Lynch, 1955). However, in some cases, auxin may increase abscission at certain concentrations as excess levels of auxin stimulate ethylene production (Abeles et al., 1992). But in most crop plants, auxin strongly decreases the sensitivity of abscission zone cells to ethylene. Similar results were reported in cherry, plum (Addicott and Lynch, 1955), and Citrus species (Einset et al., 1979).

15.5.2 Gibberellins

Gibberellins may have a regulatory function, albeit this is not yet obvious. In a controlled setting, it has been suggested that they may have an impact on vegetative and regenerative behavior. The application of GA3 has increased orange productivity as it decreases the possibility of bud drop by mobilizing nutrients to grow organs and has an impact on bud longevity (Almeida et al., 2004). According to reports, the 'BloodRed' sweet orange's bloom drop and natural product drop were both reduced by GA @45 mg/L treatment (Saleem et al., 2008). Application of gibberellin has been found to hasten pedicel abscission (Chatterjee, 1977). Reduction in floral drop with an application of GA_3 @ 20 ppm was reported earlier (Jawanda et al., 1974). Furthermore, in wax apples, the number of flower drop incidents was resolved when spraying with 10 ppm GA3 (Tuan and Chung-Ruey, 2013). In citrus, gibberellins spray might trigger auxin levels and nullify the action of ABA, consequently helping to delay senescence (Bisht et al., 2018).

15.5.3 Ethylene

Ethylene is mainly associated with flower and fruit drop and their role is well established. Before floral abscission, ethylene production is frequently increased. It was reported that the first biochemical change associated with abscission is an increase in ethylene production (Burdon and Sexton, 1990). However, they also failed to prove their hypothesis. In many species, endogenous ethylene controls flower shriveling and abscission, which are considered signs of flower senescence and are typically related to ethylene production (Halevy and Mayak 1981; Van Doorn, 2001). Previous physiological research has not demonstrated that ethylene is a necessary component of abscission, but it has been demonstrated that in some plant species that first experience senescence, the abscission mechanism is ethylene-independent (Lewis and Bakshi, 1968). It was reported that exogenous ethylene exposure accelerated plant inflorescence abscission (Weis et al., 1988; Heyer, 1985;

Rewinkel-Jansen, 1985). In Phalaenopsis cultivars, when the impact of ethylene inhibitor 1-MCP (1-methylcyclopropene) was studied, it was reported to diminish ethylene-induced flower bud drop (Sun et al., 2009). It suppresses senescence forms such as decreased water substance, increased membrane penetrability, and increased ABA content by anticipating ethylene activity. Similarly, additional ethylene inhibitors, such as silver thiosulphate (STS) and Amino-oxyacetic acid (AOA), which limit ethylene synthesis, reduced floral abscission to nil in some species (Cameron and Reid, 1983; Joyce, 1989; Dostal et al., 1991; Sexton et al., 1995). The effects of these ethylene inhibitors unambiguously show that endogenous ethylene regulates floral abscission. The rate of ethylene synthesis in bananas rose prior to petal abscission (Israeli and Blumenfeld, 1980). Floral abscission in citrus *spp.* was found to be delayed with AVG spray (Slpes and Einset, 1982), indicating its regulation by endogenous ethylene. Ethephon is proven to be the most effective for flower bud formation promotion (Byers, 2003), although, its use on bearing trees was limited because it also caused thinning. Endogenous levels of this hormone may also be linked to the transmission of wounded or pathogen-induced stimuli, which increase in anticipation of injury, owing to increased ACC synthesis (Taylor and Whiteclaw, 2001).

15.5.4 Abscisic Acid

Since abscisic acid impacts cell elongation and is opposed to both auxins and gibberellins, it creates favorable conditions for the differentiation of floral buds (Upreti et al., 2013). The effect of Abscisic acid and auxins (IAA) on flower and fruit drops in mangosteen trees was studied (Rai et al., 2013), and was reported that the abscised flowers had a high ABA content, low supply of photosynthate, and low IAA depicting that the excessive flowers abscission might be due to these factors. A low amount of photosynthates, as seen by the decreased sugar content, emerges from the flowerless shoots. In Shamouti orange (*Citrus sinensis*) before flower abscission, a small increase in concentration of abscisic acid was found in the petals (Goldschmidt, 1980). A similar effect has been reported in Grape (*Vitis vinifera*) flowers when treated with abscisic acid (ABA) and resulted in early abscission of the flower (Weaver and Pool, 1969). However, with the increase in exogenous ABA concentration, the general rate of ethylene production also increases, so any effect may be indirect (Abeles et al., 1992).

15.6 ROLE OF NUTRIENTS IN FLOWER ABSCISSION

A reduction in the longevity of individual flowers, bud and leaf loss, stem etiolation, and floral abscission are some of the negative impacts of plant nutrients, even though they do boost the overall growth, yield, and quality of the plants at greater levels (Nell et al., 1995). Nitrogen plays a very important role in the abscission process as a high level of it drastically causes a reduction in organ longevity. It has been reported that low calcium and high ammonium increased the chlorosis of leaves and flower senescing (Starkey and Pedersen, 1997). Although fertilizer use varies depending on plant species, growing medium, concentration, and type used, calcium sprays and fertilizer drops help to reduce the frequency of floral abscission (Nell et al., 1995).

15.7 THINNING

A fruit crop tree is unable to supply the photoassimilates to all of the growing regions, including buds, floral parts, etc. and this led to the use of thinning in many fruit crops (Untiedt and Blanke, 2001). Since most fruit crops tend to bear extra loads of flowers, the existence of well-thinning techniques is of special importance to getting quality fruit crops every year. Thinning is the practice of removing certain blossoms or clusters of flowers after they have naturally dropped in order to reduce limb breaking, increase fruit size, enrich color and quality, and boost floral initiation for the following year's harvest (Valenzuela, 1992; Westwood, 1993). Accelerating cell division and removing competing blooms and fruits during phase I of the cell-division stage early influences the potential size of the fruit. Time of thinning influences the amount of resource allocation such as sugar, starch, and nitrogen to different growth processes (McAlister and Krober, 1958). Spring frost, which leads to alternate bearing in apples, has become less important mainly due to the development of suitable chemical thinners that regulate the flower production cycle. In general, the severity of the thinning is inversely correlated with the proportion of pollinated flowers that produce fruit (Martin et al., 1961).

15.8 OBJECTIVE OF THINNING

1. To better distribute the crop load so that size increases in crops like apples, pears, plums, peaches, cherries, grapes, and nectarines.
2. To optimize the ripening process by increasing the uniformity of sunshine and air penetration inside the tree.
3. To eliminate the periodicities in the fruiting cycle by enhancing the vegetative growth and bud differentiation, and eliminates the problem of possible alternate (biennial) bearing for several fruit crops such as mango, apple, peaches, etc.
4. To increase fruit quality overall while also lowering the prevalence of pests and diseases.

15.9 METHODS

15.9.1 Hand Thinning

Earlier, it entailed removing fruit blooms from a specific spacing, but today it consists of deliberately eradicating small, weak fruit regardless of spacing with account of the intended level of size thinning (Westwood, 1993). In grapes and pears, hand thinning of floral clusters has been shown to increase fruit size (Einerson and Link, 1976; Messaoudi et al., 2009). The effect of hand thinning on fruit yield has been reported in plums, where it reduced the cumulative yield and diseased fruits; however, it increased in apples (Kosina, 2008). It is expensive, labor-intensive, and time-consuming to thin by hand. Hand-thinning peach fruitlets at 45–50 days is the typical commercial procedure in most peach-producing regions to enhance overall fruit quality and output (Ouma, 2012). In studies on the 'Royal Gala' and 'Braeburn' apple cultivars, the effects of hand thinning on fruit size were examined. Early hand thinning treatments were associated with bigger fruit sizes (McArtney et al., 1996). Despite the fact that hand thinning improves fruit quality, the amount of hand thinning is gradually declining because of labor scarcity and rising labor expenses (Veal et al., 2011). Fruit weight significantly increased with an increase in thinning interval from control to 15 cm, according to the results of the physical/hand thinning's impact on fruit quality metrics (Babu and Yadav, 2004).

15.9.2 Mechanical Thinning

Mechanical thinning can be carried out by employing a power tree shaker, a stiff-bristled brush to "sweep" off part of the young fruits, or a direct discharge of water with high pressure from a hand-operated mister at or just past bloom (Ouma, 2010). These methods are not generally recommended for apples as they can be easily bruised. However, increases in the size of fruit in the apple cultivar Royal Delicious were reported when they were thinned using a tree shaker (Menzies, 1980). Mechanical thinning procedures might be used as a last option as an alternative to chemicals. Mechanical thinning considerably reduced the number of fruits (from 153 to 97 per branch) with a simultaneous increase in fruit growth, according to a report on the influence of thinning on the source-sink relationship (Claudia et al., 2011). Solomakhin and Blanke (2010) studied the effect of mechanical thinning on the fruit quality of apple cultivars Golden Delicious Reinder' and Gala

Mondia' that were mechanically thinned with 30–77 g (300–480 rpm rotation) and 5 or 7.5 km hl vehicle speed and reported a good response on fruit size, firmness, and sweetnesss.

15.9.3 Chemical Thinning

Several chemicals are applied as blossom thinners, and they cause a temporary delay in abscission. For example, auxin, which possesses a gradient between leaf/fruity and spurs, reverses the auxin flow and induces the drop (Hand chack, 1994). Cytokinins, which are associated with cell division (Letham, 1958) cause thinning of fruits by reducing net CO_2 assimilation and net energy available for development and promoting abscission. Even in the absence of thinning, cytokinins increase apple fruit size. In both apples and pears, ABA has been shown to be an effective thinner. (Greene, 2009). Although auxins are also used for chemical thinning, they also have the intrinsic ability to promote flower bud formation (Harley et al., 1958). Generally, it has been speculated that the combined action of auxins and ethylene is mainly associated with the formation of abscission layers and consequent drop (Ouma, 2012). Therefore, auxin analogs such as NAA are used either alone, i.e., on Gala apples (Bertzchinger et al., 1999), or in combination with other fruit thinning agents such as ethyl (Schonherr et al., 2000). Chemical thinning agents work by causing abortion or inhibiting embryo growth, delaying abscission, inhibiting phloem transport to fruit, inhibiting auxin synthesis and transport by seed, and stimulating ethylene synthesis. The cause of abortion in various fruits has been summarized in Table 15.1. Some workers suggested that chemicals that inhibit photosynthesis can also thin apples and peaches (Byers et al., 1996). Among various chemical treatments used in the guava cultivar Shweta, the maximum flower drop was recorded when using ethrel 2,500 ppm followed by NAA 600 ppm (Dhillon et al., 2018). Some insecticidal carbamates, such as oxamyl are also proven to be effective in thinning apples (Marini, 2004). In general, thinning flower buds using chemicals like ethrel and hydrogen cyanamide throughout the winter period (dormant thinning) preceding bloom minimizes manual thinning expenditures while increasing prospective fruit size (Reighard and Byers, 2022).

REFERENCES

Abbott, D.L. 1984. *The Apple Tree: Physiology and Management*. London: Grower Books.

Abeles, F.B., P.W. Morgan, and M.E. Saltveit. 1992. *Ethylene in Plant Biology*, 2nd edn. San Diego, CA: Academic Press, 414pp.

Addicott, F.T. and R.S. Lynch. 1955. Physiology of abscission. *Annual Reviews of Plant Physiology* 6: 211–238.

Almeida, I., I.M. Leite, J.D. Rodrigues and E.O. Ono. 2004. Application of plant growth regulators at pre-harvest for fruit development of 'PERA' oranges. *Brazilan Archieves of Biology and Technology* 47: 658–662.

Ascough, G.D., N. Nogemane, N.P. Mtshali and J.V. Staden. 2005. Flower abscission: environmental control, internal regulation and physiological responses of plants. *South African Journal of Botany* 71: 287–301.

Ashman, T.L. and D.J. Schoen. 1994. How long should flowers live? *Nature* 371: 788–791.

Babu, K.D. and D.S. Yadav. 2004. Physical and chemical thinning of peach in subtropical North Eastern India. *Acta Horticulturae* 662: 327–331.

Becquerel, P. 1907. Sur un cas remarkable d'autonomie depedoncule floral du tabac, provoquee par le traumatisme dela corolle. *Comptes rendus de l'Académie des Sciences* Paris 145:936–937.

Bertschniger, L., W. Stadler, C. Krebs and Pfammoter. 1999. Erhohung der Wirkung SS cher neit Von Ansudungmittern; Er Fahrungen mit Ethephon and ang'e Bter Appli kation stechnik Sweiz. *Zeitschnif Obst – Weinban* 24: 580–583.

Bisht, T.S., L. Rawat, B. Chakraborty and V. Yadav. 2018. A recent advances in use of plant growth regulators (pgrs) in fruit crops – a review. *International Journal of Current Microbiology and Applied Sciences* 7: 1307–1336.

Borochov, A., H. Itzhaki and H. Spiegelstein. 1985. Effect of temperature on ethylene biosynthesis in carnation petals. *Plant Growth Regulation* 3: 159–216.

Burdon, J.N., and R. Sexton.1990. Fruit abscission and ethylene production of red raspberry cultivars. *Scientia Horticulturae* 43: 95–102.

Bushnell, J.W. 1920. The fertility and fruiting habit in Cucurbita. *Proceedings of the American Society for Horticultural* Science 15: 47–51.
Byers, R.E., D.H. Carbaugh, C.N. Presley and T.K. Wolf. 1996. The influence of low light on apple fruit abscission. Journal of the American Society for Horticultural Science 66 7–17.
Byers, R.E. 2003. Flower and fruit thinning and vegetative:fruit balance. In: D.C. Ferree and I.J. Warrington (eds.), *Apples Botany Production and Uses*. Wallingford, UK: CABI Publishing, pp. 409–436.
Byers, R.E., H.D. Carbaugh and N.C. Presley. 1990. The influence of bloom thinning and GA sprays on flower bud numbers and distribution in peach trees. *Journal of Horticulture Sciences* 65: 143–150.
Cameron, A.C. and M.S. Reid. 1983. Use of silver thiosulfate to prevent flower abscission from potted plants. *Scientia Horticulture* 19: 373–378.
Chatterjee, S. 1977. Studies on the abscission of flowers and fruits of cotton (*Gossypium barbadense* L.). *Biologia Plantarum* 19:81–87.
Cochran, H.L. 1936. Some factors influencing growth and fruit setting in the pepper (*Capsicum frutescens* L.). *Cornell Agricultural Experiment Station,* Memoir 190: 1–39.
Claudia, S., L. Damerow and M. Blanke. 2011. Regulation of source: sink relationship, fruit set, fruit growth and fruit quality in European plum (*Prunus domestica* L.)—using thinning for crop load management. *Plant Growth Regulation* 65: 335–341.
Clifford, S. C., and M. Sedgley. 1993. Pistil structure of *Banksia menziesii* R. Br. (Proteaceae) in relation to fertility. *Australian Journal of Botany* 41:481–490.
Dag, A., and S. Gazit. 2000. Mango pollinators in Israel. *Journal of Applied Horticulture* 2:39–43.
Deng, X., S.A. Weinbaum, T.M. DeJong and T.T. Muraoka. 1991. Pistillate flower abortion in 'Serr' walnut associated with reduced carbohydrate and nitrogen concentrations in wood and xylem sap. *Journal of the American Society for Horticultural Science* 116: 291–296.
Dostal, D.L., N.H. Agnew, R.J. Gladon and J.L. Weigle. 1991. Ethylene, stimulated shipping, STS, and AOA affect corolla abscission of New Guinea Impatiens. *HortScience* 26: 47–49.
Dhillon, J.S., R.S. Boora, D.S. Gill and N.K. Arora. 2018. Effect of different chemicals and hand thinning on crop regulation in guava (*Psidium guajava* l.) cv. Shweta. *Agricultural Research Journal* 55: 365–369.
Einerson, B. and G.K.K. Link. 1976. Theophrastus *de Causis Plantarum*. Cambridge, MA: (English translation) Harvard University Press.
Einset, J.W., A. Cheng and H. Elhag. 1979. Citrus tissue culture: regulation of stylar abscission in excised pistils. *Canadian Journal of Botany* 58: 1257–1261.
Erdelska, O. and M. Ovecka. 2004. Senescence of unfertilized flowers in Epiphyllum hybrids. *Biologia Plantarum* 48: 381–388.
Esau, K. 1976. *Anatomy of Seed Plants*. *Soil Science* 90:149.
Gartner, C.F. 1844. Beitrage zur Kenntniss der Befruchtung der volkommeneren GewSchse. I. Theil. Stuttgart: E. *Schweizebart Verlag.*
Greene, D.W. 2009. Effect of abscisic acid on thinning and return bloom of "Bartlett" pears. *HortScience* 44: 1128.
Goldschmidt, E.E. 1980. Abscisic acid in citrus flower organs as related to flower development and function. *Plant and Cell Physiology* 21: 193–195.
Goldway, M., R. Stern, A. Zisovich, A. Raz, G. Sapir, D. Schniederand and R. Nyska. 2012.The selfincomtability fertilization system in Rosaceae: Agricultural and genetic aspects. *Acta* Horticulturae 967: 77.
Goszczynska, D. and N. Zieslin. 1993. Abscission of flower peduncles in rose (*Rosa x hybrida*) plants and evolution of ethylene. *Journal of Plant Physiology* 142: 214–217.
Guinn, G. 1974. Abscission of cotton floral buds and bolls as influenced by factors affecting photosynthesis and respiration. *Crop Science* 14: 291–293.
Halevy, A.H. and S. Mayak. 1981. Senescence and postharvest physiology of cut flowers: part 2. *Horticultural Reviews* 3: 59–143.
Hendriks, L. 1990. Current status of environmental research on potted and bedding plants in Europe with special emphasis on temperature. *Acta Horticulturae* 272: 71–80.
Hendriks, L. 2001. Cultural factors affecting post-harvest quality of potted plants. *Acta Horticulturae* 543: 87–96.
Herbert, S.J. 1979. Density studies on lupins. I. Flower development. *Annals of Botany London* 43: 55–63.
Hands check, M. 1994. Ausdunnung Vol II APfelbaumen mit Kozentrieten Dungern Obstbau 4: 178–181.
Harley, C.P., H.H. Moonand and L.O. Regeimbal. 1958. Evidence that post-bloom apple thinning sprays of naphthalene acetic acid increase blossom-bud formation. *Proceedings of the American Society for Horticultural Science* 72: 52–56.

Henry, E.W., J.G. Valdovinos and T.E. Jensen. 1974. Peroxidases in tobacco abscission zone tissue. II. Time-course of peroxidase activity during ethylene-induced abscission. *Plant Physiology* 54: 192–196.

Heyer, L. 1985. Bud and flower drop in Begonia elatior 'Sirene' caused by ethylene and darkness. *Acta Horticulturae* 167: 387–391.

Israeli, Y. and A. Blumenfeld. 1980. Ethylene production by banana flowers. *HortScience* 15: 187–189.

Jackson, J.E. 2003. *The Biology of Apples and Pear*. Cambridge: Cambridge University Press.

Jawanda, J.S., R. Singh and R.N. Pal. 1974. The effect of growth regulators on floral bud drop in fruit characters of Thomson Seedless grape. *Vitis* 13: 215–221.

Joyce, D.C. 1989. Treatments to prevent flower abscission in Geraldton wax. *HortScience* 24: 391.

Kaska, N. 1989. Bud, flower and fruit drop in citrus and other fruit trees. In: D.J. Osborne and M.B. Jackson (eds.), *Cell Separation in Plants*. Berlin: Springer Verlag, pp. 309–321.

Khandaker, M.M., G. Faruq, M.R. Motior, M. Sofian-Azirun and A.N. Boyce. 2013.The Influence of 1-Triacontanol on the growth, flowering and quality of potted bougainvillea plants (*Bougainvillea glabra* var. Elizabeth angus) under natural conditions. *The Scientific World Journal,* 12: 308651.

Khandaker, M.M., N.S. Idris, S.Z. Ismail, A. Majrashi, A. Alebedi and N. Mat. 2016. Causes and prevention of fruit drop of *Syzygium Samarangense* (Wax Apple): a review. *Advances in Environmental* Biology 10: 112–123.

Konsens, I., M. Oflr and J. Kigel. 1991. The effect of temperature on the production and abscission of flowers and pods in snap bean. *Annals of Botany* 67: 391–399.

Kosina, J. 2008. Response of two apple cultivars to chemical fruit thinning. Proceedings of the XXVII IHC-S9 Endogenous and Exogenous Plant Bioregulators. *Acta Horticulturae* 774: 283–86.

Li, W., R. Yuan, J.K. Burns, L.W. Timmer and K. Chung. 2003. Genes for hormone biosynthesis and regulation are highly expressed in citrus flowers infected with the fungus *Colletotrichum acutatum*, causal agent of postbloom fruit drop. *Journal of the American Society for Horticultural* Science 128: 578–583.

Letham, D.S. 1958. Cultivation of apple-fruit tissue in vitro. *Nature* 182:473–474.

Lewis, L.N. and J.C. Bakhshi. 1968. Protein synthesis in abscission: the distinctiveness of the abscission zone and its response to gibberellic acid and indoleacetic acid. *Plant Physiology* 43: 359–364.

Maria, P., C. Donatella and C. Gennaro. 1997. Susceptibility of 32 walnut varieties to *Gnomonia leptostyla* and *Xanthomonas campestris* pv. juglandis. *Acta Horticulturae* 442: 379–384.

Martin, D., T.L. Lewis and J. Cerny. 1961. Jonathan spot-three factors related to incidence: ruit size, breakdown, and seed numbers. *Australian Journal of Agricultural Research* 12: 1039–1049.

Marini, R.P. 2004. Combinations of ethephon and accel for thinning 'Delicious' apple trees. *Journal of the American Society for Horticultural* Science 129: 175–181.

McAlister, D.F. and O.A. Krober. 1958. Response of soybeans to leaf and pod removal. *Agronomy Journal* 50: 674–677.

McArtney, S., J.W. Palmer and H.W. Adams. 1996. Crop loading studies with 'Royal Gala' and 'Braeburn' apples: effect of time and level of hand thinning. *New Zealand Journal of Crop and Horticultural Science* 24: 4, 401–407, DOI: 10.1080/01140671.1996.9513977

Menzies, A.R. 1980. Timing, selectivity and varietal response to mechanical thinning of apples and pears. *Journal of Horticultural Sciences* 55: 127–131.

Messaoudia, R.E., F.K. Gmili and Y. Helmy. 2009. Effect of pollination, fruit thinning and gibberellic acid application on 'Fuyu' Kaki fruit development. Proceedings of the IVth IS on Persimmon. *Acta* Horticulturae 833: 233–238.

Monterrroso, V.A. and H.C. Wien. 1990. Flower and pod abscission due to heat stress in beans. *Journal of the American Society for Horticultural Science* 115: 631–634.

Nagy Toth, E. and Z.S. Orosz-Kovacs. 2004. Effect of rootstocks on floral and pollination biological types of apple cultivars. *Acta Horticulturae* 636:387–394.

Nautiyal, P., V. Supyal, A.S. Rathore, E. Singh, A. Chaudhary, S. Rayeen, S. Pal, A. Tiwari, S. Khaniya and N. Deo. 2022. Causes and control of flower drop in fruit crops: a review. *The Pharma Innovation Journal* 11: 1165–1168.

Nell, T.A., R.T. Leonard and J.E. Barrett. 1995. Production factors affect the postproduction performance of Poinsettia—a review. *Acta Horticulturae* 405: 132–137.

Osborne, D.J. 1989. Abscission. *CRC Critical Reviews in Plant Sciences* 8, 103–129.

Ouma, G. 2010. *Flowering, Pollination and Fruit Set in Fruit Trees*. Berlin: Lambart Academic Publisher, 138pp.

Ouma, G. 2012. Fruit thinning with specific reference to citrus species : a review. *Agriculture and Biology Journal of North* America 3: 175–191.

Roberts, J.A., C.B. Schindler and G.A. Tucker. 1984. Ethylene-promoted tomato flower abscission and the possible involvement of an inhibitor. *Planta* 160: 164–167.

Racsko, J., G.B. Leite, J.L. Petri, S. Zhongfu, Y. Wang, Z. Szabo, M. Soltesz and J. Nyeki. 2007. Fruit drop: the role of inner agents and environmental factors in the drop of flowers and fruits. *International Journal of Horticultural Science* 13: 13–23.

Rai, N., R. Poerwanto, L.K. Darusman and B.S. Purwoko. 2013. Flower and fruit ABA, IAA and carbohydrate contents in relation to flower and fruit drop on Mangosteen trees. Proceedings of the 4th International Symposium on Tropical and Subtropical Fruits. *Acta Horticulturae* 975: 323–328.

Reid, M.S. 1991. Effects of low temperatures on ornamental plants. *Acta Horticulturae* 298: 215–224.

Reighard, G.L. and B.E. Byers. 2022. Peach thinning. https://www.researchgate.net/publication/268264281.

Rewinkcl-Jansen, M.J.H. 1985. Flower and bud abscission of *Streptocarpus* and the use of ethylene sensitivity inhibitors. *Acta Horticulturae* 181: 419–424.

Saleem, A.B., A.U. Malik, M.A. Pervez and A.S. Khan. 2008. Growth regulators application affects vegetative and reproductive behaviour of 'Blood Red' sweet orange. *Pakistan Journal of Botany* 40: 2115–2125.

Schonherr, J., P. Bauerand and B. Uhlig. 2000. Rates of cuticular penetration of 1-Naphthyl acetic acid (NAA) as affected by adjurants, temperature, humidity and water quality. *Plant Growth Regulation* 31: 61–74.

Singh, Z., A.U. Malik and T.L. Davenport. 2005. Fruit drop in mango. *Horticultural Reviews* 31: 111–153.

Sexton, R, A.E. Porter and S. Littlejohns. 1995. Effects of diazocyclopentadiene (DACP) and silver thiosulphate (STS) on ethylene-regulated abscission of sweet pea flowers (*Lathyrus odoratus* L.). *Annals of Botany* 75: 337–342.

Slpes, D.L. and J.W. Einset. 1982. Role of ethylene in stimulating stylar abscission in pistil explants of lemon. *Physiologia Plantarum* 56: 6–10.

Solomakhin and M.M. Blanke. 2010. Mechanical flower thinning improves the fruit quality of apples. *Journal of the Science of Food and Agriculture* 90: 735–741.

Starkey, K.R. and A.R. Pedersen. 1997. Increased levels of calcium in the nutrient solution improves the postharvest life of potted roses. *American Society for Horticultural Science* 122: 863–868.

Stephenson, A.O. 1979. An evolutionary examination of the floral display of *Catalpa speciosa*. *Evolution* 33: 1200–1209.

Stiling, P., M. Cattell, D.C. Moon, A. Rossi, B.A. Hungate, G. Hymus and B. Drake. 2002. Elevated atmospheric CO2 lowers herbivore abundance, but increases leaf abscission rates. *Global Change Biology* 8: 658–667.

Sun, Y., B. Christensen, L.B.C. Fulai and H.R.M. Wang. 2009. Effects of ethylene and 1-MCP (1-methylcyclopropene) on bud and flower drop in mini Phalaenopsis (orchid) cultivars. *Plant Growth Regulation* 59: 83–91.

Taylor, J.E. and A.C. Whitelaw. 2001. Signals in abscission. *New Phytologist* 151: 323–339.

Thomas, H., H.J. Ougham, C. Wagstaff and A.D. Stead. 2003. Defining senescence and death. *Journal of Experimental Botany* 54: 1127–1132.

Timmer, L.W. and G.E. Brown. 2000. Biology and control of anthracnose diseases of citrus. In: D. Prusky, S. Freeman and M.B. Dickman (eds.), *Colletotrichum*: *Host Specificity, Pathology, and Host-Pathogen Interaction*. St. Paul, MI: APS Press.

Tuan, N.M. and Y. Chung-Ruey. 2013. Effect of gibberellic acid and 2,4-dichlorophenoxyacetic acid on fruit development and fruit quality of wax apple. *International Journal of Biological, Biomolecular, Agricultural, Food and Biotechnological* Engineering 7: 299–305.

Untiedt, R. and M. M. Blanke. 2000. Effects of fruit thinning agents on apple tree canopy photosynthesis and dark respiration. *Plant Growth Regulation* 35: 1–9.

Upreti, K.K., Y.T.N. Reddy, P. Shivu, G.V. Bindu, H.L. Jayaram and S. Rajan. 2013. Hormonal changes in response to paclobutrazol induced early flowering in mango cv. Totapuri. *Scientia* Horticulture 150: 414–418.

Van Overbeek, J. 1959. Auxins. *The Botanical Review* 25: 269–350.

Weaver, R.J. and R.M. Pool 1969. Effect of ethrel, abscisic acid, and a morphactin on flower and berry abscission and shoot growth in *Vitis vinifera*. *Journal of the American Society for Horticultural Science* 94: 474–478.

Weis, K.G., R. Goren, G.C. Martin and B.D. Webster. 1988. Leaf and inflorescence abscission in olive. I. Regulation by ethylene and ethephon. *Botanical Gazette* 149: 391–397.

Westwood, M.N. 1993. Temperate Zone Pomology. Physiology and Culture. 3rd edition. Timber Press Inc. Portland. Oregon 523.

Wien, H.C. and Y. Zhang. 1991. Prevention of flower abscission in bell pepper. *Journal of the American Society for Horticultural Science* 116: 516–519.

Williamson, M.J. 1966. Premature abscissions and white oak acorn crops. *Forest Science* 12: 19–21.

Wyatt, R. 1980. The reproductive biology of *Asclepias tuberosa*: I. Flower number, arrangement, and fruit-set. *New* Phytologist 85: 119–131.

Valenzuela, J.R.C. 1992. Regulating blue berry (V. Ahei) crop Ph.D Dissertation, Mississippi, State Univeristy.

Van Doorn, W.G. 2001. Categories of petal senescence and abscission: a re-evaluation. *Annals of Botany* 87: 447–456.

Van Doom, W.G. and A.D. Stead. 1997. Abscission of flowers and floral parts. *Journal of Experimental Botany* 48(309): 821–837.

Veal, D., L. Damerow and M.M. Blanke. 2011. Selective mechanical thinning to regulate fruit set, improve quality and overcome alternate bearing in fruit crops. *Acta Horticulturae* 903: 775–782.

Von Mohl, H. 1860. Ober den AblOsungsprozess saftiger Pflanzenorgane. *Botanische Zeitung* 18: 273–277.

Zieslin, N. and V. Gottesman. 1983. Involvement of ethylene in abscission of flowers and petals of *Leptospermum scoparium*. *Physiologia Plantarum* 58: 114–118.

16 Fruit Drop and Parthenocarpy

Vikrant Patiyal, Neerja Rana,
Vishal Singh Rana, and Sunny Sharma

16.1 INTRODUCTION

Fruit drop is the process in which the separation of a developed fruit from a tree or a plant is caused by the formation of a separation of layers of cells on the fruit stalk due to a series of physiological and biochemical events. Natural fruit drop occurs at periodic intervals between fertilization and maturity. The extent of fruit drop depends on the fruit set's intensity and associated biochemical and physiological alterations during the active developmental season. The period of fruit drop is interdependent. In phase one of the drop, several flowers that were not pollinated are shed together with the petals of the fertilized fading flowers. The second wave is less conspicuous, followed by the third and final drop, i.e., the pre-harvest drop.

Racsko et al. (2006) emphasized the importance of studies concerning the dynamics of fruit drop during production:

- procedure, to regulate yield by interventions (enhancing fruit set, prevention of fruit drop), should be timed accurately;
- as to predict the volume of the harvest expected;
- if the cause of fruit drop were recognized, the prevention of the loss could be attempted by agricultural means such as pruning, irrigation, and nutrition to shape the rate of fruit set according to the conditions and capacity of the tree.

Fruit drop occurs in multiple waves but three main periods are the most critical ones. They are recognized as:

1. At the end of the bloom (cleaning drop)
2. June drop
3. Pre-harvest drop

Physiologically cleaning drop occurs after the beginning of the endosperm differentiation when the fruit primordium cannot furnish further amounts of resources across the fruit stem to the tree at the cost of its growth. The cleaning fruit drop has both a positive and negative relationship with the rate of the fertilized flowers (negatively) or the number of flowers being pollinated (positively) – whereas the following June drop depends on the loading capacity of the trees. As a rule, the first period of fruit drop is the most intense and the faded petals of the flower shed within 1.8 weeks. Fruit growers call it the "cleaning drop" of the trees. The flowers not being fertilized are shed or by other reasons of sterility and do not bear viable seed primordia. Suranyi and Molnar (1981) assert that the malfunctioning fruit primordia and insufficiently fertilized are also susceptible to being shed during this phase. Nevertheless, when there is a low initial density of flowers, either due to a frost spell or intentional removal of flowers, the fertilized flowers that remain can be preserved and develop into mature fruit. As a result, the occurrence of fruit drop during the early phase can be effectively avoided (Thompson, 1996).

In pome fruit species, the flowers shed that were unable to grow; in apricot, it takes 1 or 2 weeks after bloom (Suranyi and Molnar,1981), whereas, in sour and sweet cherries, 1.5–2.5 weeks may

DOI: 10.1201/9781003354055-16

elapse (Thompson, 1996). In the latter species, the cleaning fruit drop was divided into two phases: the first lasts after the shed of petals 1.5–4 weeks long, and the second between 4–6 weeks. The two phases overlap each other generally. At that time, the fruitlets attained a diameter of 4–5 mm. Racsko et al. (2006) called that period in sour cherry "initial" fruit set. The most critical period of fruit drop is this cleaning drop after a shed of petals. The inhibition of this type of fruit drop is rather difficult to accomplish by chemical means (e.g., by inducing parthenocarpy), more promising is considered the enhancement of the rate of fruit set (Suranyi and Molnar, 1981).

Whereas in the apple variety 'Golden Delicious', it was observed that the more developed the buds and flowers of the tree, the lower the intensity of the cleaning fruit drop (Racsko et al., 2006). Considering that the differentiation of fruit primordia precedes the drop by several days, the loss is decided already at the time of cell division. Consequently, we may conclude that the drop was not triggered by the development of the embryo, but rather stopped by it. Most likely, the mechanism coordinates the processes of development in favor of the remaining fruits, where the most energy-consuming embryo-development occurs in the fruits, which the tree is able to maintain up to maturity (Suranyi and Molnar, 1981).

The second phase of the fruit drop occurs in pomaceous fruits in the phase occurring 6–8th weeks after bloom, which means in the northern hemisphere in the month of June. During this period, the embryos grow vigorously and consume the endosperm tissue, which is responsible for the synthesis of auxin. Consequently the reduced auxin production leads to fruit drop, as shown in Figure 16.1. Meanwhile, the growth of embryos is completed, and a secondary endosperm is formed, which synthesizes a lot of auxins again. That is the reason for the end of June drop.

According to Racsko et al. (2006), the main cause of June drop is the difficulty in the translocation of nutrients, but at this stage of fruit development, the requirement of organic substances is culminating. In addition to the weather conditions also take part in the processes but at a lower significance. The drop is the consequence fully insufficient nutritive capacity for fully developing

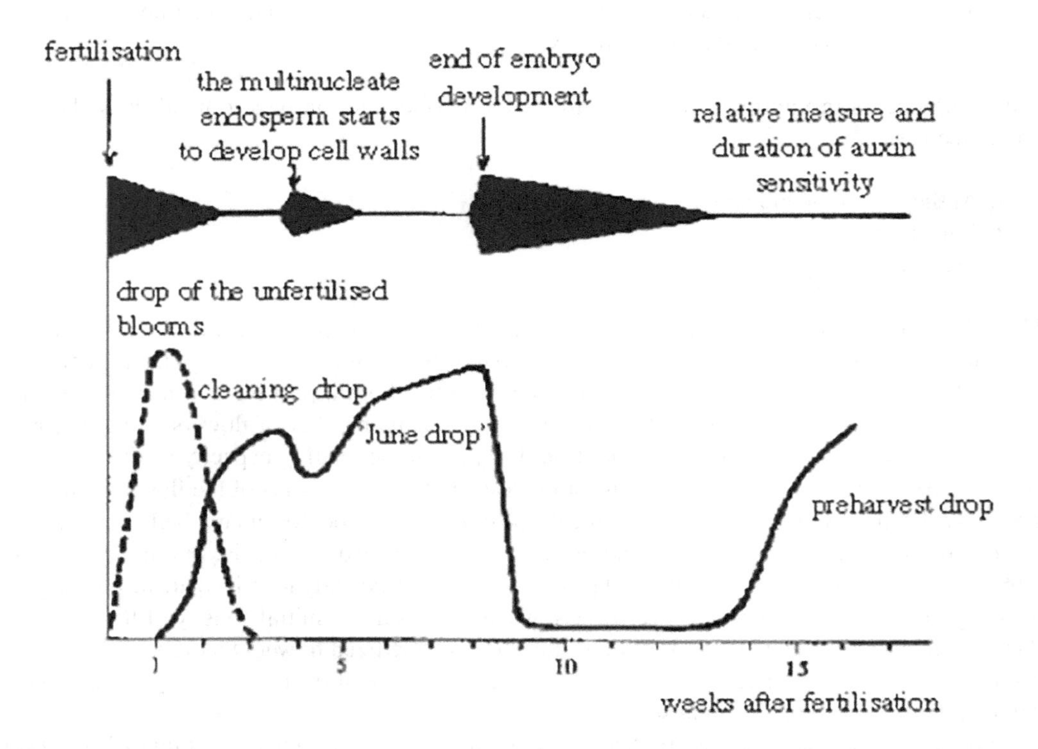

FIGURE 16.1 Relationship between hormonal activity and Fruit drop (Source: Luckwill, 1957 in: Westwood, 1993).

all fruits set after bloom. Consequently, an abundant June drop is expected every time when the leaves of the tree are damaged by frost, disease, or pest and cannot photosynthesize the due amount of organic matter. Preceding the June drop, the abscission started with the physiological processes. The dropped fruits display completely developed abscission layers, which started to form earlier as an irreversible process, but at a speed depending on the season. Therefore, the severity of the June drop is a yearly variable.

The first period of fruit drop ensues in the case of pecan nuts 14 days after pollination and lasts for 45 days at most (Smith and Romberg, 1941).

In some stone fruit varieties, the ripe fruits may drop already around the end of May. Sweet and sour cherry, the second period of fruit drop may ensue earlier than June, 6 weeks after bloom, whereas their third fruit drop finishes in June. In sweet and sour cherries, the third period of fruit drop is also called "red drop", which means that the fruits dropped started to ripen (Nyeki, 1978). More than 90% of those fruits contained an embryo, but their size was smaller than average, which means some kind of defeat (Thompson, 1996).

After June drops, the fruits begin to grow rapidly, which indicates that the auxin flow synthesized in the fruit is now slowed. As a result, many varieties face the prospect of a third phase of fruit drop. Pre-harvest fruit drop is the term for this process which uses the secondary endosperm as a source of growth factor. It is brought about by biochemical changes at the base of the fruit stem, but no cell division results in the formation of an abscission layer. As expressed by its designation, the third phase ensues near the terms of harvest. If the fruit is technically already ripe, significant financial losses could result. The ecological risks of shaking off the fruits that are only weakly anchored are impending even with cautious treatments, making it impossible to evaluate the likelihood of fruits becoming detached during this phase.

Pre-harvest fruit drop in sour cherries is uncommon and only occurs when the fruit load is abnormally high (Soltesz et al., 2003). Depending on the type and variety of the fruit, the wind might only cause pre-harvest decline (Way, 1973). Apple, pear, plum, peach, and black currant are in decreasing order of vulnerability and the most exposed species, although cherry is less vulnerable (Stosser, 2002). High fruit density per inflorescence favors pre-harvest fruit drop. The fruits may kick one another as a result of growth when the fruit stem is short (peach, apricot, and some apple kinds), mostly as other members of the same inflorescence (Soltesz, 1997). According to Suranyi and Molnar, the third stage of fruit drop is known as the "green drop" in apricots (1981). Environmental factors, as well as illnesses and pests, are acknowledged as causes. The reason for that drop in pecan, which is less significant than the first two or three drops and is brought on by embryo abortion, is clear from the shriveled and discolored tissues found inside the shell. This decline in fruit could potentially be caused by auto- or inter-incompatibility (Wood, 2000).

The fourth period of fruit drop, which occurs practically after the optimal harvest time, is rarely experienced by the fruit grower. In apples, the ethylene 5 level in the AZ tissues increases excessively after full physiological maturity leading to detachment of the fruit. Because ethylene production does not begin and is completely absent in the AZ tissues, the lime fruit remains attached to the tree for up to a month. The gooseberry is prone to dropping overripe fruits, and detached fruits retain their stem (Papp, 1984).

Unripe (green) gooseberry fruits are never dropped. Raspberries mature quickly. In fruit species with little fruit drop, the expected crop, or number of fruits, can be easily predicted after bloom, posing only the risk of unanticipated damages and technological errors (Soltesz et al., 2003). Close to harvest time, the auxin flow abruptly decreases and the abscission layer forms in apple varieties 'Berlepsch' or 'Goldenparmane,' and even a light breeze can cause heavy losses. Others, such as 'Golden Delicious' and 'Landsberger Renet,' produce auxin continuously, so picking the fruit requires a lot of energy. The detachment of fully ripe fruits is a natural phenomenon and should not be regarded as an anomaly; it is an expression of senescence, similar to that of the leaves in autumn (Soltesz, 1997).

16.2 MORPHOLOGICAL AND ANATOMICAL CHANGES

a. Formation of abscission zone at the junction of abscising organs (flower, fruit) and the parent plant body.
b. Swelling or constriction of abscission zones w. r. t other parts of the pedicle.
c. Abscission zone is 1–2 rows or up to 15 or more cell tier thickness.
d. Cells are parenchymatous, devoid of lignin and suberin, dense protoplasm, more starch, small intercellular spaces, and highly branched plasmodesmata.
e. Stele usually divides into separate bundles before it enters the zone.
f. Dissolution of middle lamella resulting in loosening of cells.
g. Weakening and breaking of vascular connections between leaves, fruits, flowers, and the parent plant.
h. Development of periderm to protect the exposed surface and cells become suberized.
i. Formation of tyloses (bladder like protrusions from xylem parenchyma cells), which block xylem vessels.

16.3 BIOCHEMICAL CHANGES

- Enhanced production of hydrolytic enzymes- cellulase, pectic enzymes, lignase which cause dissolution of the middle lamella and primary walls of the cells in the abscission zone.
- Polygalactouranase leading to the breakdown of the middle lamella.
- Ca pectate in middle lamella is hydrolyzed to pectic acid and pectate.
- Cell expansion by cellulase in the abscission zone.
- Hydrolysis of lignin by lignases.
- Auxin destruction by increased peroxidase activity.
- Active synthesis of proteins and RNA in the abscission zone.

Besides the above changes, many other processes take place to prepare the plant for this event. For instance, the vascular elements that service the organ are occluded, and the exposed fracture surfaces are protected from water loss and invasion by pathogens.

16.4 ABSCISSION ZONES

Abscission zones are present at the lower part of the fruit's pedicle and at the lower part of the leaflets. Fruits break down through various biological processes that induce the weakening of cell walls and the separation of cells from one another. The cells in the abscission zone are often smaller, more tightly packed, devoid of intercellular gaps, contain less lignin, and have gone through a longer period of cell division compared to the surrounding cells. The cells in this zone undergo more cell divisions to prepare them for subsequent abscission processes.

16.5 WALL WEAKENING

The process of abscission is triggered by growth regulator signaling that triggers physiological modifications. Cells in the abscission zone release pectinase and cellulase, which are enzymes that break down the cell walls. These enzymes weaken the structural integrity of the middle lamella and main wall that separate cells. In the abscission zone, the middle lamella, which serves as an adhesive between cells, starts to undergo dissolution. Simultaneously, the major walls in the vicinity start to expand due to alterations in the chemical makeup.

16.6 WALL CHANGES

When the connections between cell walls are loosened, the pressure of water within cells with thin walls (known as turgor pressure in parenchyma) leads these cells to increase in size. As cells enlarge, they produce shear forces by exerting pressure and tension on adjacent weaker walls. From a mechanical standpoint, fracture lines start to form between the walls of cells. The separation of cell walls can be counteracted by the deposition of obstructive substances and defensive compounds, resulting in the closure of an open wound. A robust and resilient barrier is established to safeguard the surviving tree tissues from the elements and potential threats from pests. Tyloses, suberin, lignin, and other protective substances are formed and placed on the side of the tree where the abscission zone is located.

16.7 REGULATION OF FRUIT DROP

Auxin, a key growth regulator, gets synthesized in the fruit, and then transferred gradually to the stem base via living cells. As long as auxin is efficiently delivered across the abscission zone, the cells in the abscission zone remain unresponsive. During the autumn season, as the production of auxin starts to drop and the transport rates of auxin start to decrease due to reduced availability, damage to the cells responsible for transporting auxin, and/or increased infection of live tissues by pests, the initiation of cell wall modifications occurs. The cell wall modifications progressively impede the movement of auxin and expedite the synthesis of ethylene. Minimal quantities of ethylene accelerate the process of abscission zone formation. Abscisic acid partially induces leaf dormancy by promoting ethylene synthesis and inhibiting auxin transportation. Throughout abscission, the levels of Indigenous auxin and cytokinin experience a significant reduction, while the levels of ABA and ethylene undergo a rise (Table 16.1).

16.7.1 Biological Background of Flower and Fruit Drop (Racskó et al., 2006): Physiological and Hormonal Bases:

The ovary begins to expand rapidly as a result of fertilization, which often occurs right after pollination and marks the start of fruit growth. In the meantime, the perianth's stamens, stigmata, and petals begin to fade and abscise. Following successful fertilization, the flower will undergo early modifications that indicate a fruit set (Petho, 1993). In some instances, the ovary's expansion may begin simply through pollination. Without fertilization, it is anticipated that the ovary will degenerate, followed by the flower's death and abscission. The same phenomenon occurs in fruit species with unisexual blooms, such as walnuts, where the male flowers start to deteriorate (become senescent) as soon as the pollen grains are released. An excessive amount of pollen adhering to the stigmata, where ethylene is formed, has been suggested as the cause of the strange phenomena of

TABLE 16.1
Effect of PGR on Preharvest Fruit Drops

Fruit Crop	Dose Concentration	Time of Application
In apple cv. Golden Delicious	NAA, 2,4-D at 10–20 ppm or 2,4,5-T at 20–40 ppm	one month before harvest
Plum cv. Satluj Purple	NAA @10 ppm	2nd and 4th week of April after the pit hardening stage
	Ethrel @100 ppm	4th week of March after the pit hardening stage
prunes	2,4,5-T @15–20 ppm	2 weeks after the beginning of the pit-hardening stage

the abortion of female walnut blooms (Soltesz et al., 2003). Following the flower drop, fruit drop that occurs before maturity is attributed to the breakdown of the "hormonal balance" in the developing fruits, where the growth substances (auxins, cytokinins, and gibberellins) that were active in favor of growth lost their influence against the abscisic acid (ABA) that causes abscission. The involvement of ABA was first demonstrated by Davis and Addicott (1972), and it became more prominent when the immature fruitlets dropped. A second maximum has since been noted when the fruits are almost ripe. The ABA content also varied amongst cultivars that were prone to fruit drop. Several researchers have provided evidence for the importance of ethylene in fruit abscission. IAA mostly inhibits ethylene, which is the primary controller of abscission. Ethylene induces abscission by disintegrating the central lamella of tissues, while the abscisic tissues at the bottom of the peduncle break down the cell walls of the fruit stem. Research has shown that ethylene enhances the process of de novo synthesis of the cellulase enzyme and its impact on the cell wall (Abeles, 1973). The abscisic tissue's cells are the ethylene effect's target cells. By acting locally, IAA causes the cells of a thin layer to enlarge (and soften) and separates the fruit foundation by a shearing effect. Fruit drop, according to Abeles (1973) and Abeles and Leather (1971), is caused by an increase in the cell layer's susceptibility to ethylene rather than the synthesis of ethylene (the abscisic tissue used to consist of 2–5 layers). According to Jackson and Osborne (1970), the ethylene content was rising, which predicted the natural abscission. The level of ABA (Talon et al., 1990) is increasing the frequency of abscission. Premature abscission of fruits is determined by the relative concentrations of IAA and Following bloom. Apple (Vernieri et al., 1992), litchi (Yuan and Huang, 1988), citrus (Talon et al., 1990), and cotton have all demonstrated the role of ABA in abscission. The most likely reason for the rising ABA level is a response to any stressor (drought, troubles with nutrient supply, etc.). Pozo (2001) demonstrated the combined effects of ABA and inhibitory substances like jasmonic acid on the development of citrus fruit abscission tissue as a complex fruit drop signal.

16.8 PARTHENOCARPY

Parthenocarpy is the process of developing fruit without pollination and fertilization. In the absence of pollination or any other external trigger, it is characterized as the development of seedless fruits. The phrase was first used by Noll (1902).

Parthenocarpy, in the disciplines of biology and horticulture, refers to a process of fruit development that occurs without the need for fertilization of the ovule, either by natural means or as a result of artificial manipulation. Consequently, the fruit lacks seeds. Parthenocarpy is an occasional natural mutation, although it is often regarded as a defect because it prevents sexual reproduction in plants while allowing for asexual propagation.

The ovule's fertilization often initiates the ovary's growth into a fruit (Nancy, 2015). Over fifty species, including apple, pear, grape, citrus, breadfruit, date, fig, banana, and pineapple, were classified by Gustafson (1942) as having parthenocarpy. These are usually seedless. Fruits that are parthenocarpic, are typically seedless, however this is not always the case. This is because seed lessness can be the result of embryo abortion brought on by the embryo's failure to acquire the requisite nutritional reserves, which can lead the embryo to abort. The phenomena of parthenocarpy can be viewed as the culmination of a process in which fruit growth becomes more and more independent of seed development. Some fruits rely solely on their seeds for fruit development. In strawberries, removal of the achenes (commonly called seeds) at any stage of development leads to the cessation of fruit growth. There is no free auxin in the receptacle tissue. Contrarily, the Washington Navy orange and Black Corinth grape grow despite not being fertilized, without seeds, and being entirely independent of them. Many seedless horticultural varieties of fruits are exclusively parthenocarpic. Many seeded fruits are not necessarily capable of parthenocarpy and even some single seeded fruits can be parthenocarpically set. The seeds from parthenocarpic fruits are not fully developed and they do not germinate.

16.9 TYPES

Parthenocarpy is broadly of two types

1. Natural means or genetic means
2. Artificial means (induced)

16.9.1 Natural Type of Parthenocarpy

It has been observed in several crops, such as grapes, tomatoes, mandarin, and bananas. It is further categorized into the following classes:

a. **Obligatory parthenocarpy**: Parthenocarpy caused by the suppression of pollination. Seedless fruits are consistently produced as a result.
b. **Facultative parthenocarpy**: This phenomenon arises from unfavorable conditions for pollination or fertilization, or from genetic sterility caused by continual vegetative proliferation. Examples of plants that exhibit facultative parthenocarpy include banana and pineapple.
c. **Vegetative or autonomic parthenocarpy**: Parthenocarpic fruit can be produced without the need for pollination or any other external stimulation. For example, the Washington Navel orange, Oriental persimmon, and cucumber.
d. **Stimulative parthenocarpy**: Parthenocarpy necessitates pollination or some other form of stimulation. Stimulative parthenocarpy, such as in bananas, occurs when a plant is triploid, meaning it has three sets of chromosomes due to a diploid and a tetraploid parent. As a result, these plants are unable to generate seeds. Other examples of plants that exhibit this trait are the Black Corinth grape, blackberries, and pears.
e. **Stenospermocarpy**: It refers to a phenomenon where pollination and fertilization take place, but the embryo is subsequently aborted, resulting in seedlessness. This can be observed in certain fruits, such as grapes.

16.9.2 Artificial Parthenocarpy

Seedless fruits are produced by inducing the flower with substances such as dead pollen, pollen extract, substances, or growth agents. Applying plant hormones to flowers consistently stimulates the development of parthenocarpic fruits. The auxin necessary for fruit development, typically provided by fertilized ovules, can sometimes be increased beyond the minimum level needed for fruit development due to extreme temperatures, frost, insect infestation, bark removal, and mechanical irritation to the stigma and style. The occurrence of frost and low temperatures leads to the development of parthenocarpic fruits, such as apples and pears. Certain apple cultivars may produce little, seedless apples as a consequence of insect activity. Bark ringing has been used to induce parthenocarpy in apples, grapes, and gooseberries. Several citrus fruits undergo seed loss when the style is severed at the base before the pollen tube reaches the ovary. Chemicals can be used to promote the development of fully-grown fruits. Auxins such as IAA, IPA, NAA, and PAA have been used to stimulate parthenocarpy in fruit crops (Table 16.2).

16.10 IMPORTANCE/ADVANTAGES OF PARTHENOCARPY IN FRUIT CROPS

1. This trait is valued in edible fruits with tough seeds, such as grapefruit, pineapple, banana, and orange, and is beneficial for crops that are difficult to pollinate or fertilize.
2. Parthenocarpy boosts fruit output in dioecious species like persimmon by eliminating the requirement to plant staminate trees for pollen.

TABLE 16.2
Parthenocarpy with Different Applications of Hormones

Fruit Crop	Response of Crop	Reference
Loquat	GA_3 @ 200 ppm+CPPU @ 20 ppm twice, at flowering time and between 27 and 58 days after the first treatment resulted in seedless fruits.	1. Yahata et al., 2006
Date	GA @ 25–100 ppm, once at spathe opening, and then two sprays at 4-week intervals followed by etherel @ 500 ppm spray resulted in seedless fruits of normal size	Abd-Alaal et al. (1982)
Grape	GA3 @ 100, 200, or 300 mg/L in grape cv. Triumph during flowering produced more than 20% seedless berries	Lu et al. (1997)
Cherry	NAA @ 100 ppm+GA_3 @ 5–25 ppm increased parthenocarpic fruit set.	Bhat et al. (2012)
Apple	(NOXA 500+GA_3 200+DPU 300 ppm) irrespective of the time of application (petal fall, full bloom, or 50% bloom) resulted in parthenocarpic fruits	Fortes and Petri (1983)
	NAA or Carbaryl (at bloom stage) Decrease fruit set Increase yield	Komzik (2004)
Peach	Dormex, Paclobutrazol, KNO_3 Reduce fruit set`	George and Nissan (1993)

3. Parthenocarpy makes fruits like cherries, grapes, and strawberries more appealing and waste-free to eat.
4. Enhances taste and palatability.
5. Enhances processing quality and boosts revenue in many fruit crops.
6. Extends shelf life by reducing seed ethylene production.
7. Boosts labor efficiency in food processing.
8. Environmental conditions that hinder pollination and fertilization have less impact on fruit yield.
9. Many parthenocarpic plants begin fruit set and growth before anthesis, allowing them to begin fruit production and harvest.
10. Parthenocarpy enhances fruit quality and productivity in horticulture plants valued for their fruit.

16.11 PHYSIOLOGICAL BASIS OF PARTHENOCARPY

The physiological basis of parthenocarpy is still not fully understood. It occurs most commonly among fruits with large numbers of ovules, suggesting that ovules may provide some of the chemical constituents that stimulate fruit set and fruit growth. The ovules are sites of auxin synthesis in some fruits (Nitsch, 1952), and the fact that auxin application can bring about fruit sets enhances the possibility that auxins are critical in fruit sets. However, many fruits do not respond to auxins, so there must be further components of fruit set stimulus in the plants . It has been observed that the factors favoring fruit set come from the roots of fruit trees may be interpreted and imply a role of gibberellins and cytokinins. A relationship between auxin content and the natural inclination towards seedlessness was established by Gustafson (1942) while examining the ovaries of seeded and seedless species and found that the seedless varieties have appreciably higher auxin content than the seeded varieties in citrus fruits and grapes and parthenocarpy may represent a state of auxin autotrophy in the ovary (Luckwill, 1957). Crane (1965) was able to induce parthenocarpy with auxin, gibberellin, or cytokinin and suggested their role in triggering or mobilization activities in the fruit (Figure 16.2).

Obtaining parthenocarpic fruits is a physiologically difficult process. Fruit development includes early and mature stages (Figure 16.3). Most plants have three stages of early fruit development: fruit setting, cell division, and cell expansion. The ovary decides whether to abort or continue fruit development in the first phase. Fruit size increases slowly during cell division when the cells that

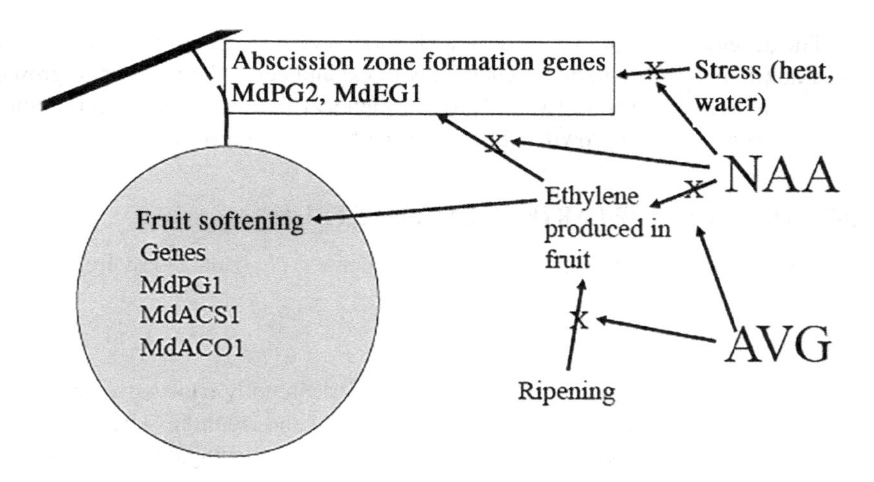

FIGURE 16.2 Proposed Model of Abscission (Rongcai Yuan, 2008).

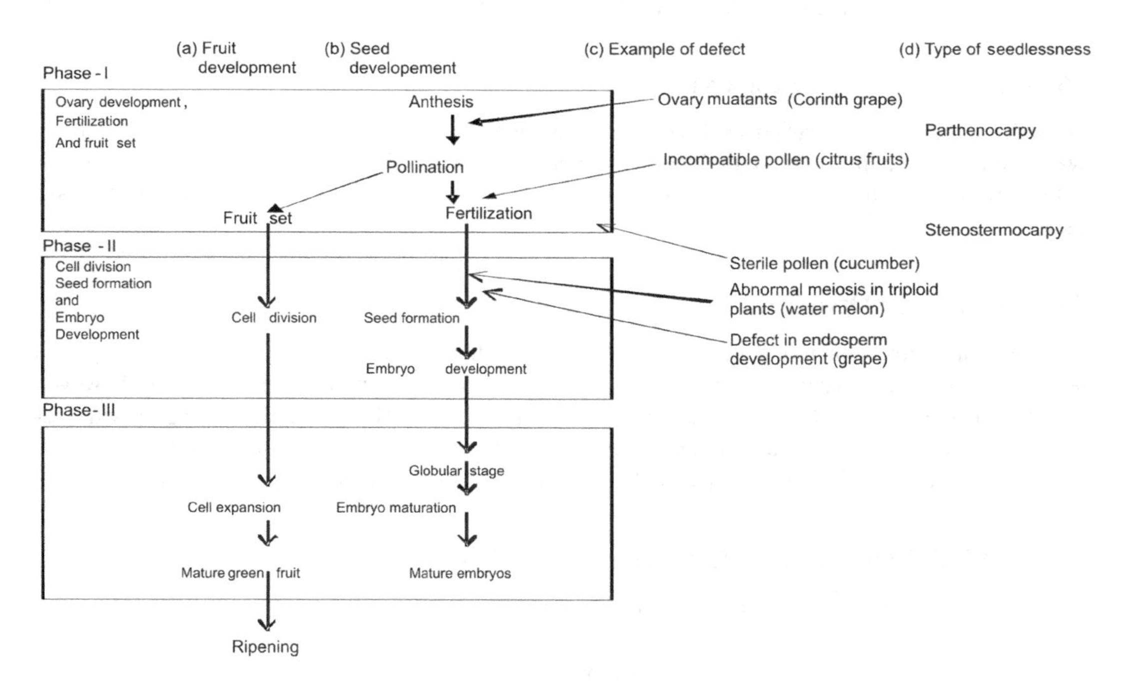

FIGURE 16.3 Fruit and seed development diagram: (a) fruit development, (b) seed emergence, and (c) parthenocarpic fruit. Events before or linked to seed development benefit fruit development (arrows between (a) and (b)). Arrows connecting (b) and (c) illustrate seed production points that seedless variants may lack.

divide are small and densely compressed, and the quantity of cells determines fruit size. After cell division stops, the fruit expands by increasing cell volume till it reaches its full size. Cell expansion frequently boosts the size of fruit by 100-fold, which correlates most to fruit size. At the end of the initial growth and development, an undeveloped fruit with mature fruit size is produced, and maturity begins.

Early fruit development includes numerous sporophyte-gametophyte pathways for communication. Pollination and fertilization influence fruit set. Exogenous gibberellins can increase the amount of auxin in the ovary of an unpollinated flower, triggering fruit setting in the absence of fertilization. The developing embryo controls the division of cells in the surrounding fruit tissue. As seeds produce auxin and other unknown chemicals to expand cells, the number of growing seeds affects fruit size and weight (Nitsch, 1952). Thus, parthenocarpic fruit production is complex since a growing seed is crucial to fruit growth. The existence of seedless crop plants dates back centuries.

16.12 DEVELOPMENT OF PARTHENOCARPIC FRUIT

Three aspects are important in the development of parthenocarpic fruit. These are:

16.12.1 Ovary Wall Development

Gardner and Marth (1939) found that parthenocarpic American holly fruits developed using indole acetic acid appeared the same as pollinated fruit at maturity and ripening. The pistillate structure of fertilized and seedless fruits was identical under a microscope during early development. There was no disordered cell proliferation like in bean plants treated with indole acetic acid (Kraus et al., 1936). Spraying holly pistils with water solutions of this growth agent causes multiple outgrowths and no unregulated cell division. Without seeds fruit produced by growth material and pollinated fruit had the same appearance, feel, taste, and carpel wall chemistry.

16.12.2 Seed Coat Development

Growth chemicals cause varied seed coat development and normal or almost normal carpel or ovary wall development in most fruits. In 80% or more of growth-set fruits, the seed coat does not develop. Why certain fruits have seed coverings, and others don't is unknown. Closely related plants vary greatly.

16.12.3 Embryo Development

Embryos have not been found in fruit set by growth chemicals, which is expected since pollen fertilizes eggs. Plants have many cases of embryo development without egg fertilization; however, growth substances cannot intimate embryos without pollination. Horticulturally, using growth factors as a substitute for pollination does not increase yield in crops where the seed (embryo) is used, such as dry cherries, though it is unclear if abscission is a primary cause of poor fruit set.

16.13 FACTORS AFFECTING PARTHENOCARPY

The factors that influence or favor the parthenocarpy in fruit crops are:

1. Environmental factors (high or low temperature).
2. Growth regulators.
3. Genetic disorder (chromosome imbalance).
4. Self-incompatibility.
5. Genetic factors like gene controlling meiosis.

16.13.1 Environmental Factors

Light, temperature, and humidity affect pollen maturation and fertilization. Poor environmental conditions hinder fertilization and fruit development. Parthenocarpic plants can grow seedless fruit under these unfavorable conditions. Parthenocarpy can increase winter and early fruit production (Acciarri et al., 2002), enabling year-round fresh horticulture products. A major challenge with facultative parthenocarpy—a genetically controlled state that produces seeds and/or seedless fruits on the same plants—is recreating the optimal environment for parthenocarpic traits. The extent of facultative parthenocarpy is affected by abiotic conditions that impact pollination, fecundity, and fruit development. Temperatures above 30°C decrease pollen viability across the majority of crops, increasing the degree of parthenocarpy. Low temperatures of less than 10°C in January-March tend to induce the abortion of growing and developing embryos that are needed to grow healthy fruits in mango cv. Haden.

16.13.2 Growth Regulators

In certain crop species, auxins, gibberellins, and cytokinin combinations induce fruit development without fertilization (Tiziana, 2009). Many physiological and genetic factors affect fruit development. Many studies have revealed that phytohormones regulate fruit set, development, and maturation at the transcriptome level in a complicated fashion. The growth regulators, including synthetic auxin, induce parthenocarpy in many fruits and vegetables. Unfortunately, this leads to higher costs of production and fruit imperfections. The genetic parthenocarpic traits, which are being introduced into berries like bananas and grapes, may mitigate these effects. Genetic parthenocarpy can be created by purposely activating ovary- or ovule-specific exogenous genes. Auxin over sensitization or accumulation in carpel tissue before anthesis activates these promoters (Carmi et al., 2003). For well-developed grape berries, GA is used. Cell growth in mesocarp tissue improves berry size and lowers seed traces. The best GA concentration and time vary by cultivar. Auxin, GAs, and cytokinins exogenously cause parthenocarpy in different plant species. Auxin and GAs are abundant in parthenocarpic citrus. GA induces cell growth with limited division, while auxin divides cells. Sprays of NAA on pineapple plants following floral primordia differentiation fully affected berries, peduncle, and axillary branch growth. Although no treatment affected fruitlet quantity, large doses of growth-regulating chemicals increased ripe fruit size and weight. This fruit growth was accompanied by huge peduncles that originated the fruit was difficult to detach. The highest quantity of NAA delayed ripening and somewhat suppressed slips and suckers, especially in the middle of the plant. Applying relatively modest amounts simultaneously stimulated latent buds, resulting in more slips per plant.

Auxin stimulates parthenocarpy, and pollen and ovary may stimulate fruit set. Pollination increases auxin synthesis in the ovary, which is low in unpollinated ovaries. The tip produces the most auxin, then the style and ovary. Some citrus and horticulture plants have high auxin concentrations for natural parthenocarpic fruit growth. Several horticultural species set fruit with GA1 or GA3 (Dorcey et al., 2009). In pollinated ovaries, gibberellin levels and GA biosynthetic gene expression are higher (Serrani et al., 2007). They set fruit well. Gibberellins cause parthenocarpy and apple fruit maturity (Molesini et al., 2009). This significant fruit-setting ability and the fact that asymmetric apple fruit growth is connected to inadequate seed development have prompted studies on endogenous GA in the endosperm. Immature apple seeds contain GA4 and GA7. Gibberellins may have a role in parthenocarpic fruit growth and development, however, GA4 and GA7 induce parthenocarpy more than GA7 in immature apple fruit seeds (Wang et al., 2009). Gibberellins' presence in additional fruit tissue and role in fruit growth is unknown. This causes parthenocarpy in various fruits. Gibberellin-induced parthenocarpy in Prunus species was first shown by Crane (1964), who showed that the presence of potassium salt of GA3 in a solution of water caused parthenocarpic fruit in peach more than almond and apricot but not cherry or plum.

Caixi et al. (2008) examined the effects of gibberellins (GAs), GA1, GA3, GA4, and GA7 with a cytokinin, CPPU, and IAA on fruit set, parthenogenesis induction, and fruit expansion in Rosaceae species like Japanese pear cv. Akibae (self-compatible) and cv. Iwate Yamanashi (seedless cultivar Also studied were Pyrus communis, Chaenomeles sinensis, Cydonia oblonga, and Malus pumila. GA4, GA7, and CPPU cause parthenocarpic fruit growth in Japanese pear; however, GA1, GA3, and IAA do not. GA4 and GA7-induced parthenocarpic fruits were smaller, harder, and ripened faster than pollinated and CPPU-induced fruits. GA4 and GA7-induced fruits had longer pedicels, a higher fruit shape index, and a small calyx protrusion. In three other Rosaceae species, CPPU, GA4, and GA7 alone or in combination with uniconazole caused parthenogenesis, but the fruit set was very low. GA1 did not promote fruit cell proliferation like the other bioactive GAs, and GA4 and GA7 were more effective than GA3 and GA1.

Cytokinins help various fruit crops develop. Cytokines applied to flowers prior to fertilization start fruit growth in some species. Auxins, GAs, and cytokinins accumulate after fertilization. Cytokines regulate cell division; hence they are linked to the first phase of fruit growth, which involves a large cell expansion. Cytokinin levels and cell division in several horticulture fruits are correlated (Srivastava and Handa, 2005). Many plants developmental processes, particularly fruit growth and development, depend on brassinosteroids (BRs). BRs boost plant development and work with auxins. The impact of naphthaleneacetic acid (NAA), gibberellic acid (GA3), CPPU, BA, DPU, and 4-PU on fruit set in unpollinated and pollinated Chinese white-flowered gourd ovaries showed that only CPPU caused parthenocarpy. CPPU could cause 100% parthenocarpy at 10–100 mg/L and was unaffected by administration time from 2 days before to 2 days after anthesis. CPPU also improved pollinated ovary fruit set and growth. Treating all or alternating ovaries enhanced fruit yield by 91.3% and 83.3%, accordingly, but reduced fruit size (Yu, 1999).

Bananas are parthenocarpic because they are sterile triploids—two sets of chromosomes from one parent and one from the other—and cannot produce seeds. Pollination, not fertilization. The banana has unfertilized ovule traces as tiny black spots. After several crosses of Laden (AAB) (BB) F1 (AB) Kadali (AA) CO I, TNAU, Coimbatore created CO I, a triploid banana. Seedlessness can be achieved by combining parthenocarpy and male sterility. Citrus cytoplasmic modifications and ploidy increases can produce sterility. Many acceptable triploids can be created by crossing diploid and tetraploid plants, such as Oroblanco, Melogold, and Winola (Spiegel and Vardi, 1992).

16.13.3 Self Incompatibility

Multi-cultivar self-incompatibility prevents seed production, resulting in seedless fruit. Self-incompatibility with male sterile plants produces seeded fruit unless they are female sterile, like barberry. By preventing ovary fertilization, Ye et al. (2009) found gametophytic self-incompatibility (SI) caused seedlessness in Mandarin cv. Wuzishatangju.

16.14 BIOTECHNOLOGICAL APPROACHES TO INDUCE PARTHENOCARPY

1. Because of the establishment of male sterility.
2. Elimination of seed coat through the production of a suicide gene.
3. Production of female sterility involves the elimination of ovules or stigma through the use of suicide genes.
4. Enhanced GA expression or heightened GA sensitivity in the ovary.

16.15 CAUSES AND INDUCTION OF PARTHENOCARPY IN FRUIT CROPS

1. **Guava**: e.g. Allahabad Seedless
   a. Aneuploidy.
   b. Poor fertilization of potential ovules.
   c. Disintegration of fertilized ovules.
2. **Grape**: e.g. Thomposon Seedless, Flame Seedless, Crimson Seedless.
   a. Reason: embryo abortion (stenospermocarpy).
3. The embryo rescue technique is employed for the purpose of generating seedless grapes. This approach involves the extraction of minuscule immature embryos, which are then cultivated artificially via tissue culture.
4. **Citrus**:
   a. Poor pollination.

   b. Self-incompatibility (mandarins).
   c. Cytoplasmic male sterlity (Satsuma).
   d. Chromosomal indiscretions (Tahiti lime, Oroblanco)
5. **Apple**:
   a. Lack of pollination.

16.16 INDUCTION OF PARTHENOCARPY IN FRUIT CROPS

The use of synthetic auxin can cause the development of an unpollinated ovary into a fruit. Synthetic auxin directly stimulates the growth of ovules, which are required for the development of parthenocarpic fruits. In apple and pear trees, auxin sprays usually result in seed abortion. At low concentrations, auxins and gibberellins have been used successfully in citrus, grapes, and pineapple to induce parthenocarpy.

16.16.1 Citrus

- Application of GA on the flowers of grapefruit (Marsh Seeded, Excelsior, and Duncan) and mandarin (Kaula, Lahore Local, and Nagpuri) at different concentrations produced parthenocarpic fruits ranging from 40% to 49%. At higher GA concentrations (500–1,000 ppm), the fruits were oblong, almost pear-shaped, and had a very rough and thick skin.
- Polyamines play an important role in fruit development in several plants, and three genes encoding aminopropyl transferases, CcSPDS, CcSPM1, and CcACL5, have been isolated from citrus. More importantly, gibberellin-induced parthenocarpic fruit set decreased CcSPDS expression in the ovaries, which was paralleled by a decrease in spermidine, while CcSPM1 and CcACL5 expression remained largely unaffected, resulting in the maintenance of spermine concentration during early fruit development. Furthermore, variations in putrescine content were mirrored by changes in the expression of one of the two putative CcODC paralogs (Trenor et al., 2010).

16.16.2 Banana

- Triploidy has been considered one of the potential causes of sterility in banana and plantain
- Genetic investigations demonstrate that numerous (at least three) complementary dominant genes (P1, P2, and P3) that are present in Musa acuminata's wild forms are responsible for the autonomous stimulation that results in fruit parthenocarpy (Simmonds, 1976).

REFERENCES

Abd-Alaal, A. F, Al-Salih, K.K., Sbabanaand, H. and Al-Salihy, G.J. 1982. Production of seedless dates by application of growth regulators. *Proceedings of the first Symposium on the date palm/ Collage of Agricultural Sciences and Food* 21:276–282.

Abeles, F.B. and Leather, G.R. 1971. Abscission: control of cellulase secretion by ethylene. *Planta* 97:87–91.

Abeles, F.B. 1973. *Ethylene in Biology*. Academic Press, New York.

Acciarri, N., Restiano, F., Vitelli, G., Perrone, D., Zottini, M., Pandolfini, T., Spena, A. and Rotino, G.L. 2002. Genetically modified parthenocarpic egg plants: improved fruit productivity under greenhouse open field cultivation. *BMC Biotechnology* 22:2–4.

Bhat, Z. A, Dhillon, W. S, Shafi, R.H.S, Rather, J. A, Mir, A. H, Shafi, W., Rashid, R., Bhat, J.A., Rather, T.R. and Wani, T.A. 2012. Influence of storage temperature on viability and in vitro germination capacity of pear (*Pyrus spp.*) pollen. *Journal of Agriculture Science* 4:128–135.

Caixi, Z., Ugyong, L. and Kinji, T. 2008. Hormonal regulation of fruit set, parthenogenesis induction and fruit expansion in Japanese pear. *Plant Growth Regulator* 55:231–240.

Carmi, N., Salts, Y. and Barg, M. 2003. Induction of parthenocarpy in tomato via specific expression of rol B gene in the ovary. *Plantanum* 217:726–735.

Crane, J.C. 1965. The chemical induction of parthenocarpy in the calimyrna fig and its physiological significance. *Plant Physiology* 37:605–610.

Crane, J.C. 1964. Growth substances in fruit setting and development. *Annual Review of Plant Physiology* 15:303–326.

Davis, L. A., and Addicott, F. T. 1972. Abscisic acid: correlations with abscission and with development in the cotton fruit. *Plant Physiology*, 49(4):644–648.

Dorcey, E., Urbez, C., Blazquez, M.A., Carbonell, J. and Perez, A.A. 2009. Fertilization dependent auxin response in ovules triggers fruit development through modulation of gibbrellin metabolism in Arabidopsis. *Plant Journal* 58:318–332.

Fortes, G.R.D. and Petri, J.L. 1983. Fruit set and parthenocarpic fruits as affected by use of Wye mixture in Golden Delicious apple cultivar. *Acta Horticulturae* 137:335–342.

Gardner, F.E. and Marth, P.C. 1939. Effectiveness of several growth substance on parthenocarpy in Hollyhock. *Botany Gaz* 101:226–229.

Gustafson, G.F. 1942. Parthenocarpy: natural and artificial. *Botanical Review* 8:599–654.

GeorgeA.P. and NissanR.J. 1993. Effect of PGR on defoliation, fruit set, retention of low chilling peach cv. Flordaprince in subtropics. *Australian Journal of Experimental Agriculture* 33(6):787–795

Jackson, M.B. and Osborne, D.J. 1970. Ethylene, a natural regulator of leaf abscission. *Nature* 225:1019–1021.

Kraus, E. J, Nellie, A.B. and Hamner, K.C. 1936. Histological reactions of bean plants to indoleacetic acid. *Botany Gaz* 98:370–420.

Komzik, M. 2004. Application of chemical appliances dedicated for chemical thinning of apples. *Acta Horticulturae et Regiotecturae* 7 (Suppl).:86–88.

Lu, J., Lamikanra, O. and Leong, S. 1997. Induction of seedlessness in 'Triumph' muscadine grape (*Vitis rotundifolia* Michx.) by applying gibberellic acid. *Horticulturae Science* 32:89–90.

Luckwill, L.C. 1957. Growth regulators in flowering and fruit development. In: Plimmer, J. R. (ed.), *Pesticide Chemistry in the 20th Century*, A.C.S. Symposium Series (1977)Vol. 11, pp. 293–304.

Molesini, B., Rotino, G. L, Pandolfini, T. and Spena, A. 2009. Expression profile analysis of early fruit development in iaaM-parthenocarpic tomato plants. *BMC Biotechcnology* 14:11–20.

Nancy, D. 2015. Scrutiny of Strategies in developing Parthenocarpic Tomato (Solanum lycopersicum)*Indian Journal of Applied Research* 5:71–74.

Nitsch, J.P. 1952. The physiology of fruit growth. *Annual Review of Plant Biology* 4:199–236.

Noll, F., 1902. Fruchtbildung ohne vorausgegangene Bestaubung (Parthenokarpie) bei der Gurke. Sitzungsber. Niederrhein. Ges. nat. Heilk. Bonn: 149–162.

Nyeki, J. 1978. Meggyfajtak gyumolcshullasa. A hullas merteke es dinamikaja. *Kertgazdasag* 31–38.

Papp, J. 1984. A koszmetenoveny leirasa. In: Papp, J. (ed.), *Bogyosgyumolcsuek*. Mezogazdasagi Kiado, Budapest, pp. 252–254.

Petho, M. 1993. *Mezôgazdasági növények élettana*. Akadémiai Kiad6, Budapest.

Pozo, L.V. 2001. Endogenous hormonal status in citrus flowers and fruitlets: relationship with postbloom fruit drop. *Scientia Horticulturae* 91:251–260.

Racsko, J, Nagy, J., Soltesz, M, Nyéki, J. and Szabo, Z. 2006. Fruit drop: I. Specific characteristics and varietal properties of fruit drop. *International Journal of Horticultural Science* 12(2):59–67.

Serrani, J.C., Fos, M., Atares, A. and Garcia, M.J.L. 2007. Effect of gibberellins and auxin on parthenocarpic fruit growth induction in thye cv . Micro tom of tomato. *Journal of Plant Growth Regulator* 26:211–221.

Simmonds, N.W. 1976. Bananas. In: Simrnonds, N.W. (ed.), *Evolution of Crop Plants*. Longman, London and New York, pp. 211–215.

Soltesz, M., Nyeki, J. and Szab6, Z. 2003. Walnut. In: Kozma, P., Nyéki, J., Soltész, M. and Szab6, Z. (eds.), *Floral Biology, Pollination and Fertilisation in Temperate Zone Fruit Species and Grape*. Akadémiai Kiad6, Budapest, pp. 451–466.

Smith, C.L. and L.D. Romberg. 1941. Pollen adherence as a criterion of the beginning of stigma receptivity in the pecan. *Proceedings of the Texan Pecan Growers Association* 21:38–45.

Soltesz, M. 1997. Termeskptodeses-ritkitas. In: Soltesz, M. (ed.) *Integralt gyümölcstermesztés*. Mezôgazda Kiad6, Budapest, pp. 309–331.

Spiegel, R. and Vardi, A. 1992. Three new selections from citrus breeding programmes. *7th Int Citrus Congress International Society Citriculture Procurement*, Acireale, Italy, Vol. 66, pp. 72–73.

Srivastav, K. and Handa, A.K. 2005. Hormonal regulation of tomato fruit development; a molecular perspective. *Journal of Plant Growth Regulator* 24:67–82.

Stosser, R. 2002. From the flower to the fruit. In: Link, H. (ed.) *Lucas' Instructions for Growing Fruit*. Eugen Ulmer GmbH Co., Stuttgart, pp. 29–37.

Suranyi, D. and Molnar, L. 1981. A kajszibarackfa elettana. In: Nyujto, F., Suranyi, D. (ed.), *Kajszibarack*. Mezogazdasagi Kiado, Bp. pp. 177–227.

Talon, M., Hedden, P. and Primo-Millo, E. 1990. Hormonal changes associated with fruit set and development in mandarins differing in their parthenocarpic ability. *Plant Physiology* 79:400–406.

Thompson, M. 1996. Flowering, pollination and Fruit set. In: Webster, A.D. and Looney, N.E. (szerk.), Cherries: Crop *Physiology, Production and U*ses. CABI Publishing, Wallingford, pp. 223–241.

Tiziana, P. 2009. Seedless fruit production by hormonal regulation in fruit set. *Nutrients* 1:168–177.

Trenor, M., Miguel, A., Juan, C. and Miguel, A.B. 2010. Expression of polyamine biosynthesis genes during parthenocarpic fruit development in *Citrus clementina. Planta* 231:1401–1411.

Vernieri, P., Tagliasacchi, A.M., Forino, L., Lanfranchi, A., Lorenzi, R. and Avanzi, S. 1992. Abscisic acid levels and cell structure in single seed tissues of shedding affected fruits of *Malus domestica* Borkh. *Journal of Plant Physiology* 140:699–706.

Wang, H., Schauer, N., Usadel, B., Frass, P., Zouine, M., Hernould, M., Latci, A., Pech, J.C., Fernie, A.R. and Bouuyzyen, M. 2009. Regulatory features underlying pollination dependent and independent tomato fruit set revealed by transcript and primary metabolite profiling. *Plant Cell* 21:1428–1452.

Way, R.D. 1973. Summer and early fall apple varieties. *Journal of Fruit Variety* 27:6–9.

Wood, B.W. 2000. Pollination characteristics of pecan trees and orchards. *Hort Technology* 10:120–126.

Yahata, S., Miwa, M., Sato, S., Ohara, H. and Matsui, H. 2006. Effects of GA3 and CPPU applications on growth and quality of seedless fruits in triploid loquat. *Horticulture Research (Japan)* 5:157–164.

Ye, W., Qin, Y., Zixing, Y., Jaime, A., Zhang, L., Wu, X., Lin, S. and Hu, G. 2009. Seedless mechanism of a new mandarin cultivar Wuzishatangju (*Citrus reticulata Blanco*). *Plant Science* 17:19–27.

Yu, J.Q. 1999. Parthenocarpy induced by N-(2-chloro-4-pyridyl)-N-phenylurea (CPPU) prevents flower abortion in Chinese white flowered gourd (*Lagenaria leucantha*). *Environment Expert Botany* 42:121–128.

Yuan, R. and Huang, H. 1988. Litchi fruit abscission: its patterns, effect of shading and relation to endogenous abscisic acid. *Scientia Horticulture* 36:281–292.

Yuan, R., and Li, J. 2008. Effect of sprayable 1-MCP, AVG, and NAA on ethylene biosynthesis, preharvest fruit drop, fruit maturity, and quality of 'Delicious' apples. *HortScience*, 43(5):1454–1460.

17 Fruit Set and Development

Shivender Thakur and Sunny Sharma

17.1 INTRODUCTION

Fruit set is a vital stage in the reproductive process of flowering plants. It involves the conversion of a dormant ovary into a developing fruit that enters its rapid development phase (Taiz and Zeiger, 2002). Pollination and fertilization are undoubtedly necessary to start the growth of fruit. In the absence of pollination, the gynoecium has a gradual development, leading to the eventual senescence and abscission of the flower (Figure 17.1). According to Ben-Cheikh et al. (1997), it appears that emasculated and unpollinated flowers ceased their development and dropped shortly after anthesis. Unpollinated ovaries' growth is mostly caused by a failure of the fruitlet's pollinated counterpart's cell division to reactivate (Ben-Cheikh et al., 1997). For an effective fruit set, it is necessary for viable pollen to undergo fertilization on the pistil, followed by the emergence of a pollen tube. Subsequently, the pollen tube must develop through the style and the ovular micropyle, ultimately delivering two sperm cells into the embryo sac. The process of fertilization involves the fusion of an egg cell with one sperm cell, while the remaining sperm cell combines with two haploid polar nuclei located within the center cell. As a result, both the embryo and the surrounding tissues can produce signals that promote the growth and development of the fruit (Youssef and Roberto, 2020). The process of fruit set is highly susceptible to climatic conditions, specifically temperatures that are either too low or too high. These temperature extremes have a significant impact on the generation of pollen and the release of anthers. High temperatures hasten anthesis and reduce the bloom period, whereas low temperatures hasten flowering and prolong the bloom period. As a result, temperature conditions may have a significant impact on pollination and fruit set. Temperature also influences bee activity, as well as the rate of development of pollen tubes, pollen viability, and fertility.

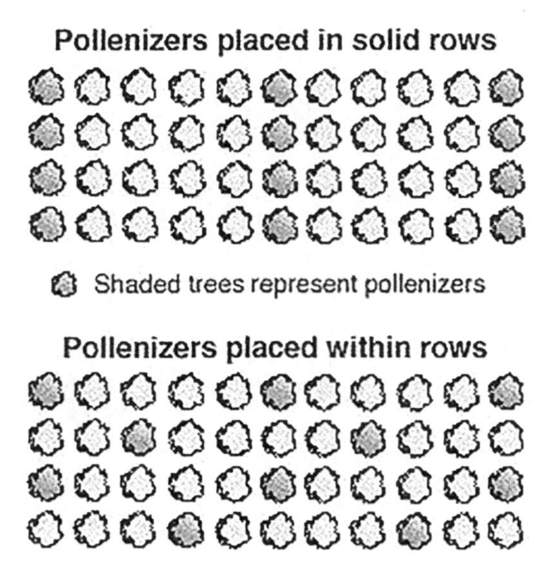

FIGURE 17.1 Control of pollination in fruit crops.

DOI: 10.1201/9781003354055-17

17.2 FACTORS AFFECTING FRUIT SET

Several factors influence the fruit set, which is classified into two groups:

17.2.1 External Factors

The following are some of the extrinsic elements that influence the fruit set:

17.2.1.1 Relative Humidity and Temperature

Temperature is the most substantial factor altering blooming and fruit sets in a variety of ways. In contrast to fruit setting, the development of flower buds and other floral components is more directly influenced by temperature in undeveloped flowers. Low temperatures during the blooming time can prevent pollen from germinating on the stigma, preventing pollen tubes from forming in the style. Certain plant species are more likely to accumulate carbohydrates, which helps to promote the development of big flower buds when temperatures fall below 20°C–25°C. Fruit set is affected by both low and high air humidity. High humidity combined with low temperatures inhibits bee activity and slows pollen release. Because of the drying of a stigmatic secretion at low humidity, pollen does not germinate (Raja et al., 2017).

17.2.1.2 Light

Light influences floral primordial differentiation through intensity, duration, and quality. Higher light intensity leads to more blooming than lesser intensity. The portions of the tree that are exposed to less light have a reduced number of blossoms or may not have any flowers at all, as this leads to a depletion of the plants' carbohydrate reserves. Photosynthesis necessitates the presence of light, and insufficient light intensity or duration leads to a reduction in the plant's carbohydrate reserves. Furthermore, low light levels stimulate fruit abscission (Shaila et al., 2019).

17.2.1.3 Wind

Wind may cause mechanical harm to flowers. It can also prematurely dry the stigmata fluid, inhibiting pollen germination. When there is a lot of wind, the fruit set on the exposed sides of the fruit trees is lower than on the other sides. The majority of fruit crops are pollinated by insects, although pollination is hampered owing to high winds. Winds that are too strong cause ovarian abortion and also dry up the stigma. Wind-pollinated crops include walnut, pecan nut, coconut, and hazelnut, among others. The wind, on the other hand, inhibits pollination in most fruit crops since they are insect-pollinated (Shaila et al., 2019).

17.2.1.4 Rain

It has a direct impact on fruit set by interfering with pollination, pollen grain germination, and staminal fertilization. Rains during the flowering season reduce productivity by washing away pollen grains, impeding pollinators, and promoting the spread of diseases, insects, and pests. For proper pollination, each fruit species requires a specified period without rain during blossoming (Shaila et al., 2019).

17.2.1.5 Tree Vigor and Age

This component, which often has a direct impact on blooming and fruit set as well as unfruitfulness, is closely related to the nutritional state of the plant or a plant's limb, location, and season (Shaila et al., 2019). In contrast, optimal vegetative growth is crucial to sustaining the crop load since it prevents fruit loss at either the early or late phases, as in the case of strawberries, walnuts, etc. Excessive vegetative growth decreases the fruitfulness of trees.

17.2.1.6 Soil Moisture Status

Another crucial factor in determining a tree's fruitfulness is the soil's moisture content at the time of flowering and fruit development. Moisture stress causes pollen abortion in pecans, whereas stress right before harvest reduces fruit set (Shaila et al., 2019). In general, reduced fruit set is caused by either extremely high or low moisture levels during flowering and fruit set.

17.2.1.7 Nutritional Status

Additionally, it has been noted that Anjou and Bosc pears produce more fruit when nitrogen levels are higher. Apple fruit set and production are increased when phosphorus levels are adequate during fruit set, whereas excess N before flowering encourages another vegetative flush and decreases fruit set (Shaila et al., 2019).

17.2.1.8 Pruning

Hormonal changes brought on by pruning alter fruitfulness. The effect of pruning varies depending on the kind, season, and quantity of pruning. Unfruitfulness can also result from trimming deciduous varieties too early or too late. It has been shown that cane-pruned grapevines produce berries and fruits better than grapes that have undergone extensive spur pruning. For large fruitset on pears, vigorous pruning is necessary.

17.2.1.9 End Season Fertility

In several fruit plants, the end-of-season fertility of ordinarily self-sterile plants is rather prevalent. Several grape types, such as Ideal, were found to be early-season self-sterile but late-season self-sterile. Figs of the San Padro class provide an eye-catching illustration of the impact of the seasons on fruit setting and unfruitfulness.

17.2.2 Internal Factors

There are several internal factors that are associated with unfruitfulness. They are further categorized into three major categories:

17.2.2.1 Evolutionary Tendencies

Cross-fertilization is essential for preserving the vitality of the species, as it aligns with evolutionary tendencies. Self-fertilization is challenging or unattainable in certain particular species. certain qualities, which are beneficial for the preservation of certain species, may, when cultivated, restrict their practicality and distribution. The evolutionary factors that contribute to infertility include imperfect and defective flowers, structural peculiarities such as dichogamy (including duo-dichogamy, hetero-dichogamy, and protogyny), stigma receptivity to facilitate germination of pollen, abortive flowers, and aborted pistils or ovules that result in infertility, as well as non-functional pollens or ovules.

17.2.2.2 Genetic Factors

Hybridity is linked to both sterility and unfruitfulness, which promotes greater crossbreeding. Incompatibility due to failure of viable pollens. Interfruitfulness and interfertility promote setting fruits and the production of seeds.

17.2.2.3 Physiological Factors

These include slow growth of pollen tubes, premature or delayed pollination leading to unfruitfulness, defective pistils caused by overbearing, drought poor nutrition, and poor pollen germination.

17.2.3 Regulation of Flowering and Fruit Setting

- Induction of flowering by the application of Gibberellic acid, which is shown to promote flowering in Apple trees, Etheral induced flowering in off-season cultivars of mango.
- At high altitudes, foliation is delayed, and bud breaks mat extend over longer periods leading to problems in cross-pollination.
- Control of blossom quality in pome and stone fruits as measured by fruit sets when hand pollination is done. Branch bending aids in improving the blossom quality as it restricts carbohydrate movement and auxin from the upper portion of the limb toward the roots.
- Although many apple varieties, particularly in warm climates, are at least partially self-fertile, bees and pollinator types are usually used to increase fruit sets. Breeding and selecting self-fertile cherry and apple cultivars or clones may minimize the difficulties of attaining sufficient pollination, particularly in chilly peripheral areas of fruit production when temperatures at blooming time are suboptimal for both bee activity and pollen tube development (Sanzol, 2007)
- Fruit set and retention in apple, pear, plum, and cherry are boosted by tipping developing shoots to limit competition from these either immediately after blooming or at any early fruitlet stage.

17.3 STAGES OF FRUIT DEVELOPMENT

While Kiwi fruit follows a triple sigmoid growth curve, most species of fruit can be represented by a sigmoid curve or a double sigmoid curve with a second burst of growth during the ripening stage. Fruit development may be broken down into four phases from a physiological and biochemical perspective. These continuous phases are split based on the main activities (Table 17.1).

Phase I: This stage comprises the development of the ovary inside the flower, blooming anthesis, and a choice to terminate the process or continue with it (i.e. rupturing the anthers to release mature pollen).

Phase II: During this phase, cell divisions happen at their fastest rate.

Phase III: During this time, cells essentially stop dividing, and cell expansion accounts for practically all of the growth. Food stores are built up during this stage, and most fruits reach their ideal size and form before the start of ripening.

Phase IV Ripening phase.

Cell division in some fruits, like the avocado, may continue long into phase III.

Fruits undergo many stages of development. After the fruit set, there is often a rapid increase in fruit volume, which is then typically followed by a time of slow development. The volume and weight of fruit increase more rapidly towards the end of the ripening process (Lasko and Goffinet, 2013). The growth resulting from cell division, cell expansion, and air space production produces a sigmoidal (S-shaped) curve. The sigmoid growth pattern can be elucidated as follows: The initial

TABLE 17.1
Stages of Fruit Development

Fruit Crop	Period of Cell Division
Ribes and Rubus	Cell division ceases at anthesis
Cherry	Ceases about 2 weeks after anthesis
Plum	4 weeks
Apple	4–5 weeks
Pear	7–9 weeks
Avocado and Strawberry	Continues up to maturity

phase is marked by fast cellular proliferation and a rise in the total cell count. The "lag" phase that follows is characterized by internal structural changes, such as the hardening of the stone (i.e. the fruit endocarp) in cherries, plums, and peaches, or the ripening of embryos in pome fruits. Several fruits, such as currants, pistachios, and seeded grapes, demonstrate a double sigmoidal growth pattern. During the early stages of stone fruit development, the pit undergoes a process called lignification, where the endocarp hardens. This causes the growth of the mesocarp (flesh) and the seed (kernel) to be suppressed. The rapid proliferation of flesh cells occurs during the final pit hardening stage before the fruit reaches a firm and mature state, after which growth decelerates and ceases.

17.4 STAGES OF FRUIT DEVELOPMENT

1. **Cell division**:
   - Predominates after bloom;
   - Typically lasts for a shorter time in smaller fruited crops.
   - In certain cases, flower thinning has extended.
   - The 2 million cells in an apple's flesh during anthesis require 21 cell number doublings.
   - Harvesting 40 million cells only requires 4.5 doublings.
   - Up to 100 days following bloom, but often within the first 2 weeks, post-anthesis cell divisions take place.
2. **Pit hardening (only with stone fruits)**: Endocarp lignification
3. **Cell enlargement**:
   - Predominates later in fruit development (and after hardening of the stone fruit's pit)
   - Starts shortly after pollination, continues during the stage of cell division, then declines until harvest
4. **Fruit maturation**: The last days (weeks) of fruit growth.

Key stages of growth of apple

Dormant: Largely dormant Fruit buds. This is the stage of overwintering (all fruits).

Silvertip: This only applies to apples. Fruit bud scales that had split apart at the tip revealed light grey tissue.

Swollen bud: Similar to the apple's silvertip stage. Fruit buds swell, and scales detach to reveal regions of tissue that are lighter in color (all fruits except apples).

Green tip: This only applies to apples. Fruit buds with the tip broken and about 1/16 inch (1–2 mm) of green visible.

Bud burst: Similar to the apple's green tip stage. Fruit buds that have been broken at the tip to reveal blooming buds include prune, pear, sweet and bitter cherries, and plum.

Tight cluster: It is applicable for apples. Mostly exposed, closely packed, and short-stemmed blossom buds.

Green cluster: only applies to prune, plum, and pears. Green, mostly divided, and with prolonged stalks, the blossom buds.

Pink cluster: It is restricted to apples and pears. Pink bloom buds in clusters cover the entire apple plant's stems. When the peach flower bud has a pink tip.

White bud: Pertains to prune, plum, sweet and tart cherries, and pear. White, divided in the cluster, and stemmed blossom buds.

17.5 FLOWER AND FRUIT DROP

Although the natural decline of fruits between fertilization and maturity is not a continuous process, most fruit species have a well-defined periodicity that is recognizable. The waves of drop are a result of biochemical and physiological changes that occur during the growth season, but the frequency and severity of subsequent waves are influenced by the intensity of the fruit set. Since the fruit was set, more fruit drop periods should be expected.

17.5.1 Types of Fruit Drop

Fruit trees often produce a great number of blooms, just a tiny portion of which are sufficient to produce a typical yield. A single inflorescence of mango, for example, can contain up to 5,000 flowers and an average of 2–3 fruits per inflorescence, resulting in a good and heavy harvest, although the actual proportion of fruit set will be significantly lower. When the fruit set exceeds what the tree can ordinarily bear to maturity, there will be fruit drop at various phases of fruit growth as the tree adjusts to its resources. Such a drop is natural and advantageous to the trees, as it prevents resource exhaustion and branch breakage caused by overbearing.

Together with the petals of the fertilized fading flowers, a large number of unpollinated flowers have disappeared during the initial phase. Usually, the second wave is less noticeable. When the initial phase is less emphasized, the opposite may take place. It's also likely that more fruits were saved during the first two drops, which made the third preharvest fruit drop more significant (Lasko and Goffinet, 2013).

17.5.2 Early Drop or Cleaning Drop

This decline happens as a result of insufficient pollination, stress from water or temperature, low nitrogen levels, and naturally occurring abscission that is presumably controlled by hormonal imbalances in the fruit. The pace at which blooms are fertilized (negatively) or the number of flowers pollinated (positively) are related to this decline, but the drop in June after that is dependent on the trees' ability to carry their burden. The initial phase of fruit drop usually proves the most potent. The unfertilized flowers shed their petals or become sterile for various causes, failing to produce viable seeds.

However, with a low initial bloom density, following a frost spell or a deliberate flower thinning, the surviving fertilized flowers may have the chance to be maintained and mature to ripe fruit, Thus the initial phase of fruit drop may be almost entirely repressed.

Given that the differentiation of fruit primordia occurs many days before the drop, the loss is determined at the time of cell division. As a result, we may conclude that the decline was caused by the embryo's growth rather than provoked by it. At the onset of hardening stones (in May), the fallen fruits still developed in the first (cell division) phase the fruitlets were dropped, i.e., 90% ('Mauritius'), 96% ('Floridian'), and 99% ('Kaimana'). The major cause was a high percentage of aberrant female flowers.

17.5.3 June Drop

Immature fruit detaches from trees with a high number of fruits in June (usually occurring in May in Florida), 1–2 months after flowering, making up approximately 10% of the total fruit loss. The reduction at this stage is mostly attributed to intra-specific rivalry among juvenile fruits for energy (carbohydrates) required for their growth and development. The June drop of citrus fruit is typically considered a natural occurrence during the fruit's maturation. However, when there is a shift coupled with high temperatures throughout the summer, it can cause the fruit to fall off at a faster rate.

Pomaceous fruit has its second period of fruit drop between the 6th and 8th week following bloom, or June in the northern hemisphere. As a result of the embryos' rapid growth and consumption of the endosperm tissue that produces auxin, fruit drop begins to occur during this time. As the embryos finish growing, a secondary endosperm forms and synthesizes a significant amount of auxin once more. That is what caused the decline towards the end of June (Singh et al., 2005). Abortion may be accepted to be caused by a lack of fertilization as it coincides with the decline of unpollinated blooms. If the trees are self-fertile, a similar decline is likewise anticipated. Low-quality pollen during this time may also lead to fruit drop. The month following pollination is when the majority of the fruit primordia in litchi are shed.

17.5.4 Pre-harvest Fruit Drop

After the June drop, robust development of the fruits begins, which suggests that the auxin flow synthesized in the fruit is slowed, and a third phase of fruit drop is threatening in many kinds. This decrease is the result of hormonal imbalance. While auxin concentration declined and promoted abscission, ethylene level rose. The secondary endosperm is also used as a source of growth factors in this process, which is known as the pre-harvest fruit drop. It is brought about by biochemical changes at the base of the fruit stem, but no cell division results in the formation of an abscission layer. The third phase begins close to the time of harvest, as indicated by its name (Arseneault and Cline, 2016).

High fruit density per inflorescence favors preharvest fruit drop. The fruits may kick one another as a result of development if the fruit stem is short (peach, apricot, and some apple kinds), mostly as other members of the same inflorescence. "Green drop" is the name given to the third stage of fruit drop in apricots.

17.6 DROP OWING TO SENESCENCE

The excessive rise in ethylene levels in the AZ tissues causes the apple to become detached after reaching complete physiological maturity. Because ethylene synthesis does not begin and is completely absent in the AZ tissues, the lime fruit can stay connected to the tree for up to a month. The overripe fruits are prone to falling from the gooseberry, and the detached fruits retain their stem. Never are green (unripe) gooseberry fruits dumped. Fruits like raspberries mature fast (Sazo and Robinson, 2013)

Fruits like raspberries mature fast. The glossy color fades, they shrivel and sand (Hippophae, strawberry), which is dissolved only at full maturity as in raspberry and blackberry, and they are already overripe after two to three days. Auxin flow abruptly decreases and the abscission layer forms in apple cultivars "Berlepsch" or "Goldenparmane" near harvest time, when even a light breeze may result in significant losses.

These three different drop types include:

1. The flowers shed and the rate of fruit set will be minimal due to insufficient fertilization. Its mechanics were covered in detail above.
2. Abscission of growing, immature fruit primordia, which may split into several stages but is never connected to senescence naturally.
3. Physiologically ripe or almost ripe fruits drop naturally, which is associated with some form of senescence.

17.7 FACTORS AFFECTING FRUIT LOSS

17.7.1 Physiological Elements

a. **Pollination and fertilization conditions**:

When a significant amount of viable pollen is assessed as being detrimental, an unusual example of "over pollination causing the abortion of female flowers" is noticed. Insufficient pollen production appears to be even more advantageous since it may lead to the creation of apomictic seeds, which might make up for the shortage.

From the perspective of the following fruit drop, the pollination and fertilization times are identical. According to Ortega et al. (2004), fruit drop is more common in almonds if pollination is carried out on the fourth or sixth day following the emasculation of flower buds, which is close to the end of the effective pollination period (EPP).

Therefore, cultivars that produce fewer seeds are more vulnerable to environmental challenges, such as water stress, low nutrition, etc., and are more likely to lose fruit. The fruit's seed content is the most crucial need for the fruit to remain on the tree. The crucial fruit seed content mostly relies on the species or variation.

17.7.2 Seed Content

From the perspective of fruit drop, the development of seeds throughout the early stages of fruit growth is of particular importance (Emery and Offord, 2019). The sites of synthesis, or the places where growth-promoting chemicals are produced, are seeds, particularly their endosperm. First, when the endosperm develops, auxin causes the fruits to begin developing rapidly. The endosperm is then consumed by the embryo at a point in fruit development that is frequently accompanied by fruit drop.

17.7.3 Competition between the Vegetative and Generative Organs

The formation of seeds during the early stages of fruit growth is particularly significant from the perspective of fruit drop. Seeds, specifically their endosperm, are the sites of synthesis, or the locations where growth-promoting compounds are created. Auxin first causes the fruits to start growing quickly as the endosperm grows. At a stage in fruit development where fruit drop is often present, the endosperm is then devoured by the embryo.

17.7.4 Competition between Different Generative Organs

It is widely recognized that the buildup of organic components in a large cluster of flower or fruit buds is not ideal, leading to a sudden and strong drop in fruit (Racsko et al, 2007). The supernumerary bloom exhibited a low fruit set rate. The flower or fruit set that initiates development sooner exhibits dominance over other sets that are growing at a slower rate. Primogenous dominance is the designated word for this phenomenon. The process of abscission is triggered by the correlative signal of dominance, which is caused by the synthesis of ethylene in senescent cells positioned distally from the relevant abscission layer. This process is influenced by both ABA and ethylene.

Ethylene is first produced at the base of the fruit stem and acts as an inhibitor of auxin's ability to regulate abscission in the abscission zone (AZ) because auxin (IAA) counteracts its effects. Given the observed significant decrease in IAA content in fruits exposed to ethylene, it is evident that all of those processes, except for the binding of IAA, are impacted by ethylene.

17.8 ENVIRONMENTAL FACTORS

17.8.1 Climatic and Meteorological Conditions

Premature fruit drop is typically brought on by a variety of circumstances, including most likely unfavorable environmental conditions. Weather conditions before, during, and after the growth and vitality of flowers, bloom, fertilization, and flower-to-fruit transition are very important. Due to the destruction of crucial conductive tissues and the subsequent fruit drop before and after fruit set, late frosts that occur in April and May in the mild climate significantly harm bloom buds, flowers, and young fruit primordia. Temperatures experienced during bloom are already affecting the embryo sac's viability and longevity, which lowers the likelihood of fruit set.

The viability of the embryo sac is also compromised by low temperatures above freezing. The damage might go undetected, render the bloom partially or completely sterile, and speed up floral wilting. Damages are also anticipated as a result of the unusually high temperatures (Table 17.2).

To determine the function of temperature in the abscission of flowers during plum bloom, a study was commissioned (Racsko et al., 2007). Flowers disappeared as a result of temperature variations from the range that the egg cells like. A significant rate of blossom drop was seen at 20°C, while at lower temperatures (5°C, 10°C, and 15°C), it was less noticeable. The outcome also varied according to the variety. Williams (1970) found that high temperatures had similar effects on apples, pears, and cranberries (Table 17.3).

TABLE 17.2
Common Names and Chemical Names of Compounds Effective at Reducing Preharvest Drop

NAA	Naphthalene acetic acid
NAAm	naphthalene acetamide
Fenoprop, 2,4,5-TP	2-(2,4,5-trichlorophenoxy)propionic acid
Daminozide, SADH	butanedioic acid mono(2,2-dimethylhydrazide)
Dicamba	3,6-dichloro-2-methoxybenzoic acid
Dichlorprop, 2,4-DP	2-(2,4-dichlorophenoxy) propanoic acid
AVG	Amino ethoxyvinyl glycine hydrochloride N-(phenylmethyl)-1H-purine-6-amine
2,4-D	(2,4-dichlorophenoxy) acetic acid
MCPB	4-(2-methyl-4-chlorophenoxy)butyric acid

TABLE 17.3
Effect of Post- and Pre-bloom Factors in Temperate Fruit Crop

	Temperate Crop
Post-harvest defoliation	Reduction in Fruit size in the following year
Spur size & position	Larger spurs bear larger fruits "King bloom" of apple produces largest fruit
Pre-blossom temperature	Low temperatures - smaller fruit (shorter growing season, greater fruit set & competition, or reduced pre-bloom cell division)
Age of bearing wood	Larger fruit on 2-year-old spurs than 1-year-old spurs
Post-Bloom Factors	
Seeds	Fruit size-dependent for first 7 weeks Aborted seeds alter fruit shape
Light & carbohydrates	Controls supply of CHOs - competition between fruit & shoot growth & between fruits The most important source of within-tree variation in fruit growth

17.9 THE INFLUENCE OF PHYTOTECHNICAL INTERVENTIONS ON FRUIT DROP

17.9.1 Fruit Thinning

One of the primary reasons for fruit drop is competition between the vegetative and generative organs as well as between the fruits, which is intended to be reduced. Fruit size, flower bud development, and the increase in stem diameter are important factors, it is preferable to thin before the June drop occurs. Premature thinning might also be harmful. Although more expensive, June is also a more significant month.

17.9.2 Irrigation Water Supply

Fruit loss is attributed to several factors, including irrigation and water supply restrictions. Watering plays a crucial function in growing environments that are dry. In addition to raising the air humidity in the canopy, the water spray also efficiently lowers the temperature of the leaves and fruits which reduces the risk of fruit loss. Substantial fruit loss might be brought on by overwatering, which causes split fruits to form close to maturity (Anonymus, 2005).

17.9.3 Nutrition

Due to the relative shortage of nutrients, an abundance of fruits is sometimes thought to cause immature fruit to drop off. The indicators of scarcity start with the blooming phase, competition amongst flowers, then fruitlets, and developing shoots. The nutritional perspective increasingly takes importance following the cleansing decline (Emery and Offord, 2019). The negative impact of too much nitrogen generates a buildup of the toxic nitrite that causes fruit drop.

17.9.4 Harvest

To control fruit drop, the timing of harvest is crucial, especially for types that are prone to preharvest fruit drop. Harvest technique is also crucial, particularly for mechanical harvest. Because of the shocks of the peduncles, a posterior fruit drop is anticipated in the event of utilizing shakers (for thinning) that operate on the vibrational principle, several weeks distant from fruit thinning.

It appears that to prevent abscission, an auxin gradient must be maintained from the organ to the plant body. A signal to abscise might be reduced auxin synthesis by an organ or reduced auxin transport. It appears that during the process, division occurs in all tissues. The secondary abscission layer's cell division and differentiation are absent in mature pedicels (those that have survived through Junedrop).

17.10 BIOTIC FACTORS

Damage from diseases and pests frequently results in the loss of flowers and fruits. It mostly depends on the organisms' life cycles, behaviors, and interactions with climatic events. As a result, we must refer to the crucial times of the growing season, when the damage's symptoms used to manifest (Sharma and Singh, 2023).

17.10.1 Diseases

The metabolism of plants infected by a pathogen used to be altered distinctively. Enzymatic alterations in the terrified tissues show up as indicators of senescence that also appear in healthy tissues and that frequently cause fruit to fall off. The metabolism of ill plants is comparable to that of a healthy but senescent plant, i.e. a senescence that is expected.

The majority of diseases that affect flowers promote and hasten the abscission (Sharma and Singh, 2023).

17.10.2 Pest

Flower parts (stamina and pistils) of fruit species (plum, apple, sour and sweet cherry, walnut, and almond) are generally threatened by the attack of maybeetle (*Melolontha hippocastani*) (*Phyllopertha horticola* and the Japanese maybeetle (*Popillia japonica*) Another purple backed proboscid beetle (*Rhynchitesbacchus)* is known all over Europe, Sibiria and Alger and feeds on apple, pear, plum, peach, apricot and almond. It also consumes floral buds and flower components. It eats the fruit stem after oviposition but does not extract the fruit. The fruit is attacked as it reaches the size of a nut, and when it shrivels, it becomes mummified and is dropped later in the summer (Sharma and Singh, 2023).

17.11 CONTROL OF FRUIT DROP

17.11.1 Horticultural Control Methods for Reducing Pre-harvest Drop

Gardner et al. (1939) provided the first report of a plant growth-controlling agent effective for minimizing preharvest drop. This article on the efficacy of plant growth regulator sprays in minimizing preharvest drop sparked a lot of interest in utilizing these chemicals as cultural aids. Many chemicals have been studied for this purpose (Marini et al., 1989), with some being proven to be successful. In the history of chemical preharvest drop control, three types of chemicals have been especially important:

1. auxins and compounds with auxin-like action,
2. SADH or daminozide, and
3. AVG, an ethylene biosynthesis inhibitor. Each of these classes is examined in the order in which they were first used.

17.11.2 Compounds with Auxin-Like Activity and Auxins

The synthetic NAA and NAAm auxins were shown to be the most efficient in preventing preharvest decline in Apple. Other auxins (IBA, IAA, and indole propionic acid) were significantly less efficient than NAA, however, they may have decreased decline slightly when compared to untreated controls (Gardner et al., 1940).

17.11.2.1 Daminozide

Daminozide was initially reported to reduce preharvest decline in 1966 (Edgerton and Hoffman). Applications from full bloom to a few days before harvest resulted in a lower decline, with the best response to treatment around 1 month before harvest. Daminozide offered preharvest drop management even in a year with a significant drop.

17.11.2.2 AVG Stands for Amino Ethoxyvinyl Glycine

The modalities of action of the numerous PGRs employed for preharvest drop control remain unknown. Because AVG is an ethylene biosynthesis inhibitor, it appears to be a viable technique for suppressing abscission. AVG works as well as or better than auxin chemicals in delaying apple fruit drop. Nonetheless, the precise function of AVG in delaying decline is unknown. AVG applied exclusively to the foliage reduced the internal ethylene levels of the fruit, showing foliar absorption of AVG and transfer of the AVG effect (Shaila et al., 2019). It is uncertain whether this is its primary (or only) mechanism of action for postponing drop.

17.11.2.3 PGR Combinations and Interactions in Preharvest Drop Control

Commercially, PGR combinations have been utilized to minimize preharvest decline, restrict the fruit-softening action of auxins, and boost fruit red color. When administered as stop drops, auxins, particularly 2,4,5-TP, soften the fruit. Maleic hydrazide can be used to decrease these effects. Making another PGR treatment to offset the unfavorable side effects of auxins incurs an investment that would only be justified if softening was a major issue (Shaila et al., 2019)

17.12 CONCLUSION

A number of the issues limiting agricultural output is low fruit set, which is caused by insufficient pollination in orchards. Pollination issues must be addressed early on to ensure optimum productivity and quality. The preferred answer to this problem is to cultivate an optimal ratio of pollinizer kinds and honey bees for pollination. Under harsh environmental circumstances, flower drop results in poor fruit set, lowering output. Thus, the usage of PGRs, notably auxin, gibberellin, and/or cytokinin (GA3 and CPPU), nutrients (Ca, Boron, K, and N) alone or in combination enhance fruit set, either through Parthenocarpic fruit set or by improving pollen viability and germination. The production of crops is influenced by pollination. Pollination issues must be addressed early on to improve yield and quality. The ideal solution to this problem is to encourage the planting of the appropriate ratio of pollinizer varieties, along with the use of honey bees for pollination.

Carbohydrate reserves appear to assist floral development but not fruit growth after bloom. The present photosynthesis of the leaves promotes post-bloom fruit development. Leaves are the source of photosynthate, hence maintaining optimal leaf:Fruit ratio is critical for producing large fruits.

REFERENCES

Arseneault, M.H., and Cline, J.A. 2016. A review of apple preharvest fruit drop and practices for horticultural management. *Sci. Hortic.* 211:40–52.

Ben-Cheikh, W., Perez-Botella, J., Tadeo, F.R., Talon, M., and Primo-Millo, E. 1997. Pollination increases gibberellins levels in developing ovaries of seeded varieties of citrus. *Plant Physiol.* 114:557–564.

Emery, N.J., and Offord, C.A. 2019. Environmental factors influencing fruit production and seed biology of the critically endangered *Persoonia pauciflora* (Proteaceae). *Folia Geobot.* 54:99–113.

Gardner, F.E., P.C. Marth, and L.P. Batjer. 1939. Spraying with plant growth substances to prevent apple fruit dropping. *Science.* 90:208–209.

Gardner, F.E., P.C. Marth, and L.P. Batjer. 1940. Spraying with plant growth substances for control of the pre-harvest drop of apples. *Proc. Amer. Soc. Hort. Sci.* 37:415–428.

Lasko, A.N., and Goffinet, M.C. 2013. Apple fruit growth. *New York Fruit Quart.* 21(1):11–14.

Ortega, E., Egea, J., and Dicenta, F. (2004). Effective pollination period in almond cultivars. *HortScience.* 39: 19–22.

Racsko, JL., GabrielP.J., Zhongfu, S., Wang, Y., and Nyeki, J. 2007. Fruit drop: The role of inner agents and environmental factors in the drop of flowers and fruits. *Int. J. Hortic. Sci.* 13.

Raja, W.H., UnNabi, S., Kumawat, K.L., and Sharma, O.C. 2017. Pre harvest fruit drop: a severe problem in apple. *Indian Farmer* 4:609–614.

Sanzol, J. 2007. Self-incompatibility and self-fruitfulness in pear cv. 'Agua de Aranjuez'. *J. Am. Soc. Horticult. Sci.* 132(2):166–171.

Sazo, M.M., and Robinson, T.L. 2013. The "Split" application strategy for pre-harvest fruit drop control in a super spindle apple orchard in western NY. *New York State Hort. Soc.* 21:21–24.

Shaila, D., Rayees, A.W., Fouzea, N., Sayeda, F., Seerat, R., Tajamul, F., and Shemoo, N. 2019. Fruit set and development: Pre-requisites and enhancement in temperate fruit crops. *J. Pharmacogn. Phytochem.* 8(2):1203–1216.

Sharma, D., Nayak, R., and Singh, J. 2023. *Insect Pests and Diseases of Temperate Fruits and Their Management.* Biotechbook, New Delhi.

Singh, Z., Malik, A.U., and Davenport, T.L. 2005. Fruit drop in mango. *Hort. Rev.* 31:111–153.

Taiz, L., and Zeiger, E. 2002. *Plant Physiology.* Sunderland: Sinauer.

Youssef, K., and Roberto, S.R. 2020. Premature apple fruit drop: associated fungal species and attempted management solutions. *Horticulturae* 6:31.

18 Pre-harvest Factors Affecting Post-harvest Fruit Quality

Anindita Roy, M. Viswanath, Chetanchidambar N. Mangalore, and K. Ravindra Kumar

18.1 INTRODUCTION

Cultivating fruits in the best post-harvest conditions actually starts very early in the farm planning process. Pre-harvest influences on postharvest quality are frequently ignored or underappreciated. However, a lot of the choices we make during crop production can have a significant impact on how well crops perform after harvest (Bamini et al., 2009). It is crucial to keep in mind that vegetable quality is not improved during the harvesting or storing operations, but only preserved afterward. Therefore, it is crucial to take into account the pre-harvest elements that enable us to maximise the quality of the vegetables entering storage (Bhagwan et al., 2013). The grower's selections regarding what and when to plant, as well as the subsequent cultivating and harvesting procedures, determine the produce's potential final market worth. In choosing crops and cultivating them, growers typically rely on their own knowledge and regional customs (Chaudhari et al., 2019). However, when consumers want assistance, they may be directed to agricultural extension officers or maybe research and development specialists within their country's national department of agriculture or a comparable entity (Black et al., 2008).

Understanding how pre-harvest factors affect fruit quality is one way to confirm the highest possible fruit quality (Chaudhari et al., 2019). A climacteric fruit, such as the mango, is typically picked when it is still green and then allowed to ripen during the marketing process, which includes storage and shipment. However, the time between harvest and eating could differ and is not always consistent. Furthermore, it is important to take into account the substantial uncertainty of pre- and post-harvest conditions, as well as the difficulty of selecting fruit at the optimal level of ripeness (Bamini et al., 2009). Nevertheless, similar to other stone fruits, the implementation of pre-harvest cultural techniques has a substantial influence on the post-harvest performance and ultimate quality of the fruit by modifying the environmental parameters that affect fruit growth (Blaikie et al., 2004; Black et al., 2008).

18.2 FRUIT QUALITY

In post-harvest technology for horticultural crops, "quality" is not fully and objectively defined, although it can be described in terms of end-use. Fresh horticulture products' "quality" refers to a collection of traits, qualities, and characteristics that offer the product value to humans for use as food (fruits, vegetables, ornaments, etc.) and gratification (Nakasone, 1999; Scalzo and Mezzetti, 2010).

Fruits with fresh market quality, such as those with good colour, firmness or tenderness, flavour, and nutritional content, are what consumers generally regard to be of high quality (Figures 18.1 and 18.2). Despite the fact that consumers choose products based on appearance and feel, edible quality is what determines whether they will be satisfied and make additional purchases (Nakasone et al., 1999; Cordenunsi et al., 2005).

18.3 PRE-HARVEST FACTORS

The quality of a horticulture crop is determined by a variety of pre-harvest procedures, which have a substantial influence on the produce's quality and post-harvest life (Lloyd and Farquhar, 2008).

 DOI: 10.1201/9781003354055-18

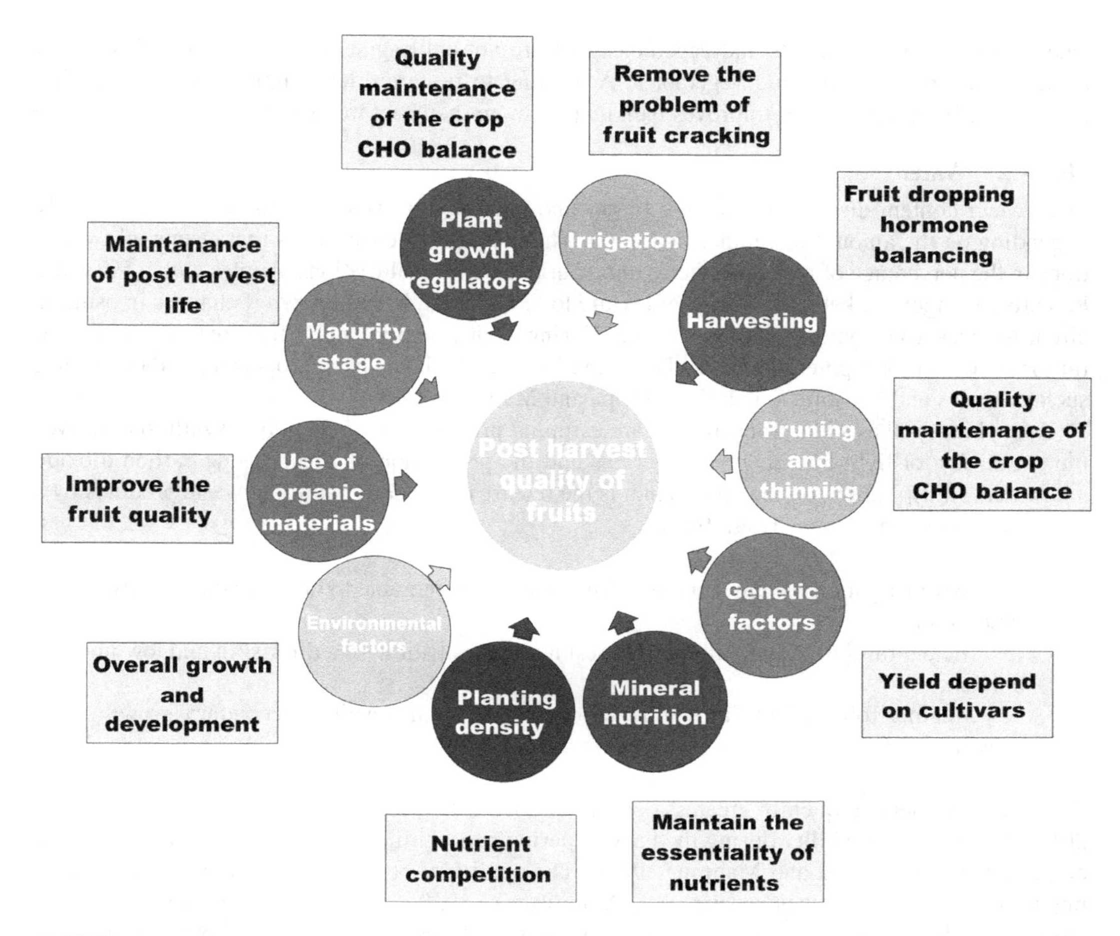

FIGURE 18.1 Pre-harvest factor affecting the post-harvest quality.

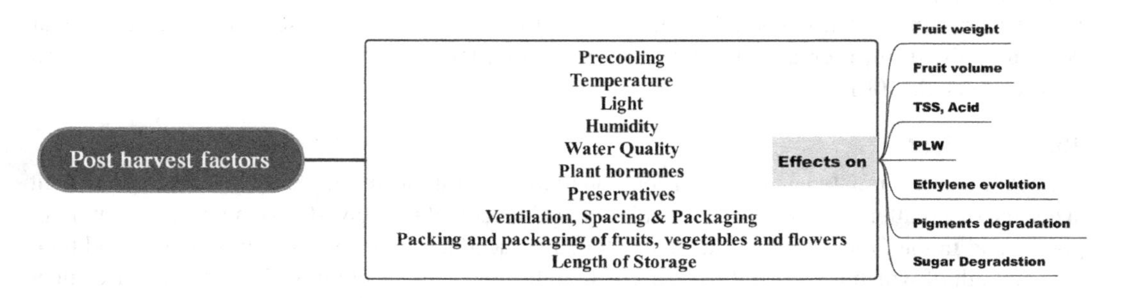

FIGURE 18.2 Post-harvest factors affecting the quality and shelf life.

Pre-harvest techniques, such as applying fertilizer, irrigating, or using chemical sprays, may interact with pre-harvest factors as shown Table 18.2 (Asrey et al., 2013; Khan and Hossain, 2015).

18.3.1 Environmental Factors

18.3.1.1 Temperature

It is widely acknowledged that the temperature during development has an impact on growth and development. Since tissue density would be larger at lower temperatures, this link is presumably true for the majority of fruits and vegetables that are not chilling sensitive, even though it is not always measured in fruits and vegetables. According to research, low-temperature conditioning prevents chilling injuries and improves fruit quality more than heat treatment.

18.3.1.2 Water

Plant water content directly affects cell turgor and plant texture is known to fluctuate noticeably depending on the amount of cellular hydration. The prevailing consensus is that even slight variations in the percentage of water can cause unfavourable textural alterations. Apples with a 5% water loss are no longer marketable because of texture loss. Although it is unknown if changes in osmotic circumstances contribute to texture changes during fruit ripening, some fruit research points to turgor alterations as a potential factor (Dalal and Brar, 2012). It has been proposed in other species, such as apples and Satsuma mandarins (Krupa and Manning, 1988).

Both photosynthesis and transpiration are essential processes in plants. Photosynthesis involves the conversion of light into chemical energies and the production of sugars using carbon dioxide and water. Transpiration, on the other hand, is the release of vapour-containing waste products by a plant. Negative impacts may result from:

- Excessive irrigation or rain can make fruits fragile, vulnerable to harm, and more susceptible to rot.
- Insufficient rainfall or irrigation might result in citrus fruit with a thick skin and low juice content.
- Dry weather followed by rain or irrigation, which might result in secondary growth or growth cracks in tomatoes or potatoes.

Water management frequently creates a conflict between harvest quality and yield (Asrey et al., 2013). The water availability during the harvest period can significantly affect the nutritional value of berry harvests (Krupa and Manning, 1988). The quality of berry harvests after harvest can be impacted by water shortages or surpluses (Maniwara et al., 2015). Severe water scarcity reduces crop productivity and quality, moderate water scarcity reduces crop productivity but may improve specific fruit quality traits, and mild water scarcity improves crop productivity but may reduce post-harvest quality (Henson, 2008). As strawberries reach maturity and ripen, the application of irrigated or rainfall that helps to alleviate water stress leads to a decrease in firmness and sugar content. This, in turn, creates a more favourable environment for physical fruit damage and rot (Stockdale et al., 2001).

18.3.1.3 Light

Shaded leaves assimilate less carbon. From top to lower positions in the canopy, "Kensington" fruit sizes and dry matter contents shrink (Thakur and Chandel, 2004). Investigation revealed that mangoes found in the lower section of the canopy had reduced levels of soluble solids content and total sugars, both of which are crucial indicators of quality that are closely linked to dry matter content (Spinardi et al., 2020; Gon Calves et al., 2021). Through the influence of light on the development of anthocyanin, another effect of light exposure on mango quality relates to its appealing characteristic, particularly the red skin colour (Khan and Hossain, 2015). Mangoes under the canopy keep their skin colour better because of the reduced exposure to sunlight. Fruit bagging, a cultural practice intended to decrease fruit disease, reduces the amount of light that reaches the skin and the area and intensity of the red colour on the peel. Fruit texture is impacted by sun radiation intensity (Thakur and Chandel, 2004; Mendlinger, 2018).

Sun scald or sunburn may occur as a result of excessive exposure to sunlight. Sunscald leads to a decrease in the quality of apple fruit. The firmness of apple fruit grown in the interior shaded areas of the apple tree was lower compared to the fruit produced in the outside sections of the tree's canopy, which received a greater amount of sunshine (Scalzo and Mezzetti, 2010). Nevertheless, excessive light levels beyond the saturation point of photosynthesis, particularly when the exposure is intense, can elevate the temperature of fruits and potentially result in fruit damage and a decrease in firmness (Nakasone, 1999). Light is essential for the proper growth of fruits and can improve their appearance (Lee and Kader, 2000). The dimensions of the fruit, the concentration of dissolved substances, and the firmness of the meat all rise in correlation with heightened light exposure. According to Cordenunsi et al. (2005), the pear soluble solid content in South Africa increased in places where there was a higher buildup of heat units, perhaps because of the elevated rates of photosynthesis and carbohydrate reserves.

18.3.1.4 Availability of Carbon

It is common knowledge that the availability of carbs has a significant impact on fruit growth (Lee and Kader, 2000). To deal with the issue of biennial bearing in mango cultivars, which is often caused by carbohydrate exhaustion, it would be beneficial to determine the foliage-to-crop ratio of a girdled stem or the crop load of a tree required to reach the desired fruit size. Multiple studies have shown that an increase in the leaf-to-fruit ratio is directly correlated with an increase in mango fruit size (Wurr et al., 1996). The association between fruit size and the fraction of meso-carp at the tree level can be observed by effectively managing cultural methods that impact crop load (Wang and Zheng, 2001).

The quantity of carbohydrates delivered to tree fruits during mango growth is determined by the size of the source, its activity, and the demand from the sink, as stated by Felzer et al. (2007). The variations in the size and productivity of the source were directly influenced by the fluctuations in the leaf count per fruit (Lloyd and Farquhar, 2008). The three main components of fruit, namely the peel, pulp, and stone, exhibit varying dry and water masses based on carbon availability (Mauzerall and Wang, 2001). Research conducted by Krupa and Manning in 1988 indicates that mango fruits increase in size when there is an abundance of nutrient sources. According to Sekse (1995), calcium concentrations in the flesh were higher when measured per unit of structural dry mass in cases where the leaf-to-fruit ratios was low.

18.3.2 Cultivation Practices

In order to produce fresh products with good yields and quality, good crop husbandry is crucial. Particularly crucial are some factors like:

18.3.2.1 Cultivar and Rootstock Genotype

By selecting the proper genotype for the environment, degradation, insect damage, and physiological diseases can be less common and less dangerous (Gon Calves et al., 2021). Breeding activities are constantly developing novel cultivars and rootstocks that have superior quality and better adaptability to various environmental conditions and agricultural pest scenarios (Spinardi et al., 2020). Some experts believe that disease resistance, notably resistance to diseases that affect postharvest quality, is the most crucial cultivar feature for fruits and vegetables (Kremer and Kohne, 2014). To effectively manage certain post-harvest illnesses, it may be essential to selectively breed plants that possess resistance to the vector responsible for transmitting the disease, including an aphid, nematode, leafhopper, or mite, rather than solely focusing on combating the pathogen itself (Dalal and Brar, 2012). Reports from Australia indicate that the rootstock of the avocado tree has a substantial impact on the amount of after-harvest rots that develop in ripe avocado fruit (Kumar et al., 2008).

18.3.2.2 Crop Load

To optimize fruit size and total output, it is necessary to find a balance between yield and fruit thinning. Research has shown that thinning the fruit can enhance its size, but it may also reduce the total amount of fruit produced (Kumar et al., 2016). Maximum profit is usually not achieved when the marketable output is at its highest due to the higher market value of larger fruits. Moreover, there is a widespread acknowledgment that the quality of fruit in high-bush blueberries is primarily determined by the ratio of fruit count to leaf count (F:L) rather than mineral nutrition (Lal et al., 2000; Somkuwar et al., 2014).

18.3.2.3 Fruit Canopy Position

According to Basak's study in 2006, peaches that were cultivated outside under a cover in low light settings had a longer shelf life and marketplace lifespan compared to peaches that were produced indoors within a canopy. Multiple studies have shown that enhancing the amount of light that reaches the grape canopy enhances the quality of the fruit. This results in higher levels of total soluble solids (TSS), aroma, anthocyanin, and total soluble phenols. However, it also leads to a drop in titratable acidity and the amount of potassium (Milic et al., 2013). Shade has the effect of reducing the number of fruits in kiwifruit, rather than affecting the weight of each fruit. It also causes a delay in the maturity of the harvest, lowers the SSC (soluble solids concentration), and accelerates the rate at which the fruit softens in preservation (Thakur and Chandel, 2004; Mendlinger, 2018).

18.3.2.4 Girdling

Four to six weeks prior to harvest, girdling (a commercial process in which the phloem of the tree or vine is removed) can improve peach and nectarine fruit size as well as advance and synchronize maturity (Yu et al., 2001). Girdling can sometimes increase fruit SSC while simultaneously increasing fruit acidity and phenolic, which can help hide the taste of the extra sugars (Sembok et al., 2016; Kudachikar et al., 2000; Maniwara et al., 2015).

18.3.2.5 Crop Rotations

Crop rotation might be a useful management strategy for lowering decay inoculums in a production field and minimising post-harvest losses (Singh et al., 2013). Rotating crops is often recommended in areas with significant vegetable production due to the potential buildup of soil-borne pathogens such as fungi, bacteria, and nematodes. Repeatedly growing the same vegetable crop can lead to hazardous levels of these pathogens (Stockdale et al., 2001).

18.3.2.6 Pruning and Thinning

Pruning has an impact on the fruit's size, colour, acidity, and sugar content, among other characteristics. Size is increased by thinning fruits like grapes, dates, peaches, and plums. Fruits' colour, acidity, and sugar content. TSS, acidity, and berry sugar levels all increased as a result of the decrease in clusters per grapevine (Anttonen and Karjalainen, 2009).

18.3.3 Mineral Nutrition

The post-harvest quality and plant nutrition are both heavily researched effects of pre-harvest interventions (Paull, 2020; Wurr et al., 1996). The postharvest life and quality of a crop can be impacted by an excess or shortage of particular components (Mendlinger, 2018). The lack of mineral nutrients is thought to be responsible for a number of physiological problems in horticultural produce (Kumar et al., 2016). Even though the exact involvement of the mineral in avoiding the condition has not been shown for the majority of disorders, these problems are reduced or prevented by pre-harvest supplies or treatments with a specific mineral (Stampar et al., 2018; Gill et al., 2012).

18.3.3.1 Nitrogen

Multiple studies have recorded the effects of different fertilizers on various crops, including the impact of potassium on tomatoes, nitrogen on oranges, and organic fertilizers on mangoes. Plant nutrition is a crucial factor that can affect both the quality and longevity of fruit once it is harvested (Kumar et al., 2008, 2016). Optimal plant function requires a sufficient and well-balanced availability of nutrients from minerals, which are often lacking in soils across the globe. Plants mostly require nitrogen, or N, and potassium (K) as their major nutrients (Yu et al., 2001; Sembok et al., 2016).

18.3.3.2 Phosphorus

The available literature lacks comprehensive coverage of the impact of phosphatic fertilisers on crop storage. When comparing the use of different amounts of phosphorus on onions, it was found that applying 100 kg/ha resulted in decreased weight loss, sprouting, and rotting. P supplementation can alter the physiological functioning of cucumbers after they are harvested by affecting the lipid chemistry of the cell membrane, the integrity of the membrane, and the respiratory metabolism. This has been demonstrated in studies conducted by Lal et al. (2000) and Somkuwar et al. (2014). Foliar fertilisation with phosphate and potassium resulted in increased sugar content (glucose, sorbitol, soluble solids) and organic acid levels (malic and citric acid) in the "Williams" pear (Henson, 2008; Maniwara et al., 2015; Bower and Cutting, 2019).

18.3.3.3 Potassium

Potassium is a key nutrient for plants, and both high and low potassium levels have been linked to aberrant metabolism. The development of an apple's bitter pit has been linked to high potassium levels (Singh et al., 2013). Low potassium affects how tomatoes ripen and prevents the production of lycopene, which slows the development of a deep red colour (Thakur and Chandel, 2004). When potassium is applied to citrus trees, the fruits' form and acidity may be impacted. The frequency of rind pitting was decreased by applying a 9% "Bonus," a potassium fertilizer, to Shamouti orange trees (Maniwara et al., 2015). When potassium is applied to Jonagold apple trees prior to harvest, the resulting fruits are larger in size and have higher soluble solids contents than control apples. Potassium and titratable acidity in "Rocha" pears have a close relationship (Cordenunsi et al., 2005; Scalzo and Mezzetti, 2010).

18.3.3.4 Calcium

When calcium salts are used as a pre-harvest spray, some illnesses, like tomato blossom-end rot, can be easily eradicated, but others, like apple bitter pit, can only be partially controlled (Table 18.1). The calcium post-harvest dip and pre-harvest foliar spray enhanced the firmness of tomato fruits (Krupa and Manning, 1988; Sembok et al., 2016). The capability of calcium to control these different

TABLE 18.1
Calcium Related Disorders of Fruits

Fruits	Disorders
Apple	Bitter Pit
	Lenticel Blotch
	Cork Spot
	Lenticel Breakdown
	Fruit Cracking
	Jonathan Spot
	Water Core
Avocado	Fruit Endpoint
Persea americana	
Cherry	Fruit Cracking
Mango	Soft nose
Pear	Cork spot
Strawberry	Leaf tip burn

systems' functions has raised the possibility that calcium may play a part in the start of the natural ripening process for fruit. Additionally, various physiological problems may be prevented or delayed by calcium (Stockdale et al., 2001; Singh et al., 2009).

Sapota's shelf life is increased by applying calcium chloride (2%) one month before harvest, followed by a post-harvest GA3 (100 ppm) dip. The shelf-life of the fruits was increased by pre-harvest spraying papaya plants with calcium nitrate or calcium chloride fifteen days before harvest, though TSS and total carotenoid concentrations fell (Spinardi et al., 2020; Gon Calves et al., 2021). The quality of the mushrooms was enhanced by the addition of calcium chloride to the irrigation water, which increased the whiteness at harvest and decreased post-harvest browning.

18.3.4 Other Minerals

Citrus fruits, for example, gained weight, size, and vitamin C due to magnesium. Grapes grow in straggly clusters when Zn levels are low. Citrus fruits with a cu deficit have an unnatural splotch that detracts from their attractiveness (Asrey et al., 2013; Khan and Hossain, 2015). According to a study conducted in Japan, organic farming has no impact on the quality of Philippine bananas after harvest. Although imported organic Robusta bananas in Britain ripened more quickly at 22°C–25°C than conventionally cultivated bananas, ripe fruits had identical TSS (Kudachikar et al., 2000; Maniwara et al., 2015).

18.3.5 Growth Regulators

18.3.5.1 Chitosan

Chitosan pre-harvest sprays greatly decreased the occurrence of fungal rot after harvest and preserved the quality of strawberries for up to 4 weeks at 3°C (Lal et al., 2000; Somkuwar et al., 2014). Anthocyanin concentration and titrable acidity showed that the fruit from plants that had been sprayed with chitosan was harder and matured more slowly (Kumar et al., 2008, 2016). It has been demonstrated that fungicide sprays in the field to control Sigatoka leaf spots minimize premature ripening (Stockdale et al., 2001; Singh et al., 2009, 2012).

18.3.5.2 Benomyl

Prevents the development of stem-end rot after harvest in oranges and peaches due to dormant infections with Dipolodia and Phomopsis. Bordeaux-based pre-harvest treatments for pest control *Alternaria* cause soft rot in capsicum and black rot in tomatoes. For instance, the quick emergence of benzimidazole group fungicide resistance in *Pencillium* species clearly suggests that pre-harvest applications of thick fungicides would be undesirable, especially if the same fungicide was being used to control *Pencillium* after harvest (Kumar et al., 2008, 2016; Dalal and Brar, 2012).

18.3.5.3 Polyamines

A study on the effects of putrescine treatments, one of the polyamines, found that pre-harvest putrescine spray was more effective than post-harvest dip in enhancing fruit ripening, quality, and shelf-life of "Kensington Pride" mangoes. The study conducted by Thakur and Chandel in 2004 found that the storage period of 20 days at a temperature of 13°C and a relative humidity of 85%. Pre-harvest treatment of fruit resulted in increased firmness, total soluble solids (TSS), and reduced fruit rot. However, fruit treated with both approaches exhibited decreased acidity, total sugars, and non-reducing sugars compared to the control group (Mendlinger, 2018). Overall, the use of pre-harvest putrescine spray was found to be more successful compared to post-harvest dip, as supported by studies conducted by Henson (2008), Maniwara et al. (2015), and Bower and Cutting (2019).

18.3.5.4 Daminozide

It is also known as Alar, B9, or B995, when applied to apples at a rate of 2,500 mg/L, caused the skin to turn a deeper shade of red and made the fruit firmer than when not sprayed (Spinardi et al., 2020;

Gon Calves et al., 2021). The ethylene production of apples at 15°C was postponed by 3 days by pre-harvest treatment of alar. Fruits were less vulnerable to ethylene during storage, and it also lowered ethylene production by 30%. In many nations, daunozide has been taken off the market (Basak, 2006; Milic et al., 2013).

18.3.5.5 Gibberellic Acid and Cytokinin

Various citrus fruits have been shown to exhibit delayed softening, colour development, and rind problems when GA3 was applied prior to harvest (Table 18.2). Although the GA3 pre-harvest treatment reduced fruit weight loss in storage after harvest, the findings were mixed when it came to fruit degradation (Gill et al., 2012; Canli et al., 2015; Mendlinger, 2018; Stampar et al., 2018).

TABLE 18.2
Summarized form of Pre-harvest Factors Affecting Post-harvest Life of Fruits

Factors	Quality Affected	Reference
	A. Environmental	
Temperature	High temperatures have an impact on things like ripeness, colour, sugar, acidity, etc., and decreases the quality of things like citrus, etc., while improving the quality of things like grapes, melons, tomatoes, etc.	Bartley (1988)
Light	Essential for the synthesis of anthocyanins. Fruit exposed to sunlight lose weight, develop a thinner peel, produce less juice and acids, and have a higher TSS than fruits grown in shade, such as citrus, mango, etc.	Panse and Sukhatme (1985)
Rainfall	Grape, date, litchi, lime, lemon, tomato, and other fruits and vegetables are susceptible to this cause of cracking.	Childers (1975)
Wind	Causes citrus fruits, such as citrus, to bruise, scrape, and develop a corky scar.	Beckman (2000)
Humidity	High humidity decreases the colour and TSS and increases the acidity of fruits and vegetables like citrus, grapes, tomatoes, etc., yet it is necessary for improved banana, litchi, and pineapple quality.	Augustin et al. (1978)
	B. Cultural Factors	
Mineral Nutrition		
Nitrogen	Elevated levels of nitrogen (N) lead to a decrease in the quantity of ascorbic acid, the ratio of total soluble solids (TSS) to acid, and the preservation of quality. However, it results in an increase in riboflavin, thiamine, and carotene levels in citrus fruits.	Adedeji et al. (1992)
Phosphorus	Numerous fruits lose size, weight, and vitamin C when their phosphorus levels are high. Citrus fruits, for example, look bad when it is lacking.	Ghosh and Mitra (2004)
Potassium	Increase sugars, vitamin C, size, and weight. Uneven ripening is caused by its absence.	Grosser et al. (2000)
Calcium	Increased firmness of several fruits, including tomato, apple, mango, guava, etc. Check for physiological issues in a variety of fruits.	Hearn (1984)
Magnesium	Increase in fruit weight, size, and vitamin C content, especially in citrus fruits.	Koltunow et al. (2000)
Zinc	Grape clusters get straggly due to deficiency.	Lee (2002)
Boron	Gummy staining of the albedo and browning of the flesh in citrus. Fruit becomes brittle and distorted.	Matus et al. (2008)
Copper	Deficiency affects the look of citrus fruits by producing an uneven splotch.	Matus et al. (2008)
Growth Regulators		
Auxins	Loquat (2, 4, 5-TP), grape (IAA), mandarins (NAA), and mango TSS sizes have all increased (2, 4-D).	Akaninwor and Sodje (2005)

(Continued)

TABLE 18.2 *(Continued)*
Different Fruit Crops with Various Cause of Fruit Drop

Factors	Quality Affected	Reference
Gibberellic acid	Grape, apricot, strawberry, and other fruit size and weight are increased by pre-harvest treatment. Causes parthenocarpic fruit, such as tomatoes, guavas, and grapes.	Cheirsilp and Umsakul (2008)
Cytokinin	Parthenocarpic fruit is caused by BA and PBA.	Emaga et al. (2008)
Ethylene	The levels of anthocyanin (found in grapes, plums, apples, chillies, and brinjal), carotenoids (found in mangoes, guavas, papayas, citrus fruits, tomatoes, etc.), ascorbic acid, and TSS increase, while the levels of tannin (found in grapes, dates, etc.) fall and acidity are all consequences of the application of ethephon on crops such as grapes, mangoes, tomatoes, and others.	Kanazawa and Sakakibara (2000)
Growth retardant	Alar (B9) is applied prior to harvest to improve fruit colour in apples, cherries, apricots, and other fruits, and MH prevents onion bulb sprouting.	Juárez, et al. (2004)
Rootstock	Oranges, mandarins, and lemons of the highest quality are produced in the citrus industry by the rootstocks Troyer and Carrizo (Citrange).	Aurore et al. (2009)
Irrigation	High acidity is brought on by excessive irrigation, while a lack of moisture affects fruit size, juice	Fageria et al. (2007)
Pruning	It impacts the dimensions, hue, acidity, and sugar content of fruit like apples, phalsa, ber, and grapes, among others.	Milic et al. (2013)
Thinning	Thinning in grapes, dates, peaches, plums, etc. enhances fruit size. The characteristics of fruits include their colour, acidity level, and sugar content.	Drury et al. (1999)
Girdling	It has an impact on the sugar, colour, and size of grape berries.	Lee (2002)
Bunch covering	Banana bananas with greater colour and quality were obtained when pre-harvest bunches were covered in plastic bags.	Ladanyia and Ladaniya (2010)
Variety	Size, shape, colour, and chemical makeup vary amongst varieties. The most crucial traits of the variety are productivity, brightness, and high-keeping properties.	Ladanyia and Ladaniya (2010)
Diseases and pests	Anthracnose of mango, papaya, crown rot of banana, stem end rot of citrus, lenticel rot of apple, and brown rot of peaches are examples of quiescent infections that can be prevented by pre-harvest administration of systematic fungicides.	Kayesh et al. (2013)

18.4 CONCLUSION

The accumulation of water and dry matter, as well as biochemical and mineral components, during fruit development and storage, are all impacted by pre-harvest variables. To adopt cultural practices that will produce high-quality fruits and to define optimal postharvest procedures that will take fruit production conditions into account, for example, to improve final fruit quality traits such as size, colour, taste, nutritional value, and flavour is also to build an integrated approach that links the two categories, having knowledge of and then being able to control changes in fruit quality in response to environmental conditions may be essential. It is necessary to assess how preharvest factors affect fruit quality up until consumer acceptance and consumption.

REFERENCES

Adedeji J, Hartman TG, Lech J, Ho C-T. Characterization of glycosidically bound aroma compounds in the African mango (Mangifera indica L.). *J. Agric. Food Chem.* 1992;40:659–661. doi: 10. 1021/jf00016a028

Akaninwor JO, Sodje M. The effect of storage on the nutrient composition of some Nigerian food stuffs: banana and plantain. *J. Appl. Environ. Manage.* 2005; 9:9–11.

Anttonen MJ, Karjalainen RO. Evaluation of means to increase content of bioactive phenolic compounds in soft fruits. *Acta Hort.* 2009;839:309–314.

Asrey R, Patel VB, Barman K, Pal RK. Pruning affects fruit yield and postharvest quality in mango (*Mangifera indica* L.) cv. Amrapali. *Fruits* 2013;68:367–380.

Augustin J, Johnson SR, Teitzel C, True RH, Hogan JM, Toma RB, et al. Changes in the nutrient composition of potatoes during home preparation: 2. Vitamins. *Am. J. Potato Res.* 1978;55:653–662. doi: 10. 1007/ BF02852138

Aurore G, Parfait B, Fahrasmane L. Bananas-raw materials for making processed food products. *Trends Food Sci. Technol.* 2009;20:78–91.

Bamini T, Parthiban S, Manivanam MI. Effect of pruning and paclobutrazol on yield and quality of mango (*Mangiftra indica* var. Neelam). *Asia Sci.* 2009;4(1&2):65–68.

Bartley JP. Volatile flavours of Australian tropical fruits. *Biomed. Environ. Mass Spectrom.* 1988;16:201–205. doi: 10. 1002/bms. 1200160136

Basak. The effect of fruitlet thinning on fruit quality parameters in the apple cultivar 'Gala'. *J. Fruit Ornamental Plant Res.* 2006;14(2):143–150.

Beckman C.H. Phenolic-storing cells: keys to programmed cell death and periderm formation in wilt disease resistance and in general defense responses in plants? *Physiol. Mol. Plant Pathol.* 2000;57:101–110. doi: 10. 1006/pmpp.2000.0287

Bhagwan A, Vanajalatha K, Sarkar SK, Girwani A, Misra AK. Standardization of dose and time of soil application of cultar on flowering and yield in mango cv. Banganapalli. *J. Eco-friendly Agric.* 2013;8(1): 39–43.

Black RE, Allen AH, Bhutta ZA, Caulfield LE, Onis M, Ezzati M, Mathers C, Rivera J. Maternal and child undernutrition: global and regional exposures and health consequences. *Lancet* 2008;371:243–260.

Blaikie SJ, KulkarniVJ, Muller WJ. Effects of morphactin and paclobutrazol flowering treatments on shoot and root phenology in mango cv. Kensington Pride. *Sci. Hortic.* 2004;101:51–68.

Bower JP, Cutting JGM. Some factors affecting post-harvest quality in avocado fruit. *S. Afr. Avocado Growers' Assn. Yrbk.* 2019;10:143–146.

Canli FA, Sahin M, Ercisli S, Yilmaz O, Temurtas N, Pektas M. Harvest and postharvest quality of sweet cherry are improved by pre-harvest benzyladenine and benzyladenine plus gibberellin applications. *J. Appl. Bot. Food Qual.* 2015;88:255–258.

Chaudhari AU, Krishna B, Soman P, Balasubrahmaniam VR. Development of a package for intensive cultivation of mango using Ultra-HighDensity Planting (UHDP), drip and fertigation technologies for higher productivity. *Int. J. Agric. Sci.* 2019;11(23):9280–9284.

Cheirsilp, B, Umsakul K. Processing of banana-based wine product using pectinase and alpha-amylase. *J. Food Process Eng.* 2008;31:78–90.

Childers NF. *Modern Fruit Science*. Horticulture Publication, Rutgers University Nichol Avenue, New Brunswick, NJ. 1975; pp. 976.

Cordenunsi BR, Genovese MI, Oliveira do Nascimento JR, Aymotov Hassimotto NM, José dos Santos R, Lajolo FM. Effects of temperature on the chemical composition and antioxidant activity of three strawberry cultivars. *Food Chem.* 2005;91(1):113–121.

Dalal RPS, Brar JS. Relationship of trunk cross-sectional area with growth, yield, quality and leaf nutrient status in Kinnow mandarin. *Indian J. Hort.* 2012;69(1):111–113.

Drury R, Donnison I, Bird CR, Seymour GB. Chlorophyll catabolism and gene expression in the peel of ripening banana fruits. *Physiol. Plantar.* 1999;107:32–38.

Emaga TH, Andrianaivo RH, Wathelet B, Tchango JT, Paquot M. Effects of the stage of maturation and varieties on the chemical composition of banana and plantain peels. *Food Chem.* 2007;103:590–600.

Fageria MS, Lal G, Dhaka RS, Choudhary, MR. Studies on postharvest management of ber cv. Umran. *Indian J. Hortic.* 2007;64:469–471.

Felzer BS, Cronin T, Reilly JM, Melillo JM, Wang X. Impacts of ozone on trees and crops. *Compters Rendus Geosci.* 2007;339:784–798.

Ghosh DK, Mitra S. Postharvest studies on some local genotypes of ber (*Z. mauritiana* Lamk.) grown in West Bengal. *Indian J. Hortic.* 2004;61:211–214.

Gill J, Dhillon WS, Gill PPS, Singh N. Fruit set and quality improvement studies on semi-soft pear cv. Punjab Beauty. *Indian J. Hort.* 2012;69(1):39–44.

Gon Calves B, Moutinho-Pereira J, Santos A, Silva AP, Bacelar E, Correia C, Rosa E. Scion-rootstock interaction affects the physiology and fruit quality of sweet cherry. *Tree Physiol.* 2021;26:93–104.

Grosser JW, Ollitrault P, Olivares-Fuster O. Somatic hybridization in Citrus: an effective tool to facilitate cultivar improvement. *In Vitro Cell. Dev. Biol. – Plant* 2000;36:434–449.

Hearn CJ. Development of seedless orange, citrus sinensis, cultivar Pineapple and grapefruit, Citrus paradisi, cultivars through seed irradiation. *J. Am. Soc. Hort. Sci.* 1984;109:270–273.

Henson R. *The Rough Guide to Climate Change*. 2nd end. Penguin Books, London. 2008; pp. 384.

Kanazawa K, Sakakibara H. High content of dopamine, a strong antioxidant, in 'Cavendish' banana. *J. Agric. Food Chem.* 2000;48:844–848.

Khan MSI, Hossain, AKMA. Effect of pruning on growth, yield and quality of ber. *Acta Hort. (ISHS)* 2015;321:684–690.

Koltunow AM, Vivian-Smith A, Sykes, SR. Molecular and conventional breeding strategies for seedless citrus. *Acta Hortic.* 2000;535:169–174.

Kremer-Köhne S, Köhne JS. Yield and fruit quality of Fuerte and Hass on clonal rootstocks. *S. Afr. Avocado Growers' Assn. Yrbk.* 2014;15:69.

Krupa SV, Manning WJ. Atmospheric ozone: Formation and effects on vegetation. *Environ. Poll.* 1988;50: 101–137.

Kudachikar VB, Ramana KVR, Eiperson WE. Pre and post-harvest factors influencing the shelf life of ber (*Ziziphus mauritiana* Lamk.) – a review. *Indian Food Packer.* 2000;1:81–90.

Kumar D, Pandey V, Anjaneyulu K, Nath V. Relationship of trunk cross-sectional area with fruit yield, quality and leaf nutrients status in Allahabad Safeda guava (*Psidium guajava*). *Indian J. Agric. Sci.* 2008;78:337–339.

Kumar D, Singh DB, Srivastava KK, Singh SR, Zargar KA. Performance of apricot varieties/genotypes in north western Himalayan region of India. *SAARC J. Agri.* 2016;14(2):107–116.

Ladanyia M, Ladaniya M. *Citrus fruit: Biology, Technology and Evaluation*. Academic press, San Diego, CA. 2010.

Lal S, Tiwari JP, Mishra KK. Effect of plant spacing and pruning intensity on fruit yield and quality of guava. *Prog. Hort.* 2000;32:20–25.

Lee KS, Kader AA. Pre-harvest and postharvest factors influencing vitamin-C content of horticultural crops. *Postharv. Biol. Technol.* 2000;20:207–220.

Lee HS. 2002. Characterization of major anthocyanins and the color of red-fleshed bud blood orange (Citrus sinensis). *J. Agric. Food Chem.* 50:1243–1246.

Lloyd J, Farquhar GD. Effects of rising temperatures and [CO_2] on the physiology of tropical forest trees. *Philos. Trans. R. Soc. Biol. Sci.* 2008;363:1811–1817.

Maniwara P, Boonyakiat D, Poonlarp PB, Natwichai J, Nakano K. Changes of postharvest quality in passion fruit (Passiflora edulis Sims) under modified atmosphere packaging conditions. *Int. Food Res. J.* 2015;22(4):1596–1606.

Matus JT, Aquea F, Arce-Johnson P. 2008. Analysis of the grape MYB R2R3 subfamily reveals expanded wine quality-related clades and conserved gene structure organization across Vitis and Arabidopsis genomes. *BMC Plant Biol.*, 8:83.

Mauzerall DL, Wang X. Protecting agricultural crops from the effects of tropospheric ozone exposure: Reconciling science and standard setting in the United States, Europe, and Asia. *Annu. Rev. Energy Environ.* 2001;26:237–268.

Mendlinger S. Effect of increasing plant density and salinity on yield and fruit quality in muskmelon. *Sci. Hortic.* 2018;57(1–2):41–49.

Milic B, Keserovic Z, Magazin N, Doric M. Effects of BA and NAA on thinning and fruit quality of apple cultivars 'Golden delicious' and 'Red delicious'. *Acta Hortic.* 2013;981:46.

Nakasone HY, Paull RE. *Tropical Fruits*. CABI Publishing, Wallingford. 1999; pp. 45.

Panse VG, Sukhatme PV. *Statistical Methods for Agricultural Workers*. Indian Council of Agricultural Research (ICAR), New Delhi. 1985.

Scalzo J, Mezzetti B. Biotechnology and breeding for enhancing the nutritional value of berry fruit. *Biotechnology in Functional Foods and Nutraceuticals*. 2010. ISBN 978-1-4200-8711-6(H); 978-1-4200-8712-3(P).

Sembok WZW, Hamzah Y, Loqman NA. Effect of plant growth regulators on postharvest quality of Banana (Musa sp. AAA B.). *J. Trop. Plant Physiol.* 2016;8:52–60.

Singh AK, Singh Sanjay S, Rao A. Influence of organic and inorganic nutrient source on soil properties and quality of anola in hot semi –arid ecosystem. *Indian J. Hort.* 2012;69(1):50–54.

Singh AP, Luthria D, Wilson T, Vorsa N, Singh V, Banuelos GS, Pasakde S. Polyphenols content and antioxidant capacity of eggplant pulp. *Food Chem.* 2009;114:955–961.

Singh SK, Singh RS, Awasthi OP. Influence of pre and post-harvest treatments on shelf life and quality attributes of ber fruits. *Indian J. Hort.* 2013;70(4):610–613.

Somkuwar RG, Samarth RR, Prerna I, Navale S. Effect on cluster thinning on bunch yield, berry quality, and biochemical changes in local clone of table grape cv. Jumbo seedless (Nana purple). *Indian J. Hort.* 2014;71(2):184–189.

Spinardi AM, Visai C, Bertazza G. Effect of rootstock on fruit quality of two sweet cherry cultivars. *ISHS Acta Horticulturae 667. IV International Cherry Symposium.* 2020; pp. 54–59. ISBN 978-90-660.

Stampar F, Hudina M, Usenik V, Dolenc K, Zadravec P. Influence of planting densities on vegetative and generative growth and fruit quality of Apple (Malus domestica Bork.). *Acta Hort. (ISHS)* 2018;513:349–356.

Stockdale EA, Lampkin NH, Hovi M, Keatinge R, Lennartsson EKM, Macdonald DW, Padel S, Tattersall FH, Wolfe MS, Watson CA. Agronomic and environmental implications of organic farming systems. *Adv. Agron.* 2001;70:261–327.

Thakur A, Chandel JS. Effect of thinning on fruit yield, size and quality of kiwifruit cv. Allison. *Acta Hortic.* 2004;662:53.

Wang SY, Zheng W. Effect of plant temperatute on antioxident capacity in strawberry. *J. Agric. Food. Chem.* 2001;48:140–146.

Wurr DCE, Fellows JR, Phelps K. Investigating trends in vegetable crop response to increasing temperature associated with climate change. *Sci. Hortic.* 1996;66(3–4):255–263.

Yu JQ, Li Y, Qian YR, Zhu ZJ. Cell division and cell enlargement in fruit of Lagenaria leucantha as influenced by pollination and plant growth substances. *Plant Growth Regul.* 2001;33:117–122.

19 Fruit Maturity, Ripening, and Storage

Susmita Das and Sunny Sharma

19.1 INTRODUCTION

Fruits provide vital components, such as vitamins, antioxidants, polyphenols, and minerals, that substantially enhance human health. Approximately 20%–40% of fruit is wasted during the entire process, from harvesting to reaching the consumer. Around 2% of the resources are allocated for processing, while more than 25% are lost as a result of ineffective handling and storage (Aggarwal et al., 2018). To evaluate the quality of fresh fruit, distributors, traders, and customers use the visual appeal of the fruit as one of the main criteria. Fruit appearance is the most basic factor in the underlying purchase, while flavor and texture may become increasingly important in the ensuing purchase. Commercially important fruits consist of a combination of tissues, including the expanded ovary, seeds, receptacles (apples, strawberries), bracts, and peduncles (pineapples). Fruits may develop from a single ovary of a flower, as in simple fruits (apples, bananas, mangos, peaches, plums, tomatoes, etc.), from ovaries of several flowers, as in aggregate fruits (strawberries), or from ovaries of different flowers, as in compound fruits (pineapples). These plants come from both monocotyledonous and dicotyledonous families. Even though fruits differ enormously in origin, structure, composition, and climate adaptability, they exhibit considerable similarities in their maturation and ripening patterns. In climacteric fruits, physiological maturity refers to ripening that continues even after they are detached from the plant. Climacteric fruit, such as mango, can be harvested when its quality is acceptable to consumers. Fruits are considered consumer mature when they are suitable for consumption or other intended uses. Consumer maturity for climacteric fruits comes after harvest maturity, whereas for non-climacteric fruits, both phases occur simultaneously. Fruit maturity is not a measurement of culinary excellence when it comes to climacteric fruits since mature bananas remain green and inedible prior to adequate ripening or when consumer maturity is accomplished (Rajkumar et al., 2012). The lifespan of the fruit crops can be divided into two parts; preharvest and postharvest stages. Postharvest operations begin with harvesting at the orchard and continue in the storage/packing house and marketing processes. Several factors affect the quality of fresh produce after harvest, but one of the most important is harvest maturity. To ensure repeat purchases and consumer satisfaction, it is essential to harvest fruit crops at the most appropriate maturity stage.

19.2 FRUIT MATURITY

The nutritional value and storage life of harvested fruits is significantly impacted by maturity, which may also have an influence on how they are handled, traded in, and transported (Vanoli a Buccheri, 2012). An important factor influencing postharvest life is harvest maturity, which also affects product quality, such as appearance, texture, taste, and flavor. "Maturity" means "fully developed". "Mature" comes from the Latin word "maturus," which means "maturation." Once the maturation process is complete, the fruit is ready to be harvested. Maturation usually occurs when the fruit is on the tree or plant, but sometimes it continues after the fruits have been detached from the tree (Figures 19.1).

DOI: 10.1201/9781003354055-19

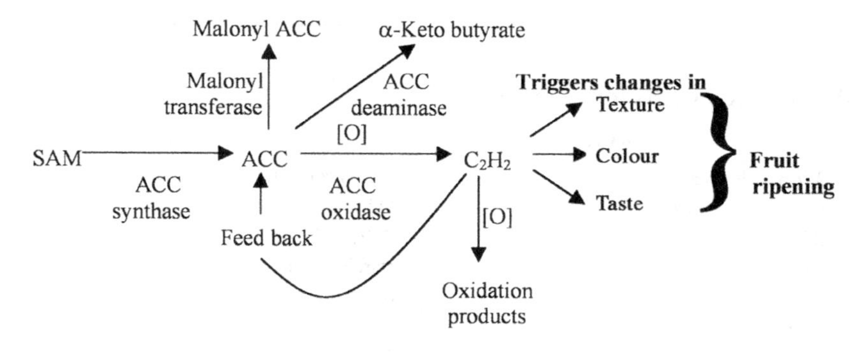

FIGURE 19.1 Different stages of fruit.

The maturity of a fruit indicates its readiness for harvesting. Despite the fact that it is not yet acceptable for consumption, the fruit's edible portion is completely matured in size and shape at this time. Most of the time, ripening begins after or during maturity, at which point the produce is ready for consumption. The word "maturation" refers to a range of phases, including immature, mature, completely mature, and over-mature. Horticultural crops' quality and nutritional value change according to their stages of maturity. The ripeness of fruit at the time of harvest is a crucial determinant of its storage life and quality. Fruits will be unevenly ripened and of low quality if harvested too early. Furthermore, delayed harvesting can increase the susceptibility of fruits to decay, resulting in greater postharvest losses. Deviations in picking optimum mature fruit can negatively affect the quality of the fruit over its shelf life. Maturity indices are useful for trade regulation, marketing strategy, and the efficient use of labor and resources. Generally, fruit quality is measured by its appearance, such as its shape, size, and surface condition. It is especially important to consider color when determining the efficiency of fruit (Raut and Bora, 2016). Color was found to be the most dominant factor for classifying the plums according to maturity level (Kaur et. al., 2018). The more mature the fruit, the more likely it is to soften early and be susceptible to mechanical handling and damage, thereby reducing the fruit's shelf life (Doerflinger et al., 2015; Rathore et al., 2012). Nevertheless, such fruit will be incredibly tasty and juicy. Under prolonged cold storage conditions, immature fruits lose their flavor and are prone to internal breakdown and wooliness due to their hard texture and low flavor (Doerflinger et al., 2015).

Based on maturity, fruits and vegetables can be categorized into:

- **Physiological maturity**: Physiological maturity is achieved whenever a vegetable or fruit accomplishes an adequate level of growth while still connected to a tree or plant. Horticultural goods are typically harvested at the stage of physiological maturity.
- **Horticultural maturity**: A fruit or vegetable reaches horticultural maturity when it has attained the level of development required to meet customer demands (Table 19.1).
- **Commercial maturity**: The market necessitates the presence of plant organs in a specific condition. It often has minimal correlation with physiological maturation and might manifest at any point during the developmental process.

One of the key variables to determine when a fruit should be picked is the maturity index. This assists with marketing flexibility and enables consumers food that is appropriate for consumption. The majority of horticultural crops develop according to a pattern that commences with the bud initiation stage, continues through the development stage, and concludes with death. The more durable development stage involves senescence, ripening, maturation (physiological maturity), and growth (Doerflinger et al., 2015). Numerous guiding principles must be followed during the harvesting of fruits:

TABLE 19.1
Harvest Maturity Standards for Fruit Crops

Fruit	Physical	Chemical
Apple	DFFB, color size	Starch content; firmness as measured by pressure tester
Avocado	Skin color, firmness	Oil content
Banana	Skin color Angularity	Pulp/peel ratio starch content Starch to sugar conversion, ethylene production
Cherry	Color, firmness	Sugar content, acid content
Citrus	Color break of the skin from green to orange, size	Sugar/acid ratio, TSS
Grape	Color, firmness, easy separation of berries,	Sugar content, acid content
Kiwi	Firmness, color	SSC (soluble solids TSS >6.2 Deg. Brix concentration), acid content
Mango	Color: green color; lenticels should be clear Others: specific gravity, days from fruit set	Starch content, flesh color
Orange	Color, firmness	TSS (total soluble solids), acid content
Papaya	Color turning stage green to yellowish	Jelliness on the seed, seed color
Peach	Firmness, ground color	Sugar content, acid content
Pear	Firmness, color	Starch content, ethylene production
Pineapple	Color, firmness	TSS (total soluble solids), acid content
Plum	Firmness, skin color	Sugar content, acid content
Strawberry	Color, size, firmness	Sugar content, titratable acidity
Watermelon	Ground spot color, sound	SSC (soluble solids concentration), sugar content

- Fruits should be harvested during the peak growth period.
- Fruit should be harvested when the taste or appearance of the fruit is acceptable.
- Fruit should be harvested when it reaches the market size
- The harvesting process should ensure minimal mechanical damage, resulting in fruit with adequate shelf life.

In fruit crops, maturity indices can be determined by several methods. Most of the indices differ based on the type of fruit. Common methods for determining fruit maturity are:

- **Visual method**: Frequently, visual clues are employed as a means of evaluating the ripeness of fruit. Prominent visual cues encompass alterations in hue, surface quality, and overall manifestation. As an illustration, the ripening process of bananas involves a transition from a green hue to a yellow coloration, while tomatoes undergo a transformation from a firm and glossy texture to a softer and more matte surface. The visual method is mostly based on appearance, such as skin color, shape, size, and weight of the fruit, the presence of mature leaves, the completeness of the fruit, and the drying of the plant structure.
- **Physical method**: Physical approaches encompass the evaluation of the basic physical characteristics exhibited by fruits. One potential approach involves quantifying hardness through the utilization of a penetrometer, while another method is evaluating the texture of the fruit by means of tactile input. In general, ripe fruits exhibit a softer and less hard texture compared to their unripe counterparts. For example, firmness, specific gravity, formation of abscission layer, etc.
- **Chemical method**: Chemical methodologies encompass the examination of chemical constituents in order to ascertain the state of ripeness. As an example, the measurement of sugar content via a refractometer can serve as an indicator of the grade of ripeness.

The sugar levels of fruits frequently increase as they undergo ripening. For example, soluble solids content (SSC), titratable acidity (TA), TSS:TA ratio, and oil content.

- **Computation method**: The computation approaches employed in fruit ripening research entail the utilization of mathematical models and algorithms to forecast the ripening process by considering diverse parameters, including temperature, humidity, and gas concentrations within storage facilities. These models have the potential to facilitate the effective management of the ripening process and storage conditions pertaining to fruits. For example: days from flower bloom (DFFB), Growing degree days until optimum fruit growth.
- **Physiological method**: The physiological approach to evaluating fruit ripeness involves examining alterations in the fruit's internal physiological mechanisms. One example of a naturally occurring plant hormone involved in the process of fruit ripening is ethylene gas. The monitoring of ethylene production can be utilized as a means to ascertain the specific stage of ripeness. Other expels are respiration, aroma development, and Pigments.

19.3 FRUIT RIPENING

Fruit ripening is a complex process involving physical, biochemical, and sensory changes that lead to the development of soft, edible fruit with desirable characteristics. These processes are genetically determined, coordinated, and irreversible. Fruits are collected after they have reached their optimal level of ripeness. Even when separated from its parent plant, it retains the catalytic machinery to sustain an autonomous existence. Fruits that have been harvested can be categorized as either climacteric or non-climacteric, depending on their respiratory pattern and the production of ethylene throughout the ripening process. Fruit ripening can exhibit a wide range of variations. A climacteric fruit can be matured off the parent plant once it has been harvested at full maturity. During the initial phases of ripening, the rates of respiration and production of ethylene are small. However, during the climacteric peak, they increase significantly. Non-climacteric fruits do not continue to mature after being separated from the parent plant. In addition, these fruits have a low production of natural ethylene and do not show any response to ethylene treatment from the environment. Non-climacteric fruits experience a steady decrease in respiration patterns and ethylene production as they ripen (Gamage and Rehman, 1999).

Several biochemical changes that are vital to the sensory quality of the fruit occur when it ripens, including those involving its color, sugar content, acidity, texture, and aroma volatile constituents. Throughout the process of fruit ripening, various significant biochemical changes take place. These include heightened respiration, breakdown of chlorophyll, production of carotenoids, anthocyanins, and essential oils, development of flavor components, enhanced activity of enzymes that break down cell walls, and increased production of ethylene. The color change that occurs during fruit ripening is caused by the decomposition of the chlorophyll and breaking down of the photosynthetic assimilates, the production of different kinds of anthocyanins and their buildup in vacuoles, and the buildup of carotenoids. Throughout the process of fruit ripening, certain volatile molecules like ocimene and myrcene are generated, while bitter principles, antioxidants, tannins, and related compounds break down (Lizada, 1993). The increase in flavor is mostly attributed to gluconeogenesis, the breakdown of polysaccharides such as starch, reduction in acidity, buildup of sugars and organic acids, and the maintenance of a balanced ratio between sugars and acids. During fruit ripening, several metabolic changes occur. These include an increase in the production and release of the ripening hormone, ethylene, which is an increase in respiration facilitated by specific enzymes found in mitochondria, particularly oxidases, and the creation of new enzymes that catalyze ripening-specific transformations. Disrupted cell structure encompasses various elements such as cell wall thickness, permeability of the plasma membrane, hydration of the cell wall, degradation of the cell's structural integrity, and a rise in intracellular content. The citation for this information is from Redgwell et al. in 1997. The physiological processes associated with senescence, which result in membrane breakdown and cell death, occur towards the later stages of ripening. In this particular context, fruit ripening might be perceived as the initiation of a programmed cell death process.

The process of maturity and ripening is followed by numerous biochemical alterations. These include:

- Changes in carbohydrate composition, which result in sugar accumulation and increasing fruit sweetness.
- An alteration in the color of fruit skin occurs due to changes in pigments such as chlorophyll, carotenoid, anthocyanin, and betalain.
- Changes in fruit texture and flesh softening due to cell wall degradation.
- A formation of volatile scent molecules.
- The disappearance of astringent compounds and acidity.
- In climacteric fruits and vegetables, there appears to be an increase in ethylene production and respiration rate.

In particular, in climacteric fruits, the phytohormone ethylene possesses a tendency to initiate a number of cell metabolic processes, including ripening and senescence. According to their potential for involvement in a program of higher ethylene production and a corresponding rise in respiration rate at the initial phase of ripening, fruits are usually classified into several categories. Apples, peaches, and bananas are a few examples of climacteric fruits that go through this transition; non-climacteric fruits, on the other hand, don't release as much ethylene, which include citrus, grape, and strawberry. First, the enzyme ACC synthase transforms S-adenosylmethionine (SAM) into 1-aminocyclopropane carboxylic acid (ACC). Multiple ACC synthase genes are activated at the start of fruit ripening, increasing the production of ACC. The rate of ethylene biosynthesis is typically governed by the ACC synthase activity. ACC oxidase then converts ACC to ethylene.

The usage of ethylene biosynthesis inhibits or would be beneficial to delay fruit ripening. There are several compounds that inhibit ethylene production. Ethylene production can be reduced by:

- Use of ethylene absorbent e.g. KMn04, AgNO3 etc.
- Use of anti-ethylene compounds, e.g., AVG (Aminoethoxyvinyl glycine), AOA (Aminoethoxy acetic acid), Cobalt, AIB (à-amino isobutyric acid), EDTA.
- Use of radical quenching agents (Sodium benzoate, n-pyrogllate, silver ions, etc.).

1-Methylcyclopropene (1-MCP), an inhibitor of ethylene action, is commonly employed as a postharvest treatment to regulate the ripening and softening process of many fruits. The application of 1-MCP had a notable impact on fruit ripening by reducing respiration and ethylene generation. The application of 1-MCP treatment can extend the shelf life of fruits for a duration of up to 5 weeks by preserving higher levels of ascorbic acid, phenolic content, and fruit firmness (Lim et al., 2016). In banana fruit, concentrations of 0.1–0.5 ml 1-MCP inhibit the action of ethylene (Jiang et al., 2004).

19.4 STORAGE

Biological functions, including respiration and transpiration, continue to occur in fruits and vegetables after harvesting, which means they only keep their freshness over as long as their usual metabolism continues. To sustain these operations, the packing materials must be appropriate. Properly or scientifically packing fresh fruits and vegetables reduces wastage by minimizing mechanical damage, contamination, moisture loss, and other unwanted physiological modifications and pathological deterioration during storage, transit, and marketing. Harvesting fruits after they have reached the appropriate level of ripeness might increase the amount of time they can be stored without spoiling. The main objective of storage is to regulate the transpiration rate, respiration rate, diseases, and pest infestations, and maintain the food in its most consumable state. The main objectives of the store include the following:

- Slow down the biological activity of fruits and vegetables.
- Slow down the growth of different microorganisms.
- Reduce transpirational losses to avoid different undesirable changes (sprouting, greening, rotting, toughening, etc.) in fruits and vegetables.

For fruits and vegetables, a range of storage techniques have been implemented, including ground storage, ambient storage, refrigerated storage, air-cooled transpiration rate storage, and atmosphere storage, depending on the type of commodity being stored and the span of time it is meant to be retained (Rathore et al., 2012).

19.4.1 Controlled Atmosphere (CA) Storage

The incorporation of controlled atmosphere (CA) storage technology by the fruit and vegetable sector proved to be its most triumphant advancement. In order to achieve Controlled environment (CA) storage, it is common to employ a combination of refrigeration together with a storage environment that has low levels of oxygen (O_2) and high levels of carbon dioxide (CO_2) (Dilley, 2006). Within a CA storage facility, the gas composition within a food storage room is constantly monitored and fine-tuned to maintain the optimal concentration with utmost precision. Because CA storage requires a large initial investment and is costly to run, it is better suited for foods like apples, kiwifruit, and pears that can be kept for a long time. There are certain operations for the production and maintenance of CA, which involve the removal of excess CO_2, the addition of air to replace the O_2 depleted by means of respiration, the removal of C_2H_4, as well as the addition of CO_2. Based on the horticultural produce that is being stored and the specific storage requirements for each product, the proper functions and technology for developing and maintaining CA need to be determined. The ripening process of 'Pedro Sato' guava fruit was significantly delayed in atmospheres with lower O_2 concentrations (1 and 5 kPa). As a result, low-oxygen atmospheres have been found to be effective for the prolonged storage of guava fruit.

19.4.2 Hypobaric Storage

Hypobaric storage is an analogous approach that entails confining the food in an atmosphere with precisely controlled pressure, humidity, and air temperature, as well as changing the air in the storage environment at controlled intervals (Burg, 2004). No gases other than air are needed, unlike CA storage and modified atmosphere storage. The O_2 concentration is directly proportional to the total pressure inside the hypobaric chamber, so it's important to determine that pressure. Hypobaric or low-pressure storage acts by diminishing pressure by vacuuming air outside of the container with a vacuum pump. In fruit and vegetable markets, this approach is typically used to minimize the partial pressure of O_2. In order to achieve the required low pressure within the store, air intake and evacuation from it should be balanced. It is feasible to regulate the O_2 level in a hypobaric store precisely and easily by assessing the pressure inside the store utilizing a vacuum gauge.

19.4.3 Modified Atmospheric Storage

Storage of fruits in the packaging of polymer materials with selective permeability for gases is called modified atmospheric storage. As a way to safeguard fresh items from losing quality and to extend the time that products of acceptable quality may be marketed, modified atmospheric storage is predominantly utilized. Modified atmosphere packaging (MAP) is a method that involves placing plant goods in a package and allowing the environment inside the package to interact with the goods, the packages, and the external atmosphere. Initially, the fruit's respiration was utilized to decrease oxygen levels and increase carbon dioxide levels in sealed storage. However, the implementation of sustained modification of the atmosphere using various techniques such as combustion devices, air separators, liquid nitrogen gas flushing, and carbon dioxide scrubbing offered additional advantages in terms of quicker 'pull down' times and improved control over the atmospheric conditions (Ben-Yehoshua et al., 2005). The main attribute that differentiates MAP from CA is that, in comparison to CA, with MAP, active involvement by humans ceases at the moment of sealing. Additionally, MAP is frequently utilized on lesser amounts of product than CA. Khan and Singh (2008) reported that 1-MCP combined with MAP extended the storage life up to 7 weeks followed by 8 days of ripening of plum fruit. A combination of 1-methyl cyclopropene (1-MCP) and MAP improves the storability of 'Harbiye' persimmon fruits (1°C and 90% RH) (Oz, 2011).

19.4.4 Zero Energy Cool Chambers

A zero-energy cool chamber is a sustainable solution for storing perishable goods without relying on external energy sources like electricity. This innovative structure operates on passive cooling principles, utilizing natural processes to maintain low temperatures within the chamber.

The design of a zero-energy cool chamber focuses on several key features. First, it incorporates high levels of insulation in the walls, ceiling, and floor to minimize heat transfer between the interior and exterior environments. Additionally, the chamber utilizes materials with high thermal mass, such as stone or concrete, to absorb and store heat during the day and release it slowly at night, helping to maintain stable temperatures. Strategic ventilation openings facilitate airflow and allow for the exchange of cooler air from outside with warmer air inside, while external shading devices reduce direct sunlight exposure to minimize heat gain. Some designs may also include mechanisms for controlling humidity levels to prevent moisture buildup and spoilage.

Operationally, the cool chamber takes advantage of natural cooling cycles. During cooler periods or at night, the chamber is ventilated to allow cold air to enter and replace warmer air inside. Natural convection helps to circulate the cooler air throughout the chamber, displacing warmer air and maintaining uniform temperatures. In some cases, evaporative cooling techniques, such as wetting the floor or using water-soaked materials, may be employed to further lower temperatures. Zero-energy cool chambers are particularly beneficial in off-grid or remote areas where access to electricity is limited or unreliable. By harnessing natural processes and design principles, they offer a sustainable and cost-effective solution for preserving perishable goods while minimizing environmental impact.

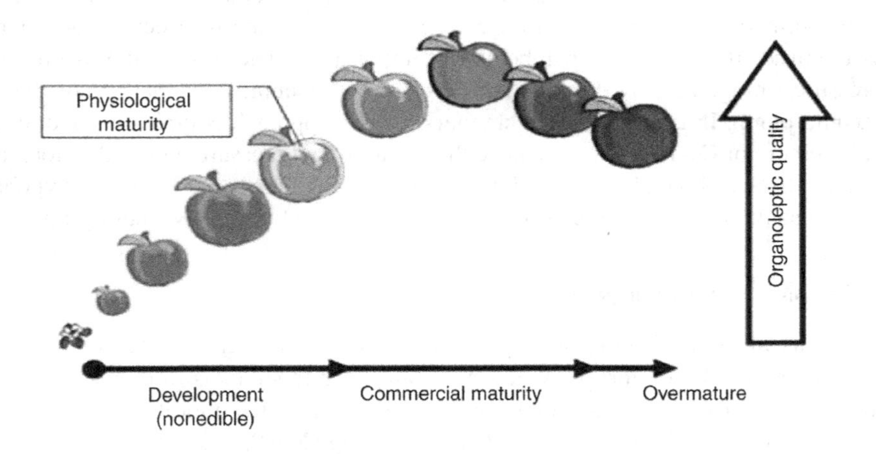

FIGURE 19.2 Ethylene biosynthesis pathway (Prasad et al., 2018).

REFERENCES

Aggarwal, S., Mohite, A., and Sharma, N. 2018. The maturity and ripeness phenomenon with regard to the physiology of fruits and vegetagles: a review. *Bulletin of the Transilvania University of Brasov*, **11**:77–88.

Ben-Yehoshua, S., Beaudry, R.M., Fishman, S., Jayanty, S., and Mir, N. 2005. Modified atmosphere packaging and controlled atmosphere storage. In *Environmentally Friendly Technologies for Agricultural Produce Quality.* Ben-Yehoshua, S. (Ed.). Taylor and Francis Group LLC, Boca Raton, FL, pp. 51–73.

Burg, S.P. 2004. *Postharvest Physiology and Hypobaric Storage of Fresh Produce*. CAB International, Wallingford, UK.

Dilley, D.R., 2006. Development of controlled atmosphere storage technologies. *Stewart Postharvest Review*, **6** (5):1–8.

Doerflinger, F.C., Rickard, B.J., Nock, J.F., and Watkins, C.B. 2015. An economic analysis of harvest timing to manage the physiological storage disorder firm flesh browning in 'Empire' apples. *Postharvest Biology and Technology*, **107**:1–8.

Jiang, Y., Joyce, D.C., Jiang, W., and Lu, W., 2004. Effects of chilling temperatures on ethylene binding by banana fruit. *Plant Growth Regulation*, **43**(2):109–15.

Kaur, H., Sawhney, B.K., and Jawandha, S.K. 2018. Evaluation of plum fruit maturity by image processing techniques. *Journal of Food Science and Technology*, **55**:3008–15.

Lim, S., Han, S.H., Kim, J., Lee, H.J., Lee, J.G., and Lee, E.J. 2016. Inhibition of hardy kiwifruit (Actinidiaaruguta) ripening by 1-methylcyclopropene during cold storage and anticancer properties of the fruit extract. *Food Chemistry*, **190**:150–7.

Lizada, C. 1993. Mango. In *Biochemistry of Fruit Ripening*. Seymour, G.B., Taylor, J.E., and Tucker, G.A. (Eds.) Chapman and Hall, London, pp. 255–271.

Oz, A.T., 2011. Combined effects of 1-methyl cyclopropene (1-MCP) and modified atmosphere packaging (MAP) on different ripening stages of persimmon fruit during storage. *African Journal of Biotechnology*, **10**(4):807–14.

Prasad, K., Jacob, S., & Siddiqui, M. W. (2018). Fruit maturity, harvesting, and quality standards. In Preharvest modulation of postharvest fruit and vegetable quality (pp. 41–69). Academic Press.

Rajkumar, P., Wang, N., Elmasry, G., Raghavan, G.S.V., and Gariepy, Y. 2012. Studies on banana fruit quality and maturity stages using hyperspectral imaging. *Journal of Food Engineering*. https://doi.org/10.1016/j.jfoodeng.2011.05.002.

Rathore, N.S., Mathur, G.K., and Chasta, S.S., 2012. *Postharvest Management and Processing of Fruits and Vegetables*. ICAR, New Delhi.

Redgwell, R.J., MacRae, E.A., Hallet, I., Fischer, M., Perry, J., and Harker, R. 1997. In vivo and in vitro swelling of cell walls during fruit ripening. *Planta*, **203**:162–73.

Vanoli, M., and Buccheri, M. 2012. Overview of the methods for assessing harvest maturity. *Stewart Postharvest Review*. https://doi.org/10.2212/spr.2012.1.4.

20 Molecular Approaches in Fruit Crop Growth Regulation

Komaljeet Gill, Shagun Sharma, and Pankaj Kumar

20.1 INTRODUCTION

Crop regulation is the pillar for reliable and quality crops. The process of growth regulation encompasses an intricate series of connected activities that encompass cell division, expansion, and multiple tiers of genetic regulation. Conventional science has a variety of limitations that are particularly restrictive in fruit crops when it comes to studying plant growth regulation. The method of conducting scientific research has changed significantly since the development of molecular biology. Recently, the research of growth regulation has been transformed by the availability of molecular techniques and resources, which assist in the evaluation of the genotype and its relationship to the phenotype, primarily for complex traits. Genes that determine qualities in fruit crops are being examined through massive gene sequencing, transcription studies, and protein expression. Molecular approaches are often used for different gene activities, particularly interactions between them and how these interactions are regulated (Vasistha et al., 2020; Nautiyal et al., 2022). Several molecular approaches, like the omics approach, molecular markers, transgenic, genome editing, RNAi, etc., are used to modulate fruit crop growth regulation (Figure 20.1).

20.2 IMPORTANCE AND NEED OF MOLECULAR APPROACHES IN CROP GROWTH REGULATION

- The objective of this study is to examine the impact of both simple and complex genes on the regulation of crop growth.
- Additionally, the study aims to investigate the interplay between genes and the surrounding environment.

20.3 OMICS APPROACH

With the emergence of sequencing technology, the scientific community is nowadays capable of discovering hidden information effectively and inexpensively at the genome, transcriptome, and epigenome levels. Various major fruit crops have completed genome-wide sequencing including grapevine (487Mbp), papaya (372Mbp), apple (742Mbp) (Velasco et al., 2010), woodland strawberry (240Mbp) (Shulaev et al., 2011), tomato (760Mbp) (Sato et al., 2012), banana (523Mbp) (D'hont et al., 2012), watermelon (425Mbp) (Guo et al., 2013), sweet orange (367Mbp) (Xu et al., 2013), pear (512Mbp) (Wu et al., 2013), peach (220Mbp) (Verde, 2013), kiwifruit (758Mbp) (Huang et al., 2013), jujube (444Mbp) (Liu et al., 2014), pineapple (526Mbp) (Ming etal., 2015), pomegranate (336 Mbp) (Qin et al., 2017), longan (445Mbp) (Lin et al., 2017), red bayberry (313Mbp) (Jia et al., 2019), and avocado (446Mbp) (https://www. ncbi.nlm.nih.gov/biosample/7709230) (Figure20.2).

Whole genome sequencing (WGS) technologies for these fruit crops primarily use second-generation sequencing methods, including the Illumina/Solexa Genome analyzer, the Roche/454 platform, and the ABI/Solid technology. Second-generation sequencing technologies are cost-effective, highly efficient, and cover a wider area of the genome than first-generation sequencing technologies. The estimated gene numbers of these 17 fruit crops varied from 14,777 to 57,386. The largest and

DOI: 10.1201/9781003354055-20

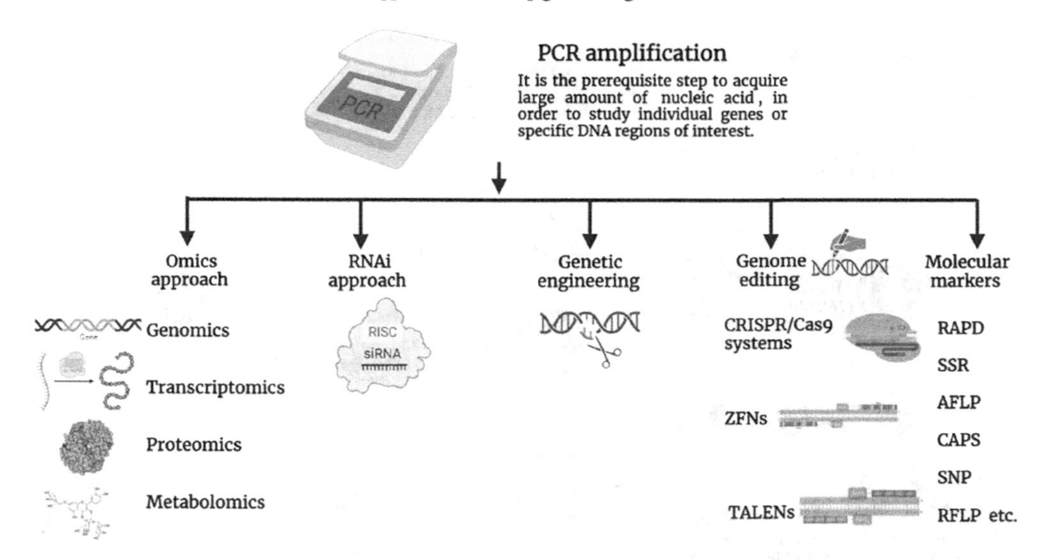

FIGURE 20.1 Schematic representation of different molecular approaches used for crop growth regulation.

the smallest estimated genome sizes, respectively, were reported in tomatoes and peaches. In recent times, there has been a comprehensive examination of the genomic data pertaining to fruit crops, resulting in a substantial accumulation of resources and expertise in this field. This phenomenon has been enabled by the improvement in sequencing efficiency and the reduction in sequencing expenses. These sources of information serve as vital references for further studies of biotic and abiotic stress resistance mechanisms of fruit crops (Li et al., 2019).

The presence of a nucleotide binding site (NBS) in the proteins encoded by NBS genes was observed, indicating its function as a cellular detector for pathogen effectors (Jones et al., 2006; Li et al., 2019). The genes associated with the illness constitute a substantial portion of the grapevine genome. The genome series data by Velasco et al. (2007) identified 341 NBS genes in grapevine. Several resistance genes, which encode signaling pathways involved in the plant's response to disease, have been identified within the grapevine genome. The set of proteins mentioned consists of PAD4, RAR1, NDR1, COI1, EDS1, NPR1, MPK4, JAR1, ETR1, and EIN2, as shown by Grant and Lamb (2006) and Velasco et al. (2007). The grapevine genome has facilitated the identification of specific sequences of DNA that encode pathogenesis-related proteins (PRs), such as PR-1, PR-2, PR-3, PDF1, PDF2, and PR5. The discovery of powdery mildew resistance-related genes PEN1, PEN2, and PEN3 was made based on the analysis of grapevine genome sequence data (Van et al., 2006; Velasco et al., 2007). In a study conducted by Velasco et al. (2007), it was shown that there are eight genes (MLO1–MLO8) that exhibit complete similarity to the MLO gene family, known for its ability to confer resistance against mildew. A distinct research group conducted transcription of seven grapevine MLO genes, which demonstrated the expression of grapevine MLO genes in reaction to biotic stress. The potential of the encoded protein to enhance resistance against mildew infections in plants suggests that the MLO grapevine genes could serve as functional genes to bolster resistance against powdery mildew in grapevines and other fruit crops (Chen et al., 2006; Feechan et al., 2008; Li et al., 2019). A total of 992 NBS genes have been identified within the apple genome, which are responsible for encoding resistance (R) proteins. Nevertheless, it has been observed that the papaya fruit (57) possesses a lower number of NBS genes compared to other fruits, as indicated by studies conducted by Ming et al. (2008) and Velasco et al. (2010). During the course of plant evolution, certain NBS genes have undergone tandem duplications, resulting in their non-random distribution across the chromosomes. Based on comprehensive studies and whole

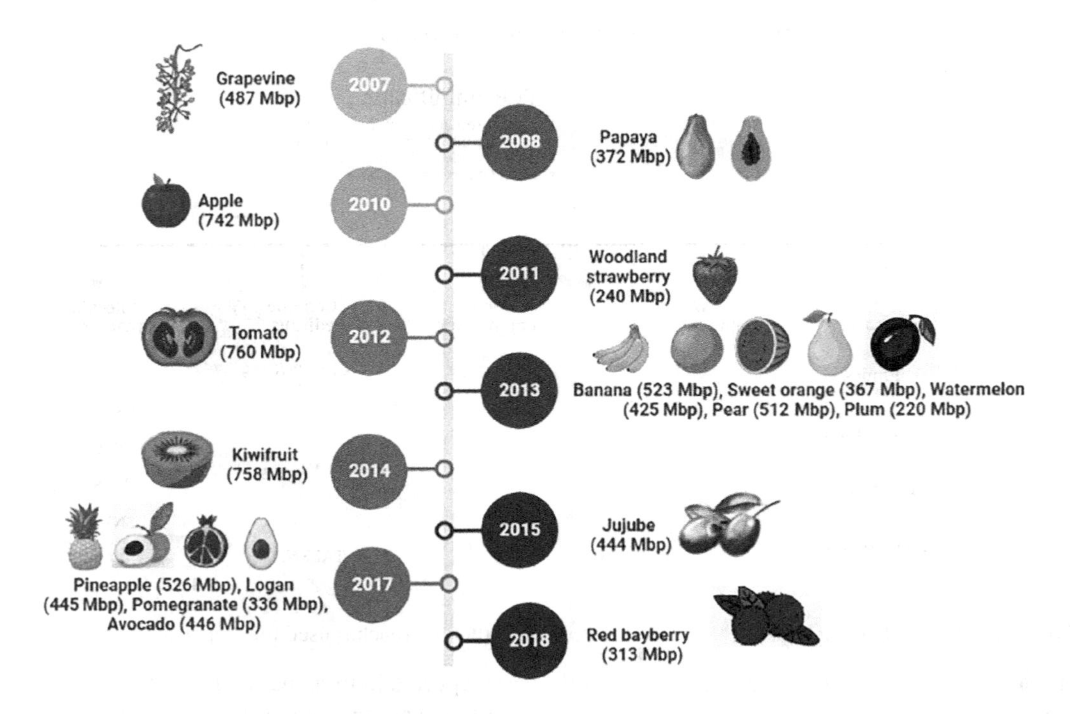

FIGURE 20.2 A timeline depicting the genome sequencing of different fruit crops.

genome sequencing, the following numbers of NBS genes have been discovered in various plant species: 341 in grapevine (Velasco et al., 2007), 251 in tomato (Santo et al., 2012), 44 in watermelon (Guo et al., 2013), 96 in kiwifruit (Huang et al., 2013), and 594 in longan (Lin et al., 2017). The apple genome consists of various disease-resistance genes, such as EDS1, RAR, NPR1, ETR1, MPK4, PR1, PR2, PR3, and PR5 (Velasco et al., 2010). According to Qin et al. (2017), the genome of the pomegranate has a total of 995 genes responsible for encoding protein kinases, along with 710 R genes. The jujube genome analysis revealed the identification of 839 resistance (R) genes. Notably, around 16% of these genes were found to be distributed across chromosome 9. The R genes have been classified into seven distinct groups, namely CC-NBS-LRR, CC-NBS, LRR-RLK, NBS-LRR, NBS, TIR-NBS-LRR, and TIR-NBS, as documented by Liu et al. (2014).

Both the genomes of the papaya and grapevine species encompass genes belonging to the lipoxygenase (LOX)family, which have been found to have a significant role in plant defense mechanisms and resistance against pests. In the genomes of papaya and grapevine, there were 15 and 18*LOX* genes, respectively (Li et al., 2019). Numerous biotic stresses, such as insect or pathogen attacks, have an impact on the growth and development of fruit crops as well as fruit yield and quality. Fruit crops have varying levels of resilience to various biotic stresses. The adoption of diverse methodologies is crucial in acquiring comprehensive knowledge regarding signal transduction, as well as the physiological and biochemical mechanisms behind the response of fruit crops to biotic stress. This knowledge is vital for enhancing the resilience of these crops against such stressors. Omic methods for acquiring a massive amount of information include transcriptomics, proteomics, and metabolomics. The genomes of several fruit cultivars have undergone extensive sequencing due to the development of sequencing technologies. According to Li et al. (2019), the utilization of advanced sequencing technologies, such as new or next-generation sequencing, holds promise for the comprehensive sequencing of a wider range of fruit crops in the near future.

20.4 MOLECULAR MARKERS

Molecular markers are 'gene tags' that are strongly connected to genes. The use of DNA markers has different uses in fruit crop progress particularly in areas of genetic diversity and their assessment, identification of parents, cultivar sand variety, disease tolerance, high-resolution development of genetic linkage groups, identification of QTLs, hybrid identification, sex differentiation, etc. (Sonwane, 2018; Goswami et al., 2022).

20.4.1 Genetic Diversity Assessment of Fruit Crops

There are numerous research articles on the practice of DNA markers to determine the genomic relatedness of various horticulture crop species and to evaluate genetic diversity among them. This has broad implications, particularly for woody perennials that are challenging to breed. A study conducted on the mandarin germplasm discovered in the Northeast Himalayas revealed a significant degree of diversity within the citrus species, as indicated by the use of RAPD markers. In China, SSR markers were employed to elucidate the genetic variety present in mandarin landraces, as well as wild races of sweet oranges, mandarins, grapefruit, lemon, and citranges (Lassois et al., 2016; Goswami et al., 2022).

20.4.2 Varietal Identification

Varietal identification refers to the practice of DNA fingerprinting. Molecular markers can produce patterns that are specific to each genotype, whether they are used singly or in groups. The patterns generated, irrespective of the method employed (PCR or hybridization with single, multiple, or repeating sequences), are commonly denoted as genetic fingerprints (Dhyani et al., 2015; Nwosisi et al., 2019; Goswami et al., 2022).

20.4.3 Disease Diagnostics

The utilization of molecular markers has facilitated the development of accurate and dependable diagnostic techniques for pathogen identification. Furthermore, genes derived from diverse sources such as microbes, plants, and animals have been employed by researchers to enhance the production of disease-resistant plants (Islam et al., 2019; Goswami et al., 2022). There have been several viral diseases among them Banana bunchy top virus (BBTV), Banana bract mosaic virus (BBrMV), Banana streak virus (BSV), Papaya ringspot virus (PRSV), Apple mosaic virus (ApMV, Citrus tristeza virus (CTV), etc., are considered important. An outbreak of BBTV has destroyed a large area under banana cultivation. Another severe viral disease that attacks the banana fruit crop is BBrMV. To prevent the disease from spreading, especially to a new location, early and precise virus diagnosis is essential (Sonwane, 2018).

20.4.4 Construction of Linkage and QTL Maps

The utilization of DNA markers in agricultural research primarily involves the development of linkage maps for diverse crop species. To locate chromosomal regions containing quantitative traits and single gene traits (regulated by a single gene), linkage maps are utilized. The combined activity of multiple genes results in the development of significant heritable traits. Such traits are commonly described as polygenic or quantitative. Numerous agronomic characteristics shown by plant species, such as drought resistance, yield, and maturation time, are subject to quantitative inheritance. The genetic locus associated with these characteristics has been often referred to as the quantitative trait loci (QTLs). The key aspect that enables both the characterization and identification of a quantitative trait loci (QTL) is its linkage to a recognized marker locus that segregates as per Mendelian ratios. This potential is made possible by molecular markers, which enable the identification, mapping, and assessment of the effects of the genes underlying quantitative traits (Gulsen et al., 2011; Goswami et al., 2022).

20.4.5 Marker-Assisted Selection

The potential enhancement of the use of indirect selection in crop improvement can be achieved by the utilization of molecular markers, hence increasing its importance and efficacy. Marker-assisted selection (MAS) enables the breeder to make better and early decisions regarding future selection while studying fewer plant species. Once marker information is available, breeding for disease-resistant behavior can be achieved without a pathogen, resulting in an additional advantage. Previously, markers were mainly developed for monogenic traits controlled by single genes, but currently, markers are being developed for traits controlled by multigene or polygenes (Goswami et al., 2022). In previous studies, it was hypothesized that markers that are directly associated with genes or quantitative trait loci (QTLs) in core QTL mapping might be effectively employed in marker-assisted selection (MAS) without any intermediary steps. Molecular marker-assisted breeding (MAB), alternatively referred to as molecular-assisted breeding, entails the manipulation and enhancement of plant characteristics through the utilization of genotypic assays facilitated by molecular biotechnologies, namely molecular markers, in conjunction with linkage maps and genomics (Jiang, 2013). The aforementioned nomenclature is employed to distinguish between many recent breeding methodologies, including genome-wide selection (GWS) or genomic selection (GS), marker-assisted selection (MAS), marker-assisted backcrossing (MABC), and marker-assisted recurrent selection (MARS) (Ribaut et al., 2010). MAB has been widely used in multiple crop species and is recognized as a potent tool for the genetic improvement of crop plants (Diaz et al., 2011; Xu, 2012; Jiang, 2013; Judith et al., 2021). However, MAB has not yet been extremely effective in comparison to the resources used and the expectations set.

20.4.6 Marker-Assisted Gene Pyramiding

Molecular markers have the capability to facilitate gene pyramiding by effectively identifying and locating many genes within plants, wherein the phenotypic manifestations of such genes present considerable challenges in terms of segregation. The integration of several disease resistance genes (also known as the integration of qualitative resistance genes) into a single genotype is the most common use of pyramiding. Because pathogens often outgrow single-gene resistance over time due to the introduction of new plant pathogen strains, the goal of the study is to develop "durable" or stable disease resistance. According to some research, combining several genes (that work against certain disease strains) can result in long-lasting resistance. In the past, it was challenging to stack resistance genes together because their phenotypes were frequently similar. It is simple to estimate the amount of resistance genes present in each plant using linked DNA markers. The potential to use molecular markers to improve fruit crops genetically will depend on close collaboration between plant breeders and biotechnologists, the availability of qualified labor, and considerable financial investment in research (Bhat et al., 2010;Li et al., 2019; Goswami et al., 2022).

20.5 RNA INTERFERENCE

The study of modern biology, especially biotechnology, presents several advantages in comparison to conventional methods of crop improvement. RNA Interference (RNAi) has emerged as a significant biotechnological tool in the field of fruit crop improvement. RNA interference (RNAi) is a method of post-transcriptional gene silencing (PTGS) that is evolutionarily conserved. It involves the degradation of mRNA sequences in a sequence-specific manner, which is triggered by the presence of double-stranded RNA (dsRNA). RNA interference (RNAi)-mediated gene silencing has emerged as a highly effective approach for elucidating gene function and improving agronomic characteristics in crops. This technique involves the targeted suppression of pathogen/pest genes and plant genes associated with desired trait enhancement (Jain et al., 2018). Due to the general exponential growth in available genomic and transcriptome sequence databases, extremely precise targeting dsRNAs may now be synthesized, reducing the possibility of off-target effects or silencing in non-target individuals (Christiaens et al., 2018). Normally, there is ssRNA, but by chance, if dsRNA is produced inside the cell, it will result in the activation of the RNAi pathway. dsRNA acts as a precursor for miRNA, shRNA, and siRNA activation. miRNA, siRNA, and shRNA play a central role in the RNA interference pathway. A few examples of RNA interference techniques used in the modification of fruit crops are mentioned in Table 20.1.

TABLE 20.1
Modification in Fruit Crops *via* RNA Interference

Crop(s)	Target Gene(s)	Outcomes	Reference
Tomato	Suppression of *DET1* gene	Increase in nutritional content	Davuluri et al. (2005)
	Suppression of *polygalacturonase* (*PG*) gene	enhanced post-harvest shelf life	Krieger et al. (2008)
	Suppression of *SlNCED1* gene	Ameiolorate the coloring pigments	Sun et al. (2012)
	ACC synthase gene	Delayed ripening	Gupta et al. (2013)
	Silencing of *SlERF A1, SlERF B4, SlERF C3* and *SlERF A3* gene	Decrease the plant susceptibility to *Botrytis cinerea*	Ouyang et al. (2016)
	Downregulated the activity of ζ-carotene desaturase (ZDS)	Variegated tomato transformants with increased phytoene content	Babu et al. (2021)
	Overexpression of SlMYB75 gene	Improves the tolerance to *Botrytis cinerea* and prolongs fruit storage life	Liu et al. (2021)
	Down-regulation of *SlGRAS10*	dwarf plants with smaller leaves, internode lengths, and enhanced flavonoid accumulation	Habib et al. (2021)
	Silencing of *SLZF57* gene	Lower down the drought stress Problem	Gao et al. (2022)
Apple	*Mal d1* gene	10-fold reduction in leaf expression deprived of distressing phenotypic characteristics	Gilissen et al. (2005)
	Silencing of *MdTFL1* gene	Early flowering	Szankowski et al. (2008)
	Suppression of *MdGA20-ox* gene	Decrease the plant height	Zhao et al. (2016)
	Suppression of *AGAMOUS* gene	Promoted floral attractiveness	Klocko et al. (2016)
	Overexpression of *MdHB7* gene	Enhanced photosynthates production	Zhao et al. (2021)

(Continued)

TABLE 20.1 (*Continued*)
Modification in Fruit Crops *via* RNA Interference

Crop(s)	Target Gene(s)	Outcomes	Reference
Strawberry	Silencing of *chalcone synthase* (CHS) gene	Increases in levels of (hydroxy) cinnamoyl glucose esters.	Hoffmann et al. (2006)
	Silencing of *SEP1/2*- like (*FaMADS9*) gene	Retards the ripening and development of petal, achene, and receptacle tissues	Seymour et al. (2011)
	Down regulation of *FaPYR1* gene	Delayed ripening process	Chai et al. (2011)
	Silencing of *Flavanone 3-Hydroxylase* gene	Decreased anthocyanin content	Jiang et al. (2013)
	Down regulation and overexpression of *FaMYB1* gene	Down regulation of *FaMYB1* gene increases anthocyanin content and overexpression decrease anthocyanin content	Kadomura-Ishikawa et al. (2015)
	Down regulation and overexpression of *FaTPK1* gene	Promoted fruit ripening	Wang et al. (2018)
	Up and down regulation of *FaSnRK1α* gene	Overexpression of *FaSnRK1α* gene increases sucrose content whereas down regulation decreases sucrose content	Luo et al. (2020)
	Silencing of *FaPG1*gene	Increased fruit firmness and resistance to fungal decay	Paniagua et al. (2021)
	Silencing of *FaAKR23* gene	Decreased ascorbic acid and anthocyanin content	Wei et al. (2022)
	Overexpression of *FaSnRK1α* gene	Tolerance to waterlogging	Luo et al. (2022)
	Overexpression of -*FvemiR167b* and RNAi-*FveARF6* transgenic lines	Increased number of roots and leaves	Li et al. (2022)
Citrus	Suppression of the *callose synthase 1* gene	Resistance against *Xanthomonas citri subsp. citri* (*Xcc*)	Enrique et al. (2011)
	p25, *p20*and *p23*	Citrus tristeza virus (CTV) resistant	Soler et al. (2012)
Papaya	Down regulation of *1-aminocyclopropane-1-carboxylic acid oxidase* gene	Prolong shelf life	Sakeli et al. (2014)
Banana	*DNA-R,BBTV* viral genome	Resistance to banana bunchy top disease	Elayabalan et al. (2013) and Elayabalan et al. (2017)
	Silencing of *Foc*TR4 *ERG6/11* genes	Resistance to fusarium wilt	Dou et al. (2020)
	Silencing of*MaBAM9b* gene	Decrease in the level of starch content during postharvest fruit ripening	Liu et al. (2021)

20.6 GENOME EDITING TOOLS

Genetic engineering (GE) advancements during the past 10 years have given rise to new technologies. The process of genetically modifying the DNA of an organism by specifically inserting, deleting, or substituting genetic material at a targeted area is commonly known as genetic engineering (GE) (Zhang et al., 2017). The GE toolbox is generally composed of three types: zinc finger nucleases (ZFNs), transcription activator-like effector nucleases (TALENs), and CRISPR/Cas9 systems. These tools utilize engineered endonucleases which, when combined with a sequence-specific DNA-binding domain or RNA, induce a double-strand break in the DNA at a predetermined location. Upon the occurrence of double-strand breaks (DSBs) in DNA, biological processes for DNA repair, namely homology-directed repair (HDR) and error-prone non-homologous end joining (NHEJ), are initiated (Xiong et al., 2015). Zinc finger nucleases (ZFNs) and transcription activator-like effector nucleases (TALENs) have demonstrated their efficacy in plant systems, showcasing their potential applications. However, the emergence of CRISPR/Cas9 technology has revolutionized the field of genetic engineering (GE) by significantly enhancing the ability to enhance many plant species, including fruit crops. SgRNA and the Cas9 gene are the two main components of the engineered CRISPR/Cas9 tool (Ma et al., 2016; Zhang et al., 2017). The Type II Cas9 (SpCas9) gene, commonly employed in scientific research, is derived from Streptococcus pyogenes with a length of 4107 base pairs (Adli, 2018; Wang et al., 2016a). The Cas9 protein contains two distinct nuclear domains, namely RuvC and HNH. The nuclear domain cleaves a single strand of the target DNA, leading to the formation of a double-stranded break in the target DNA sequence. The protospacer-matching crRNA and the transactivating crRNA are combined to create the engineered non-coding RNA known as the sgRNA (tracrRNA). The CRISPR/Cas9 system's effectiveness and practicality are enhanced by the single guide RNA (sgRNA), which is derived from a single gene and facilitates the replication of the initial crRNA: tracrRNA duplex structure (Doudna and Charpentier, 2014; Hsu et al., 2014; Jinek et al., 2012). The utilization of the CRISPR/Cas9 method for the genetic modification of fruit crops is experiencing a notable surge in research initiatives. Table 20.2 provides a number of examples showcasing the application of genome editing technologies in the alteration of fruit crops.

TABLE 20.2
Genome Editing Technologies are Employed in the Manipulation of Fruit Crops

Crop(s)	Target Gene(s)	Genome Editing Tool Used	Traits	References
Sweet orange	*CsPDS*	CRISPR/Cas9	Carotenoid formation	Jia and Wang (2014)
Fig	*uidA*	ZFNs	Improve the performance of Beta-glucuronidase	Peer et al. (2014)
Apple	*uidA*	ZFNs	Ameliorated the performance of Beta-glucuronidase	Peer et al. (2014)
	DIPM-1, *DIPM-2*, and *DIPM-4*	CRISPR/Cas9	Fight against Fire blight diseases	Malnoy et al. (2016)
	Phytoene Desaturase (*PDS*)	CRISPR/Cas9	Carotenoid formation	Nishitani et al. (2016)
	MdMAPKKK1	CRISPR/Cas9	Resistance to White rot of apple	Wang et al. (2022)
'Duncan' grapefruit	*CsLOB1* gene	-	Resistance to Citrus canker	Jia et al. (2016)

(Continued)

TABLE 20.2 (*Continued*)
Genome Editing Technologies are Employed in the Manipulation of Fruit Crops

Crop(s)	Target Gene(s)	Genome Editing Tool Used	Traits	References
Citrus spp.	*Phytoene desaturase* gene	-	Target specific mutations	Zhang et al. (2017)
'Duncan' grapefruit	*CsPDS* and *Cs2g12470 gene*	SaCas9/sgRNA	No off-target mutations	Jia et al. (2017)
Grapes	*L-idonate dehydrogenase gene IdnDH*	CRISPR/Cas9	Tartaric acid formation	Ren et al. (2016)
	MLO-7	CRISPR/Cas9	Reduces the susceptibility of powdery mildew	Malnoy et al. (2016)
	VvWRKY52	CRISPR/Cas9	Reduces susceptibility to *Botrytis cinerea*	Wang et al. (2018a)
Watermelon	*ClPDS*	CRISPR/Cas9	Carotenoid biosynthesis	Tian et al. (2017)
Wanjincheng orange	The promoter region of *CsLOB1* gene	CRISPR/Cas9	Resistance to Citrus canker	Peng et al. (2017)
Woodland strawberry	*TAA1* and *ARF8*	CRISPR/Cas9	Auxin biosynthesis and signaling	Zhou et al. (2018)
Cultivated strawberry	*FaTM6*	CRISPR/Cas9	Flower development	Martin-Pizarro and Pose (2018)
Kiwifruit	*AcPDS*	CRISPR/Cas9	Carotenoid biosynthesis	Wang et al. (2018b)
Groundcherry	Self-Pruning *SP gene*	CRISPR/Cas9	Modulate sympodial growth	Lemmon et al. (2018)
Banana	*MaPDS*	CRISPR/Cas9	Carotenoid biosynthesis	Kaur et al. (2018)
Pear	*MDPS*	CRISPR/Cas9	Promote the early blooming process	Charrier et al. (2019)
Wanjincheng Orange	*CsWRKY22* gene	-	Citrus canker	Wang et al. (2019)
'Duncan' grapefruit	*CsPDS* gene	CRISPR/Cas12a (Cpf1)	High mutation rate	Jia et al. (2019)
Hongkong kumquat	-	CRISPR/Cas9	minimizing the generation process	Zhu et al. (2019)
Pummelo	*LOB1*	SpCas9	Homozygous canker resistant genotypes	Jia and Wang (2020)
Carrizo citrange and Hamlin sweet orange	*CsLOB1* gene	-	Canker resistant	Huang and Wang (2021)
Citrus spp.	-	RPA-CRISPR/Cas12a	Detection of citrus scab	Shin et al. (2021)
'Duncan' grapefruit and Carrizo citrange	*CsDMR6* gene		Canker resistant varieties	Parajuli et al. (2022)
'Duncan' Grapefruit	-	sgRNAs	Canker resistant; no off-target mutation	Jia et al. (2022)

20.7 GENETIC ENGINEERING

For many years, the fruit industry has effectively exploited breeding to create the majority of the commercial fruit varieties we see today. More recently, some of the limitations of traditional breeding have been overcome by innovative molecular breeding approaches. Currently, there are only five commercially available genetically engineered fruits. Among these, virus-resistant papaya and squash were introduced as commercial varieties 25 years ago. In more recent years, insect-resistant eggplant, non-browning apple, and pink-fleshed pineapple have received approval for commercialization. The production of these genetically engineered fruits has been steadily increasing over the past 6 years (Lobato-Gomez et al., 2021).

The development of molecular technologies, particularly the most recent wave of genome editing tools, offers chances to create novel fruit varieties more quickly. The Flavr Savr tomato, the first GE fruit product, was legalized in 1992 and released on the market in 1994 (Baranski et al., 2019). The transgenic fruit's longer shelf life and delayed fruit softening were caused by the down-regulation of a gene that initiates pectin solubilization (Kramer and Redenbaugh, 1994; Lobato-Gomez et al., 2021). Additional fruit crops that have had their features enhanced through genetic engineering have been given regulatory approval for commercialization in various regions of the world. These crops are either grown for human consumption or animal feed.

These include the apple (*Malus domestica* Borkh.) (USDA, 2020), pineapple (*Cucumis melo* L.) (Firoozbady et al., 2015), papaya (*Carica papaya* L.), plum (*Prunus domestica*) (Scorza et al., 1994), tomato (*Solanum lycopersicum*) (APHIS, 1998) and papaya (*Ananas comosus* L. Merr.) (Chen et al., 2001). The majority of transgenic fruits have been created to have delayed ripening, pest or disease resistance, or both to increase agronomic productivity. The quality attributes of more contemporary products, however, have been addressed by removing fruit browning or including new aesthetic characteristics like flesh color. Because they were not commercially feasible (such as the Flavr Savr tomato) (Bruening and Lyons, 2000; Baranski et al., 2019) or were never commercialized, certain transgenic fruit crops have been removed from the market (Melon A and B) (APHIS, 1999; Biosafety Clearing-House, 2020).

New techniques for creating superior fruit varieties have been made available by advancements in genetic engineering, notably the advent of genome editing technology. There have been several reported proof-of-concept cases involving fruit crops, and further research and commercialization of such types might have a significant socioeconomic impact. Several countries have recently revised existing laws and regulations or developed new ones to govern genome-edited plants and their products (Menz et al., 2020). This may facilitate the ability for genome-edited fruits, like all other genome-edited crops to reach the market sooner in countries with a policy supportive of genome editing (Alvarez et al., 2021).

20.7.1 Genetically Engineered Fruit Crops

To meet the needs of growers and/or consumers, genetically modified fruits have been created with distinctive agronomic traits that are frequently challenging to obtain through conventional breeding. Some of the genetically modified fruit crops are discussed below.

i. **Papaya**

The papaya ringspot virus (PRSV) was first identified in 1992 in Puna, which is the primary region for papaya production in Hawaii. The absence of Papaya Ringspot Virus (PRSV) resistance was observed in both papaya germplasm and wild Carica species, hence indicating their limited suitability as possible prospects for interspecific hybridization. Moreover, the efficacy of pesticides in managing the aphid vectors responsible for transferring the virus was found to be inadequate, resulting in the evacuation of several orchards due to the prevalence of PRSV infection (Lobato-Gómez et al., 2021). To create the transgenic papaya 'SunUp,' which is resistant to PRSV in Hawaii, a gene from a Hawaiian strain was added to the commonly cultivated 'Sunset' papaya (Fitch et al., 1992). The yellow-fleshed, PRSV-resistant "Rainbow" papaya was created by mating "SunUp" papaya with "Kapoho," a non-engineered variety. Since 1950, the PRSV virus has been a hazard to the papaya crop in China (Gonsalves and Ferreira, 2003). The Huanong No. 1 papaya, which has been genetically modified, exhibits resistance to the four primary strains of Papaya Ringspot Virus (PRSV) found in the southern regions of China. These strains are known as Ys, Vb, Sm, and Lc (Ye et al., 2010). This resistance is similar to that of the "SunUp" variety. In comparison to the original cultivar, Huanong No. 1 also yields larger fruits with thicker flesh. Some of the Huanong No. 1 papayas grown in Hainan showed symptoms like PRSV in 2012, indicating that resistance is starting to break down. A novel virus family that may be a threat to the cultivation of Huanong No. 1 papaya was discovered in Hainan and Guangdong papaya plantations after phylogenetic research (Wu et al., 2018).

ii. **Apple**

The activity of polyphenol oxidases (PPOs), oxidize phenolic substances, and induce progressive browning in soft fruits like apples, has an impact on fruit quality. When fruits are damaged, peeled, or sliced, PPOs are triggered by oxygen exposure, which causes browning. Storage in an atmosphere devoid of air, irradiation to render PPOs inactive, usage of chemical inhibitors and natural antioxidants, and enzymatic browning can be avoided (Moon et al., 2020). PPOs were silenced to develop the Arctic® apple idea (APHIS, 2012; Stowe et al., 2021). Arctic® Golden Delicious, Arctic® Granny Smith, and Arctic® Fuji are the three commercially available types of the Arctic® apple at present. Arctic® Golden Delicious and Arctic® Granny Smith's commercial harvest began in 2016, and Arctic® Fuji was available on markets in 2021, (Okanagan Specialty Fruits. Arctic Media Kit, 2023)

iii. **Pineapple**

Fruits exhibiting contrasting skin and flesh hues have been successfully generated using conventional breeding techniques (Zhang et al., 2020) as well as genetic manipulation approaches (Espley et al., 2007). The Pinkglow transgenic pineapple was developed in 2005 through genetic alteration of the carotenoid pathway, resulting in the accumulation of lycopene in the pink flesh of this fruit (Firoozbady and Young, 2015). In contrast to traditional pineapple, which is green and yellow, the Pinkglow pineapple's skin is a blend of green, yellow, orange, and red colors. In addition to regulating carotenoid accumulation, a gene linked to endogenous ethylene biosynthesis was inhibited to regulate blooming, however, this trait has not yet been examined (Firoozbady and Young, 2015; Tiwari et al., 2020).

20.8 CONCLUSION

The growth and development of fruit crops have proved successful by using molecular techniques. In particular, the study of plant growth, development, and gene functioning has been made feasible by the integration of genetics with biotechnology and molecular biology. At several levels of study, from gene sequence and expression to protein and specific traits, molecular approaches enable the synthesis of a substantial amount of information that will help understand fruit crop development and growth regulation at the gene level.

REFERENCES

Adli, Mazhar. "The CRISPR tool kit for genome editing and beyond." *Nature communications* 9, no. 1 (2018):1–13.

Alvarez, Derry, Pedro Cerda-Bennasser, Evan Stowe, Fabiola Ramirez-Torres, Teresa Capell, Amit Dhingra, and Paul Christou. "Fruit crops in the era of genome editing: closing the regulatory gap." *Plant Cell Reports* 40, no. 6 (2021):915–930.

APHIS. Interpretative rule on Okanagan Speciality Fruits petition for determination of nonregulated status: Arctic™ Apple (*Malus×domestica*). Events GD743 and GS784. Docket No. APHIS-2012-0025 (2012).

APHIS. Interpretive ruling on Monsanto petition of nonregulated status for insect resistant tomato line 5345. Docketnumber:97-114-1 (1998).

APHIS. Petition for determination of nonregulated status—Melon. Petition number98-350-01p (1999).

Babu, Merlene Ann, Ramachandran Srinivasan, Parthiban Subramanian, and Gothandam Kodiveri Muthukaliannan. "RNAi silenced ζ-carotene desaturase developed variegated tomato transformants with increased phytoene content." *Plant Growth Regulation* 93, no. 2 (2021):189–201.

Baranski, Rafal, Magdalena Klimek-Chodacka, and Aneta Lukasiewicz. "Approved genetically modified (GM) horticultural plants: a 25-year perspective." *Folia Horticulturae* 31, no. 1 (2019):3–49.

Bhat, Zahoor Ahmad, Wasakha Singh Dhillon, Rizwan Rashid, Javid Ahmad Bhat, Waseem Ali Dar, and Mohammad Yousf Ganaie. "The role of molecular markers in improvement of fruit crops." *Notulae Scientia Biologicae* 2, no. 2 (2010):22–30.

Biosafety Clearing-House. Melon A and B. Country's decision or any other communication (1999). Available at: https://bch.cbd.int/database/record.shtml?documentid=6351 [Accessed Sept 2022].

Bruening, George, and James Lyons. "The case of the Flavr savr tomato." *California Agriculture* 54, no. 4 (2000):6–7.

Chai, Ye-mao, Hai-feng Jia, Chun-li Li, Qing-hua Dong, and Yuan-yue Shen. "*FaPYR1* is involved in strawberry fruit ripening." *Journal of Experimental Botany* 62, no. 14 (2011):5079–5089.

Charrier, Aurélie, Emilie Vergne, Nicolas Dousset, Andréa Richer, Aurélien Petiteau, and Elisabeth Chevreau. "Efficient targeted mutagenesis in apple and first time edition of pear using the CRISPR-Cas9 system." *Frontiers in Plant Science* 10 (2019):40.

Chen, Guanghao, C.M. Ye, Jiandong Huang, Min Yu, and B.J. Li. "Cloning of the papaya ringspot virus (PRSV) replicase gene and generation of PRSV-resistant papayas through the introduction of the *PRSV* replicase gene." *Plant Cell Reports* 20, no. 3 (2001):272–277.

Chen, Zhongying, H. Andreas Hartmann, Ming-Jing Wu, Erin J. Friedman, Jin-Gui Chen, Matthew Pulley, Paul Schulze-Lefert, Ralph Panstruga, and Alan M. Jones. "Expression analysis of the *AtMLO* gene family encoding plant-specific seven-transmembrane domain proteins." *Plant Molecular Biology* 60, no. 4 (2006):583–597.

Christiaens, Olivier, Teodora Dzhambazova, Kaloyan Kostov, Salvatore Arpaia, Mallikarjuna Reddy Joga, Isabella Urru, Jeremy Sweet, and Guy Smagghe. "Literature review of baseline information on RNAi to support the environmental risk assessment of RNAi-based GM plants." *EFSA Supporting Publications* 15, no. 5 (2018):1424E.

D'hont, Angélique, France Denoeud, Jean-Marc Aury, Franc-Christophe Baurens, Françoise Carreel, Olivier Garsmeur, Benjamin Noel et al. "The banana (Musa acuminata) genome and the evolution of monocotyledonous plants." *Nature* 488, no. 7410 (2012):213–217.

Davuluri, Ganga Rao, Ageeth Van Tuinen, Paul D. Fraser, Alessandro Manfredonia, Robert Newman, Diane Burgess, David A. Brummell et al. "Fruit-specific RNAi-mediated suppression of *DET1* enhances carotenoid and flavonoid content in tomatoes." *Nature Biotechnology* 23, no. 7 (2005):890–895.

Dhyani, Praveen, Amit Bahukhandi, Arun Jugran, and Indra D. Bhatt. "Inter Simple Sequence Repeat (ISSR) markers based genetic characterization of selected Delicious group of apple..." *International Journal* 3, no. 2 (2015):591–598.

Diaz, Aurora, Mohamed Fergany, Gelsomina Formisano, Peio Ziarsolo, José Blanca, Zhanjun Fei, Jack E. Staub et al. "A consensus linkage map for molecular markers and Quantitative Trait Loci associated with economically important traits in melon (Cucumis meloL.)." *BMC Plant Biology* 11, no. 1 (2011):1–14.

Dou, Tongxin, Xiuhong Shao, Chunhua Hu, Siwen Liu, Ou Sheng, Fangcheng Bi, Guiming Deng et al. "Host-induced gene silencing of *Foc TR4 ERG6/11* genes exhibits superior resistance to Fusarium wilt of banana." *Plant Biotechnology Journal* 18, no. 1 (2020):11.

Doudna, Jennifer A., and Emmanuelle Charpentier. "The new frontier of genome engineering with CRISPR-Cas9." *Science* 346, no. 6213 (2014):1258096.

Elayabalan, Sivalingam, Kalaimughilan Kalaiponmani, Sreeramanan Subramaniam, Ramasamy Selvarajan, Radha Panchanathan, Ramlatha Muthuvelayoutham, Krish K. Kumar, and Ponnuswami Balasubramanian. "Development of Agrobacterium-mediated transformation of highly valued hill banana cultivar Virupakshi (AAB) for resistance to BBTV disease." *World Journal of Microbiology and Biotechnology* 29, no. 4 (2013):589–596.

Elayabalan, Sivalingam, Sreeramanan Subramaniam, and Ramasamy Selvarajan. "Construction of *BBTV* rep gene RNAi vector and evaluate the silencing mechanism through injection of Agrobacterium tumefaciens transient expression system in BBTV infected hill banana plants cv. Virupakshi (AAB)." (2017).

Enrique, Ramón, Florencia Siciliano, María Alejandra Favaro, Nadia Gerhardt, Roxana Roeschlin, Luciano Rigano, Lorena Sendin, Atilio Castagnaro, Adrian Vojnov, and María Rosa Marano. "Novel demonstration of RNAi in citrus reveals importance of citrus callose synthase in defence against Xanthomonas citri subsp. citri." *Plant Biotechnology Journal* 9, no. 3 (2011):394–407.

Espley, Richard V., Roger P. Hellens, Jo Putterill, David E. Stevenson, Sumathi Kutty-Amma, and Andrew C. Allan. "Red colouration in apple fruit is due to the activity of the MYB transcription factor, *MdMYB10*." *The Plant Journal* 49, no. 3 (2007):414–427.

Feechan, Angela, Angelica M. Jermakow, Laurent Torregrosa, Ralph Panstruga, and Ian B. Dry. "Identification of grapevine *MLO* gene candidates involved in susceptibility to powdery mildew." *Functional Plant Biology* 35, no. 12 (2008):1255–1266.

Firoozbady, Ebrahim, and Thomas R. Young. "Pineapple plant named 'Rosé'." U. S. Patent Application13/507, 101, filed August 4, 2015.

Fitch, Maureen M.M., Richard M. Manshardt, Dennis Gonsalves, Jerry L. Slightom, and John C. Sanford. "Virus resistant papaya plants derived from tissues bombarded with the coat protein gene of papaya ringspot virus." *Bio/technology* 10, no. 11 (1992):1466–1472.

Gao, Yongfeng, Jikai Liu, Yongfu Chen, Hai Tang, Yang Wang, Yongmei He, Yongbin Ou, Xiaochun Sun, Songhu Wang, and Yinan Yao. "Tomato *SlAN11* regulates flavonoid biosynthesis and seed dormancy by interaction with bHLH proteins but not with MYB proteins." *Horticulture Research* 5 (2018):1–18.

Gilissen, Luud J.W.J., Suzanne T.H.P. Bolhaar, Catarina I. Matos, Gerard J.A. Rouwendal, Marjan J. Boone, Frans A. Krens, Laurian Zuidmeer et al. "Silencing the major apple allergen *Mal d 1* by using the RNA interference approach." *Journal of Allergy and Clinical Immunology* 115, no. 2 (2005):364–369.

Gonsalves, Dennis, and Steve Ferreira. "Transgenic papaya: a case for managing risks of papaya ringspot virus in Hawaii." *Plant Health Progress* 4, no. 1 (2003):17.

Goswami, Manika, Kaushal Attri, and Isha Goswami. "Applications of molecular markers in fruit crops: A review." *International Journal of Economic Plants* 9, no. 2 (2022):121–126.

Grant, Murray, and Chris Lamb. "Systemic immunity." *Current Opinion in Plant Biology* 9, no. 4 (2006):414–420.

Gulsen, Osman, Aydin Uzun, Ubeyit Seday, and Gucer Kafa. "QTL analysis and regression model for estimating fruit setting in young citrus trees based on molecular markers." *Scientia Horticulturae* 130, no. 2 (2011):418–424.

Guo, Shaogui, Jianguo Zhang, Honghe Sun, Jerome Salse, William J. Lucas, Haiying Zhang, Yi Zheng et al. "The draft genome of watermelon (*Citrullus lanatus*) and resequencing of 20 diverse accessions." *Nature Genetics* 45, no. 1 (2013):51–58.

Gupta, Aarti, Ram Krishna Pal, and Manchikatla Venkat Rajam. "Delayed ripening and improved fruit processing quality in tomato by RNAi-mediated silencing of three homologs of *1-aminopropane-1-carboxylate synthase* gene." *Journal of Plant Physiology* 170, no. 11 (2013):987–995.

Habib, Sidra, Yee Yee Lwin, and Ning Li. "Down-regulation of *SlGRAS10* in tomato confers abiotic stress tolerance." *Genes* 12, no. 5 (2021):623.

Hoffmann, Thomas, Gregor Kalinowski, and Wilfried Schwab. "RNAi-induced silencing of gene expression in strawberry fruit (*Fragaria× ananassa*) by agroinfiltration: a rapid assay for gene function analysis." *The Plant Journal* 48, no. 5 (2006):818–826.

Hsu, Patrick D., Eric S. Lander, and Feng Zhang. "Development and applications of CRISPR-Cas9 for genome engineering." *Cell* 157, no. 6 (2014):1262–1278.

Huang, Shengxiong, Jian Ding, Dejing Deng, Wei Tang, Honghe Sun, Dongyuan Liu, Lei Zhang et al. "Draft genome of the kiwifruit *Actinidia chinensis*." *Nature communications* 4, no. 1 (2013):1–9.

Huang, Xiaoen, Yuanchun Wang, and Nian Wang. "Highly Efficient Generation of Canker-Resistant Sweet Orange Enabled by an Improved CRISPR/Cas9 System." *Frontiers in Plant Science* 12(2021).

Islam, Md Rafiqul, Mohammad Rashed Hossain, Hoy-Taek Kim, Denison Michael Immanuel Jesse, Md Abuyusuf, Hee-Jeong Jung, Jong-In Park, and Ill-Sup Nou. "Development of molecular markers for detection of *Acidovorax citrulli* strains causing bacterial fruit blotch disease in melon." *International Journal of Molecular Sciences* 20, no. 11 (2019):2715.

Jain, Pradeep Kumar, Ramcharan Bhattacharya, Deshika Kohli, Raghavendra Aminedi, and Pawan Kumar Agrawal. "RNAi for resistance against biotic stresses in crop plants." In*Biotechnologies of Crop Improvement, Volume*2, pp. 67–112. Springer, Cham, 2018.

Jia, Hongge, and Nian Wang. "Generation of homozygous canker-resistant citrus in the T0 generation using CRISPR-SpCas9p." *Plant Biotechnology Journal* 18, no. 10 (2020): 1990.

Jia, Hongge, and Nian Wang. "Targeted genome editing of sweet orange using Cas9/sgRNA." *PLoS One* 9, no. 4 (2014): e93806.

Jia, Hongge, Vladimir Orbović, and Nian Wang. "CRISPR-LbCas12a-mediated modification of citrus." *Plant Biotechnology Journal* 17, no. 10 (2019):1928–1937.

Jia, Hongge, Vladimir Orbovic, Jeffrey B. Jones, and Nian Wang. "Modification of the PthA4 effector binding elements in Type I *Cs LOB 1* promoter using Cas9/sg RNA to produce transgenic Duncan grapefruit alleviating XccΔpthA4: dCs LOB 1. 3 infection." *Plant Biotechnology Journal* 14, no. 5 (2016):1291–1301.

Jia, Hongge, Yunzeng Zhang, Vladimir Orbović, Jin Xu, Frank F. White, Jeffrey B. Jones, and Nian Wang. "Genome editing of the disease susceptibility gene *Cs LOB 1* in citrus confers resistance to citrus canker." *Plant Biotechnology Journal* 15, no. 7 (2017):817–823.

Jia, Hui-Min, Hui-Juan Jia, Qing-Le Cai, Yan Wang, Hai-Bo Zhao, Wei-Fei Yang, Guo-Yun Wang et al. "The red bayberry genome and genetic basis of sex determination." *Plant Biotechnology Journal* 17, no. 2 (2019):397–409.

Jiang, Fei, Jia-Yi Wang, Hai-Feng Jia, Wen-Suo Jia, Hong-Qing Wang, and Min Xiao. "RNAi-mediated silencing of the flavanone *3-hydroxylase* gene and its effect on flavonoid biosynthesis in strawberry fruit." *Journal of Plant Growth Regulation* 32, no. 1 (2013):182–190.

Jiang, Guo-Liang. "Molecular markers and marker-assisted breeding in plants." *Plant Breeding from Laboratories to Fields* 3(2013):45–83.

Jinek, Martin, Krzysztof Chylinski, Ines Fonfara, Michael Hauer, Jennifer A. Doudna, and Emmanuelle Charpentier. "A programmable dual-RNA–guided DNA endonuclease in adaptive bacterial immunity." *Science* 337, no. 6096 (2012):816–821.

Jones, Jonathan D.G., and Jeffery L. Dangl. "The plant immune system." *Nature* 444, no. 7117 (2006):323–329.

Judith, de la Cruz Marcial, Villegas Monter Angel, and Cruz Izquierdo Serafn. "Identification of zygotic and nucellar seedlings, originated by the largest embryo in mango seeds cv Ataulfo, using simple-sequence repeat (SSR)." *African Journal of Agricultural Research* 17, no. 5 (2021):794–801.

Kadomura-Ishikawa, Yasuko., Katsuyuki. Miyawaki, Akira Takahashi, and Sumihare Noji. "RNAi-mediated silencing and overexpression of the *FaMYB1* gene and its effect on anthocyanin accumulation in strawberry fruit." *Biologia Plantarum* 59, no. 4 (2015):677–685.

Kaur, Navneet, Anshu Alok, Navjot Kaur, Pankaj Pandey, Praveen Awasthi, and Siddharth Tiwari. "CRISPR/Cas9-mediated efficient editing in phytoene desaturase (*PDS*) demonstrates precise manipulation in banana cv. Rasthali genome." *Functional &Integrative Genomics* 18, no. 1 (2018):89–99.

Kaur, Navneet, Praveen Awasthi, and Siddharth Tiwari. "Fruit crops improvement using CRISPR/Cas9 system." In Vijai Singh and Pawan K. Dhar *Genome Engineering via CRISPR-Cas9 System*, pp. 131–145. Academic Press, 2020.

Klocko, Amy L., Ewa Borejsza-Wysocka, Amy M. Brunner, Olga Shevchenko, Herb Aldwinckle, and Steven H. Strauss. "Transgenic suppression of *AGAMOUS* genes in apple reduces fertility and increases floral attractiveness." *PLoS One* 11, no. 8 (2016): e0159421.

Kramer, Matthew G., and Keith Redenbaugh. "Commercialization of a tomato with an antisense *polygalacturonase* gene: The Flavr savr tomato story." *Euphytica* 79, no. 3 (1994):293–297.

Krieger, Elysia K., Edwards Allen, Larry A. Gilbertson, James K. Roberts, William Hiatt, and Rick A. Sanders. "The Flavr Savr tomato, an early example of RNAi technology." *HortScience* 43, no. 3 (2008):962–964.

Lassois, Ludivine, Caroline Denancé, Elisa Ravon, Arnaud Guyader, Rémi Guisnel, Laurence Hibrand-Saint-Oyant, Charles Poncet, Pauline Lasserre-Zuber, Laurence Feugey, and Charles-Eric Durel. "Genetic diversity, population structure, parentage analysis, and construction of core collections in the French apple germplasm based on SSR markers." *Plant Molecular Biology Reporter* 34, no. 4 (2016):827–844.

Lemmon, Zachary H., Nathan T. Reem, Justin Dalrymple, Sebastian Soyk, Kerry E. Swartwood, Daniel Rodriguez-Leal, Joyce Van Eck, and Zachary B. Lippman. "Rapid improvement of domestication traits in an orphan crop by genome editing." *Nature Plants* 4, no. 10 (2018):766–770.

Li, Tian-Yu, Shao-Xi Wang, Xiao-Guang Tang, Xiang-Xiang Dong, and He Li. "The *FvemiR167b-FveARF6* module increases the number of roots and leaves in woodland strawberry." *Scientia Horticulturae* 293 (2022):110692.

Li, Tong, Ya-Hui Wang, Jie-Xia Liu, Kai Feng, Zhi-Sheng Xu, and Ai-Sheng Xiong. "Advances in genomic, transcriptomic, proteomic, and metabolomic approaches to study biotic stress in fruit crops." *Critical Reviews in Biotechnology* 39, no. 5 (2019):680–692.

Li, Xiaojing, Jie Ye, Shoaib Munir, Tao Yang, Weifang Chen, Genzhong Liu, Wei Zheng, and Yuyang Zhang. "Biosynthetic gene pyramiding leads to ascorbate accumulation with enhanced oxidative stress tolerance in tomato." *International Journal of Molecular Sciences* 20, no. 7 (2019):1558.

Lin, Yuling, Jiumeng Min, Ruilian Lai, Zhangyan Wu, Yukun Chen, Lili Yu, Chunzhen Cheng et al. "Genome-wide sequencing of longan (*Dimocarpus longan* Lour.) provides insights into molecular basis of its polyphenol-rich characteristics." *Gigascience* 6, no. 5 (2017): gix023.

Liu, Meng-Jun, Jin Zhao, Qing-Le Cai, Guo-Cheng Liu, Jiu-Rui Wang, Zhi-Hui Zhao, Ping Liu et al. "The complex jujube genome provides insights into fruit tree biology." *Nature Communications* 5, no. 1 (2014):1–12.

Liu, Mengting, Meng Li, Yudi Wang, Jingyi Wang, Hongxia Miao, Zhuo Wang, Biyu Xu, Xinguo Li, Zhiqiang Jin, and Juhua Liu. "Transient virus-induced gene silencing of *MaBAM9b* efficiently suppressed starch degradation during postharvest banana fruit ripening." *Plant Biotechnology Reports* 15, no. 4 (2021):527–536.

Liu, Mengyu, Zhen Zhang, Zhixuan Xu, Lina Wang, Chunhua Chen, and Zhonghai Ren. "Overexpression of *SlMYB75* enhances resistance to Botrytis cinerea and prolongs fruit storage life in tomato." *Plant Cell Reports* 40, no. 1 (2021):43–58.

Lobato-Gómez, Maria, Seanna Hewitt, Teresa Capell, Paul Christou, Amit Dhingra, and Patricia Sarai Giron-Calva. "Transgenic and genome-edited fruits: background, constraints, benefits, and commercial opportunities." *Horticulture Research* 8(2021):1–16.

Luo, Jingjing, Futian Peng, Shuhui Zhang, Yuansong Xiao, and Yafei Zhang. "The protein kinase *FaSnRK1α* regulates sucrose accumulation in strawberry fruits." *Plant Physiology and Biochemistry* 151(2020):369–377.

Luo, Jingjing, Wenying Yu, Yuansong Xiao, Yafei Zhang, and Futian Peng. "Strawberry *FaSnRK1α* Regulates Anaerobic Respiratory Metabolism under Waterlogging." *International Journal of Molecular Sciences* 23, no. 9 (2022):4914.

Ma, Xingliang, Qinlong Zhu, Yuanling Chen, and Yao-Guang Liu. "CRISPR/Cas9 platforms for genome editing in plants: developments and applications." *Molecular Plant* 9, no. 7 (2016):961–974.

Malnoy, Mickael, Roberto Viola, Min-Hee Jung, Ok-Jae Koo, Seokjoong Kim, Jin-Soo Kim, Riccardo Velasco, and Chidananda Nagamangala Kanchiswamy. "DNA-free genetically edited grapevine and apple protoplast using CRISPR/Cas9 ribonucleoproteins." *Frontiers in Plant Science* 7 (2016): 1904.

Martín-Pizarro, Carmen, and David Posé. "Genome editing as a tool for fruit ripening manipulation." *Frontiers in Plant Science* 9 (2018):1415.

Menz, Jochen, Dominik Modrzejewski, Frank Hartung, Ralf Wilhelm, and Thorben Sprink. "Genome edited crops touch the market: a view on the global development and regulatory environment." *Frontiers in Plant Science* 11(2020):586027.

Ming, Ray, Robert VanBuren, Ching Man Wai, Haibao Tang, Michael C. Schatz, John E. Bowers, Eric Lyons et al. "The pineapple genome and the evolution of CAM photosynthesis." *Nature Genetics* 47, no. 12 (2015):1435–1442.

Ming, Ray, Shaobin Hou, Yun Feng, Qingyi Yu, Alexandre Dionne-Laporte, Jimmy H. Saw, Pavel Senin et al. "The draft genome of the transgenic tropical fruit tree papaya (*Carica papaya* Linnaeus)." *Nature* 452, no. 7190 (2008):991–996.

Moon, Kyoung Mi, Eun-Bin Kwon, Bonggi Lee, and Choon Young Kim. "Recent trends in controlling the enzymatic browning of fruit and vegetable products." *Molecules* 25, no. 12 (2020):2754.

Nautiyal, Pankaj, Gaurav Papnai, Khusboo Agrawal, Saugat Khaniya, Ankit Semwal, Shivangi Pandey, Ankita Pokhriyal, Ishu Nautiyal, Archit Pokhriyal, and Bibisha Karki. "Molecular approaches in plant growth regulation of fruit crops." *International Journal of Agriculture Sciences* 14, no. 2 (2022):11073–11076.

Nishitani, Chikako, Narumi Hirai, Sadao Komori, Masato Wada, Kazuma Okada, Keishi Osakabe, Toshiya Yamamoto, and Yuriko Osakabe. "Efficient genome editing in apple using a CRISPR/Cas9 system." *Scientific Reports* 6, no. 1 (2016):1–8.

Nwosisi, Sochinwechi, Kripa Dhakal, Dilip Nandwani, Joshua Ibukun Raji, Sarada Krishnan, and Yoel Beovides-García. "Genetic diversity in vegetable and fruit crops." In*Genetic Diversity in Horticultural Plants*, pp. 87–125. Springer, Cham, 2019.

Okanagan Specialty Fruits. Arctic apples: media kit. (2023). Available at: https://arcticapples.com/wp-content/uploads/2023/04/OSF-Media-Kit.pdf

Ouyang, Zhigang, Shixia Liu, Lihong Huang, Yongbo Hong, Xiaohui Li, Lei Huang, Yafen Zhang, Huijuan Zhang, Dayong Li, and Fengming Song. "Tomato *SlERF.* A1 SlERF. B4, SlERF. C3 *and* SlERF. *A3*, members of B3 group of ERF family, are required for resistance to Botrytis cinerea." *Frontiers in Plant Science* 7 (2016):1964.

Paniagua, Candelas, Cristina Sánchez-Raya, Rosario Blanco-Portales, Jose A. Mercado, Elena Palomo-Ríos, and Sara Posé. "Silencing of *FaPG1*, a Fruit Specific Polygalacturonase Gene, Decreased Strawberry Fruit Fungal Decay during Postharvest." In *Biology and Life Sciences Forum*, vol. 11, no. 1, p. 96. MDPI, 2021.

Parajuli, Saroj, Heqiang Huo, Fred G. Gmitter, Yongping Duan, Feng Luo, and Zhanao Deng. "Editing the *CsDMR6* gene in citrus results in resistance to the bacterial disease citrus canker." *Horticulture Research* 9 (2022): uhac082, 1–3.

Peer, Reut, Gil Rivlin, Sara Golobovitch, Moshe Lapidot, Amit Gal-On, Alexander Vainstein, Tzvi Tzfira, and Moshe A. Flaishman. "Targeted mutagenesis using zinc-finger nucleases in perennial fruit trees." *Planta* 241, no. 4 (2015):941–951.

Peng, Aihong, Shanchun Chen, Tiangang Lei, Lanzhen Xu, Yongrui He, Liu Wu, Lixiao Yao, and Xiuping Zou. "Engineering canker-resistant plants through CRISPR/Cas9-targeted editing of the susceptibility gene *Cs LOB 1* promoter in citrus." *Plant Biotechnology Journal* 15, no. 12 (2017):1509–1519.

Qin, Gaihua, Chunyan Xu, Ray Ming, Haibao Tang, Romain Guyot, Elena M. Kramer, Yudong Hu et al. "The pomegranate (*Punica granatum* L.) genome and the genomics of punicalagin biosynthesis." *The Plant Journal* 91, no. 6 (2017):1108–1128.

Ren, Chong, Xianju Liu, Zhan Zhang, Yi Wang, Wei Duan, Shaohua Li, and Zhenchang Liang. "CRISPR/Cas9-mediated efficient targeted mutagenesis in Chardonnay (*Vitis vinifera* L.)." *Scientific Reports* 6, no. 1 (2016):1–9.

Ribaut, J.M., M.C. De Vicente, and X. Delannay. "Molecular breeding in developing countries: challenges and perspectives." *Current Opinion in Plant Biology* 13, no. 2 (2010):213–218.

Sato, Shusei, Satoshi Tabata, Hideki Hirakawa, Erika Asamizu, Kenta Shirasawa, Sachiko Isobe, Takakazu Kaneko et al. "The tomato genome sequence provides insights into fleshy fruit evolution." *Nature* 485, no. 7400 (2012):635–641.

Scorza, Ralph, Michel Ravelonandro, Ann M. Callahan, John M. Cordts, Marc Fuchs, Jean Dunez, and Dennis Gonsalves. "Transgenic plums (*Prunus domestica* L.) express the plum pox virus coat protein gene." *Plant Cell Reports* 14, no. 1 (1994):18–22.

Sekeli, Rogayah, Janna Ong Abdullah, Parameswari Namasivayam, Pauziah Muda, Umi Kalsom Abu Bakar, Wee Chien Yeong, and Vilasini Pillai. "RNA Interference of *1-Aminocyclopropane-1-carboxylic Acid Oxidase* (*ACO 1* and *ACO 2*) Genes Expression Prolongs the Shelf Life of Eksotika (*Carica papaya* L.) Papaya Fruit." *Molecules* 19, no. 6 (2014):8350–8362.

Seymour, Graham B., Carol D. Ryder, Volkan Cevik, John P. Hammond, Alexandra Popovich, Graham J. King, Julia Vrebalov, James J. Giovannoni, and Kenneth Manning. "A *SEPALLATA* gene is involved in the development and ripening of strawberry (*Fragaria× ananassa* Duch.) fruit, a non-climacteric tissue." *Journal of Experimental Botany* 62, no. 3 (2011):1179–1188.

Shin, Kihye, Soon-Hwa Kwon, Seong-Chan Lee, and Young-Eel Moon. "Sensitive and rapid detection of citrus scab using an RPA-CRISPR/Cas12a system combined with a lateral flow assay." *Plants* 10, no. 10 (2021):2132.

Shulaev, Vladimir, Daniel J. Sargent, Ross N. Crowhurst, Todd C. Mockler, Otto Folkerts, Arthur L. Delcher, Pankaj Jaiswal et al. "The genome of woodland strawberry (*Fragaria vesca*)." *Nature Genetics* 43, no. 2 (2011):109–116.

Soler, Nuria, Montserrat Plomer, Carmen Fagoaga, Pedro Moreno, Luis Navarro, Ricardo Flores, and Leandro Pena. "Transformation of Mexican lime with an intron-hairpin construct expressing untranslatable versions of the genes coding for the three silencing suppressors of Citrus tristeza virus confers complete resistance to the virus." *Plant Biotechnology Journal* 10, no. 5 (2012):597–608.

Sonawane, Madhuri Shrikant. "Applications of Molecular Characterization in Fruit Crops." *International Archive of Applied Sciences and Technology* 8, no. 5 (2017):88–95.

Stowe, Evan, and Amit Dhingra. "Development of the Arctic® apple." *Plant Breeding Reviews* 44 (2021):273–296.

Sun, Liang, Bing Yuan, Mei Zhang, Ling Wang, Mengmeng Cui, Qi Wang, and Ping Leng. "Fruit-specific RNAi-mediated suppression of *SlNCED1* increases both lycopene and β-carotene contents in tomato fruit." *Journal of Experimental Botany* 63, no. 8 (2012):3097–3108.

Szankowski, I., S. Waidmann, A. El-Din Saad Omar, H. Flachowsky, C. Hättasch, and M-V. Hanke. "RNAi-silencing of *MdTFL1* induces early flowering in apple." In *International Symposium on Biotechnology of Fruit Species:* Biotechfruit 2008839, Dresden, Germany, pp. 633–636. 2008.

Tian, Shouwei, Linjian Jiang, Qiang Gao, Jie Zhang, Mei Zong, Haiying Zhang, Yi Ren et al. "Efficient CRISPR/Cas9-based gene knockout in watermelon." *Plant Cell Reports* 36, no. 3 (2017):399–406.

USDA. USDA announces deregulation of non-browning apples. (2015). Available at: https://content.govdelivery.com/accounts/USDAAPHIS/bulletins/f1008d [Accessed June 2020]

Van Loon, Leendert C., Martijn Rep, and Corné M.J. Pieterse. "Significance of inducible defense-related proteins in infected plants." *Annual Review of Phytopathology* 44(2006):135–162.

Vasistha, Neeraj Kumar, Shivani Thakur, Hitesh Kumar, and Jitendra Kumar. "Molecular Approach for Crop Improvement." In: Kumar Amarjeet, Birendra Prasad Anil Kumar, Editors *Classical And Molecular Approaches In Plant Breeding*, Narendra Publishing House, New Delhi, P. 282-303 (2020)

Velasco, Riccardo, Andrey Zharkikh, Jason Affourtit, Amit Dhingra, Alessandro Cestaro, Ananth Kalyanaraman, Paolo Fontana et al. "The genome of the domesticated apple (*Malus× domestica* Borkh.)." *Nature Genetics* 42, no. 10 (2010):833–839.

Velasco, Riccardo, Andrey Zharkikh, Michela Troggio, Dustin A. Cartwright, Alessandro Cestaro, Dmitry Pruss, Massimo Pindo et al. "A high quality draft consensus sequence of the genome of a heterozygous grapevine variety." *PLoS One* 2, no. 12 (2007): e1326.

Verde, Ignazio, Albert G. Abbott, Simone Scalabrin, Sook Jung, Shengqiang Shu, Fabio Marroni, Tatyana Zhebentyayeva et al. "The high-quality draft genome of peach (*Prunus persica*) identifies unique patterns of genetic diversity, domestication and genome evolution." *Nature Genetics* 45, no. 5 (2013):487–494.

Wang, Lijuan, Shanchun Chen, Aihong Peng, Zhu Xie, Yongrui He, and Xiuping Zou. "CRISPR/Cas9-mediated editing of *CsWRKY22* reduces susceptibility to *Xanthomonas citri* subsp. citri in Wanjincheng orange (*Citrus sinensis* (L.) Osbeck)." *Plant Biotechnology Reports* 13, no. 5 (2019):501–510.

Wang, Nan, Yingshuang Liu, Chaohua Dong, Yugang Zhang, and Suhua Bai. "*MdMAPKKK1* Regulates Apple Resistance to *Botryosphaeria dothidea* by Interacting with *MdBSK1*." *International Journal of Molecular Sciences* 23, no. 8 (2022):4415.

Wang, Shufang, Miaoyu Song, Jiaxuan Guo, Yun Huang, Fangfang Zhang, Cheng Xu, Yinghui Xiao, and Lusheng Zhang. "The potassium channel *Fa TPK 1* plays a critical role in fruit quality formation in strawberry (*Fragaria*× *ananassa*)." *Plant Biotechnology Journal* 16, no. 3 (2018):737–748.

Wang, Xianhang, Mingxing Tu, Dejun Wang, Jianwei Liu, Yajuan Li, Zhi Li, Yuejin Wang, and Xiping Wang. "CRISPR/Cas9-mediated efficient targeted mutagenesis in grape in the first generation." *Plant Biotechnology Journal* 16, no. 4 (2018a):844–855.

Wang, Zupeng, Shuaibin Wang, Dawei Li, Qiong Zhang, Li Li, Caihong Zhong, Yifei Liu, and Hongwen Huang. "Optimized paired-sgRNA/Cas9 cloning and expression cassette triggers high-efficiency multiplex genome editing in kiwifruit." *Plant Biotechnology Journal* 16, no. 8 (2018b):1424–1433.

Wei, Lingzhi, Huabo Liu, Yang Ni, Jing Dong, Chuanfei Zhong, Rui Sun, Shuangtao Li et al. "*FaAKR23* Modulates Ascorbic Acid and Anthocyanin Accumulation in Strawberry (*Fragaria*× *ananassa*) Fruits." *Antioxidants* 11, no. 9 (2022):1828.

Wu, Jun, Zhiwen Wang, Zebin Shi, Shu Zhang, Ray Ming, Shilin Zhu, Muhammad Awais Khan et al. "The genome of the pear (*Pyrus bretschneideri* Rehd.)." *Genome Research* 23, no. 2 (2013):396–408.

Wu, Zilin, Cuiping Mo, Shuguang Zhang, and Huaping Li. "Characterization of papaya ringspot virus isolates infecting transgenic papaya 'Huanong No. 1'in South China." *Scientific Reports* 8, no. 1 (2018):1–11.

Xiong, Jin-Song, Jing Ding, and Yi Li. "Genome-editing technologies and their potential application in horticultural crop breeding." *Horticulture Research* 2(2015):1–10.

Xu, Qiang, Ling-Ling Chen, Xiaoan Ruan, Dijun Chen, Andan Zhu, Chunli Chen, Denis Bertrand et al. "The draft genome of sweet orange (*Citrus sinensis*)." *Nature Genetics* 45, no. 1 (2013):59–66.

Xu, Yunbi, Zhi-Kang Li, and Michael J. Thomson. "Molecular breeding in plants: moving into the mainstream." *Molecular Breeding* 29, no. 4 (2012):831–832.

Ye, Changming, and Huaping Li. "20 Years of transgenic research in China for resistance to Papaya ringspot virus." *Transgenic Plant Journal* 4 (2010):58–63.

Zhang, Fei, Chantal LeBlanc, Vivian F. Irish, and Yannick Jacob. "Rapid and efficient CRISPR/Cas9 gene editing in citrus using the YAO promoter." *Plant Cell Reports* 36, no. 12 (2017):1883–1887.

Zhang, Hui, Jinshan Zhang, Zhaobo Lang, José Ramón Botella, and Jian-Kang Zhu. "Genome editing—principles and applications for functional genomics research and crop improvement." *Critical Reviews in Plant Sciences* 36, no. 4 (2017):291–309.

Zhang, Jie, Honghe Sun, Shaogui Guo, Yi Ren, Maoying Li, Jinfang Wang, Haiying Zhang, Guoyi Gong, and Yong Xu. "Decreased protein abundance of lycopene β-cyclase contributes to red flesh in domesticated watermelon." *Plant Physiology* 183, no. 3 (2020):1171–1183.

Zhao, Kai, Feng Zhang, Yi Yang, Yue Ma, Yuexue Liu, He Li, Hongyan Dai, and Zhihong Zhang. "Modification of plant height via RNAi suppression of *MdGA20-ox* gene expression in apple." *Journal of the American Society for Horticultural Science* 141, no. 3 (2016):242–248.

Zhao, Shuang, Haibo Wang, Xumei Jia, Hanbing Gao, Ke Mao, and Fengwang Ma. "The HD-Zip I transcription factor *MdHB7*-like confers tolerance to salinity in transgenic apple (Malus domestica)." *Physiologia Plantarum* 172, no. 3 (2021):1452–1464.

Zhou, Junhui, Guoming Wang, and Zhongchi Liu. "Efficient genome editing of wild strawberry genes, vector development and validation." *Plant Biotechnology Journal* 16, no. 11 (2018):1868–1877.

Zhu, Chenqiao, Xiongjie Zheng, Yue Huang, Junli Ye, Peng Chen, Chenglei Zhang, Fei Zhao et al. "Genome sequencing and CRISPR/Cas9 gene editing of an early flowering Mini-Citrus (*Fortunella hindsii*)." *Plant Biotechnology Journal* 17, no. 11 (2019):2199–2210.

Index

Note: **Bold** page numbers refer to tables and *italic* page numbers refer to figures.

9781032406367